Adapting to the Atmosphere Conference 2014

Journal of Physics: Conference Series Volume 595

Durham, United Kingdom
15 – 18 September 2014

ISBN: 978-1-5108-0252-0
ISSN: 1742-6588

Printed from e-media with permission by:

Curran Associates, Inc.
57 Morehouse Lane
Red Hook, NY 12571

Some format issues inherent in the e-media version may also appear in this print version.

Printed by Curran Associates, Inc. (2015)

Institute of Physics
Dirac House, Temple Back
Bristol BS1 6BE UK

Phone: 44 1 17 929 7481
Fax: 44 1 17 920 0979

techtracking@iop.org

Additional copies of this publication are available from:

Curran Associates, Inc.
57 Morehouse Lane
Red Hook, NY 12571 USA
Phone: 845-758-0400
Fax: 845-758-2634
Email: curran@proceedings.com
Web: www.proceedings.com

TABLE OF CONTENTS

Journal of Physics: Conference Series **595** (2015) 011001 doi:10.1088/1742-6596/595/1/011001

Adapting to the Atmosphere: Editorial

These proceedings, for the "Adapting to the Atmosphere" conference, Durham University, 2014, address the field of optical turbulence profiling of the Earth's atmosphere for astronomical observatory site selection, and the applications of Adaptive Optics and Optical Astronomical Instrumentation for small observatories as well as the future Extremely Large Telescope scales.

This conference concentrated on the future of the community in an era of change. Optical turbulence profiling and remote sensing is required for the next generation of sophisticated adaptive optics systems and enables many of the proposed instruments and technologies which are necessary for the operation of large to extremely large telescopes. The successful design and operation of these complex systems demands an increasingly detailed understanding of the nature of atmospheric optical turbulence, as well as improvements in our ability to monitor and forecast its properties.

This conference has the endorsement of the International Astronomical Union (working group on site testing instrumentation) and has received funding from the European Southern Observatory, the Thirty Meter Telescope and the Optical Infrared Co-ordination Network (OPTICON).

Peer review statement

All papers published in this volume of *Journal of Physics: Conference Series* have been peer reviewed through processes administered by the proceedings Editors. Reviews were conducted by expert referees to the professional and scientific standards expected of a proceedings journal published by IOP Publishing.

Adapting to the Atmosphere Conference 2014 IOP Publishing
Journal of Physics: Conference Series **595** (2015) 012001 doi:10.1088/1742-6596/595/1/012001

On choosing layer profiles in atmospheric tomography

Günter Auzinger
industrial mathematics institute of JKU Linz, Austria
E-mail: auzinger@mathconsult.co.at

Abstract.
In control of the AO system of an ELT, most algorithms for atmospheric tomography are based on assuming atmospheric turbulence to occur only within a certain number of horizontal layers. The number of these layers is in general significantly smaller than the number of turbulent layers in the real atmosphere. We investigate the question, how the choice of the number and heights of the hypothetical layers (as parameters of the reconstruction algorithm) influences the resulting quality of the tomographic reconstruction.

1. Introduction

In this article we investigate the influence of a chosen reconstruction layer profile, especially the heights of its layers, on the quality of the solution to the atmospheric tomography problem which arises e.g. during operation of an ELT in the operating mode LTAO. The tests were performed in the context of the project "Mathematical algorithms and software for E-ELT Adaptive Optics". We utilize the Kaczmarz algorithm (see [2]) for tomographic reconstruction and use the simulation tool MOST for evaluation. We will describe a simple *compression* algorithm for generating an appropriate reconstruction layer profile from a given atmosphere model in Section 2 and a simple method for subsequent *optimization* in Section 3. The resulting profiles are compared numerically with respect to the center Strehl ratio in Section 4, finally the importance of informations about the atmosphere is investigated in Section 5 via comparison to 'ignorant' reconstruction profiles that are constructed without any knowledge about the atmosphere model.

The contents of this article are essentially extracted from the author's PhD thesis [1], which is still work in progress.

2. Compression of an atmospheric turbulence profile

Before we can formulate the compression algorithm, we have to define *Voronoi intervals*, which are needed for determining disjoint intervals around given points on an axis: Let $0 \le h_1 < \cdots < h_n$ be a sequence of ascending real numbers. Then the Voronoi intervals $I_1, \ldots, I_n$ are defined as

$$I_1 := \left[0, \frac{h_1 + h_2}{2}\right]; \forall 1 < k < n : I_k := \left[\frac{h_{k-1} + h_k}{2}, \frac{h_k + h_{k+1}}{2}\right]; I_n := \left[\frac{h_{n-1} + h_n}{2}, 2h_n\right].$$

We now assume an N-layer model of atmospheric turbulence given in form of heights $H_1, \ldots, H_N$ and C_n^2-values $C_1, \ldots, C_N$ (in addition, wind speeds and directions are given, but we do not use this information explicitly). In order to derive an appropriate reconstruction layer profile P with a chosen number of layers L, we suggest the following algorithm:

(i) Set the temporary number of layers $\tilde{L} := L$

(ii) Create a geometric sequence $h_2, \dots, h_{\tilde{L}}$ with $h_2 = \min\{H_n \neq 0\}$ and $h_{\tilde{L}} = H_N$ and add a ground layer $h_1 := 0$.

(iii) For each $k = 1, \dots, \tilde{L}$, set up a Voronoi interval I_k around h_k and set $c_k := \sum_{n:\, H_n \in I_k} C_n$.

(iv) If the number of values $c_k \neq 0$ is less than L, set $\tilde{L} := \tilde{L} + 1$ and repeat from step (ii).

(v) Select those indices k, for which $c_k \neq 0$, resulting in an L-layer profile P.

Some examples for the compression algorithm can be seen in Figure 1: Starting from a 26-layer model for the atmosphere provided by ESO, the compression algorithm was performed for 6, 10, 15 and 20 layers. Note that in all plots the C_n^2-values are scaled by dividing through the length of the corresponding Voronoi-interval in order to represent something looking more like a continuous function, which makes the similarities more visible.

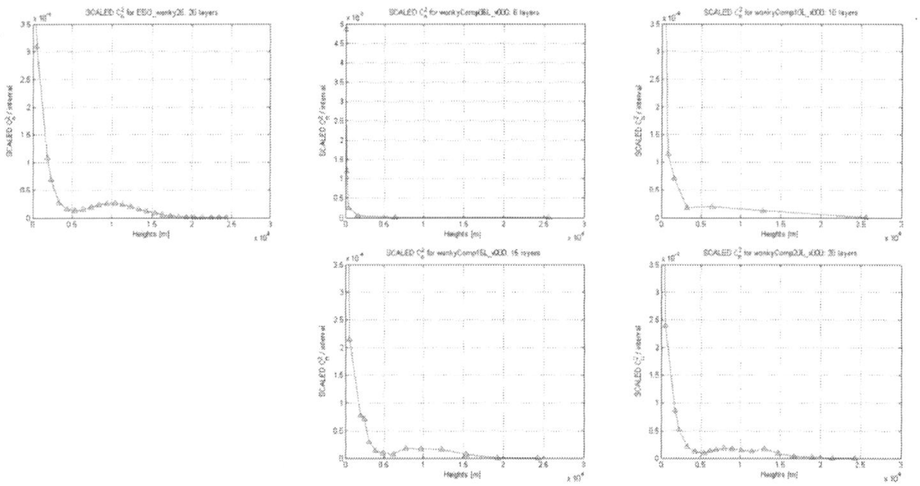

Figure 1. Full atmosphere model (top left) and compressed reconstruction layer profiles for 6, 10, 15 and 20 layers. The absolute values C_n^2 are divided by the lengths of the corresponding Voronoi-intervals.

3. Optimization of reconstruction layer profiles

During test runs in MOST, we used a very simple method for optimizing the profiles gained by compression of the atmosphere model. It is essentially a heuristic line search algorithm along canonical search directions. For application of the following algorithm, in each optimization step the Kaczmarz method was run in parallel for the same atmosphere dynamics, but with different reconstruction layer profiles over several hundreds of system time steps.

(i) For all $k = 2, \dots, L$, try a new profile with h_k increased by a fixed amount (e.g. 20%) and choose k^* such that variation of h_{k^*} has most significant impact.

(ii) Try several values for h_{k^*}, choose the optimum and assign it to h_{k^*}.

(iii) Optimize one value c_{j^*} analogously to steps (i),(ii), keeping the complete profile scaled to $\sum_{j=1}^{L} c_j = 1$.

(iv) If overall quality could be increased, perform another iteration beginning with step (i).

An example for performing step (i) can be seen in Figure 2: The same 26-layer atmosphere model is used as for Figure 1, and the compressed 6-layer profile as starting value. The curves show the the performance of the compressed profile (blue) and profiles modified by increasing the heights of layers 2 (green), 3 (red), 4 (cyan), 5 (magenta) and 6 (yellow) about 20%. The evolution of the SE center Strehl ratio clearly indicates that variation of the height of layer number 5 (magenta curve) has most direct impact. Hence we examine more variations of this layer in step (ii): h_5 is raised by 10, 20, 30, 40, 60 and 80 %, the results are shown in Figure 3 and suggest to increase h_5 by 50%.

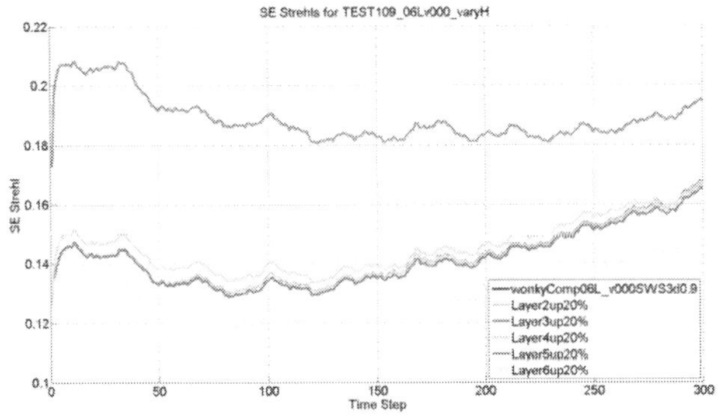

Figure 2. Step (i) of optimization: The 6 layer profile, all heights separately raised by 20% - clearly, raising layer $k^* = 5$ has the largest impact on the reconstruction quality.

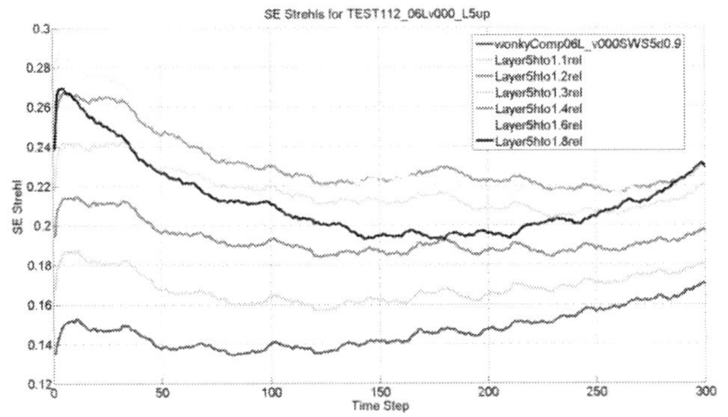

Figure 3. Step (ii) of optimization: Layer 5 is raised by several amounts

Step (iii) does not bring any significant benefit in this situation - in general it seemed throughout all these tests that variation of the heights h_k has much more impact on the quality than variation of the C_n^2-values c_j. However, we show the successive quality gain for the 10 layer profile, where optimization seems finished (settlement in a maximum) after 4 iterations - these can be seen from Figure 4: The blue curve shows the SE Strehl ratio evolution for the compressed 10 layer profile, the successively increasing quality is shown by the green, red, cyan and magenta curves. Quality gained by usage of the full 26 layer profile is shown for comparison, using 2 (yellow) and 5 (black) Kaczmarz iterations.

4. Comparison of the resulting profiles

The results with the 10 layer profiles indicate that it is not possible to reach the quality of the full 26 layer profile by usage of 10 layers. The experiments were repeated with 15 and

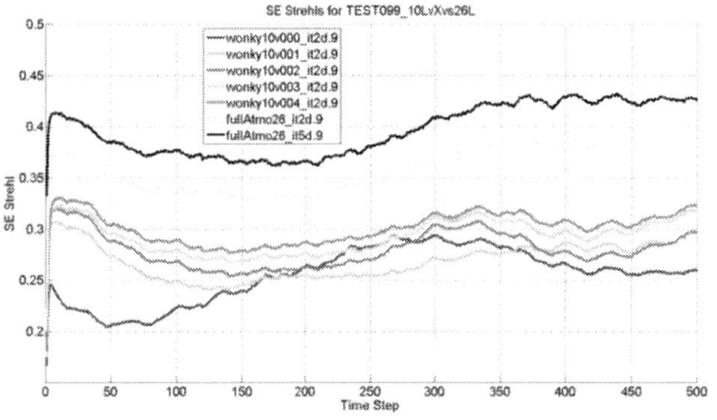

Figure 4. Original compressed 10 layer profile (blue) vs. optimization iterates 1-4. Yellow and black curves: Usage of the full 26-layer profile from the atmosphere model, for 2 and 5 Kaczmarz iterations, respectively.

20 layer profiles - in these cases the optimization routine could not significantly increase the reconstruction quality, i.e., compression alone seems to produce optimal profiles. A summary of all results for the 26 layer atmosphere is shown in Figure 5:

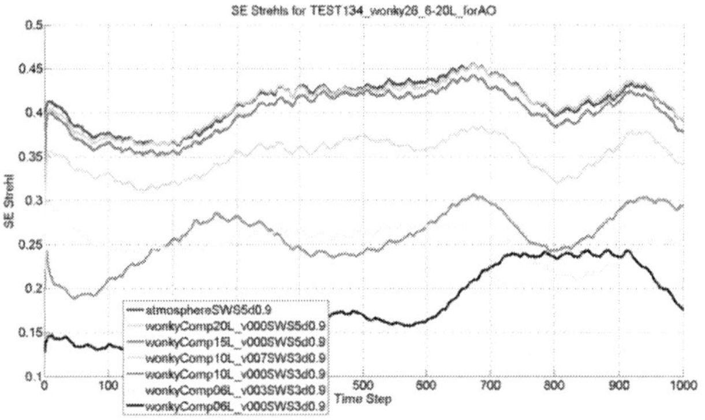

Figure 5. Performance of compressed and optimized profiles for 6, 10, 15 and 20 layers, using a 26 layer atmosphere model

The black curve results from the compressed 6 layer profile, the yellow curve from the same profile optimized by 3 iterations. The magenta and cyan curves result from the 10 layer profiles, original compressed version and after 7 optimization steps, respectively. Center Strehl ratios for the compressed 15 and 20 layer profiles are shown by the red and green curve, the blue curve results from reconstruction on the complete 26 layer profile. The curves give evidence for the following conclusions:

- For a sufficiently large number of reconstruction layers (in our case ≥ 15), usage of the compressed profile is sufficient and no optimization is needed.
- The corresponding quality in that case is not significantly less than reconstruction on the full profile.
- When not enough (in our case ≤ 10) reconstruction layers are applied, the quality of the full 26 layer profile cannot be reached, but still optimization can yield a significant increase of quality (Gain from the black curve to the yellow one for 6 layers, from magenta to cyan for 10 layers).
- For instance, the 6 layer profile after 3 optimization steps yields results comparable to the compressed 10 layer profile without optimization (yellow and magenta curve). Thus, in this case optimization reduces the amount of data to be processed without significant loss of quality.

In the following section we will briefly summarize some more optimization tests based on initial profiles that are constructed without any knowledge of the atmosphere model.

5. Layer optimization without knowledge of the atmosphere

In order to avoid too much computational time, the subsequent tests were performed on smaller amounts of data: We used a 9 layer atmosphere model provided by ESO (shown in Figure 6) and tried 'ignorant' reconstruction profiles with 3, 5 and 7 layers. These reconstruction layer profiles were simply chosen as arithmetic sequences for the heights, $c_1 = 1/2$ and all other c_k equal and such that the sum over all C_n^2-values is 1. Note that this arbitrary setting does not use any information about the atmosphere.

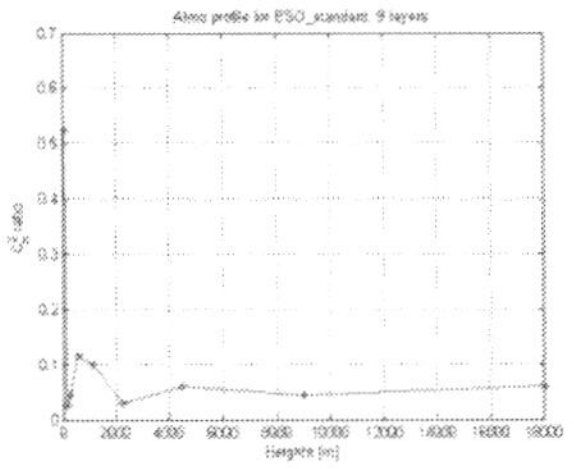

Figure 6. The 9 layer atmosphere profile defined by ESO: absolute values C_n^2 over layer height h.

The results for these profiles and compressed versions, as well as optimized versions of both, can be seen from the plots in Figures 7 and 8, all of them showing the atmosphere model used as reconstruction layer profile as blue curve for comparison:

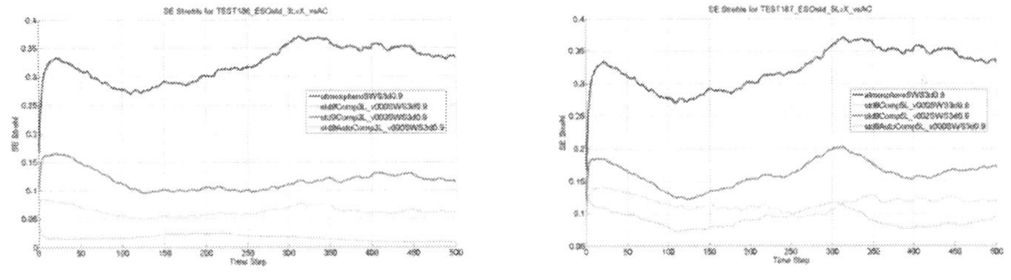

Figure 7. Comparison of different optimization iterates vs. automatic compression for 3 and 5 layers

The two plots in Figure 7 show results for the 3 layer profile (left) and the 5 layer profile (right), respectively. In both cases, the green curves show the center SE Strehl ratio evolution for the 'ignorant' profiles, the red curves result from these profiles after sufficiently many optimization steps (left: 3, right: 2). The cyan curves show the performance of the profiles resulting from compression of the given atmosphere model - in both cases, optimization did not yield any increase of quality. This means that the compressed profiles seem to be stuck in a local optimum and optimization cannot reach the global one, whereas optimization of the 'ignorant' profiles can reach higher quality. Hence, in this cases the profiles generated without any knowledge of the atmosphere turn out to be better suited as starting values for the optimization routine.

Figure 8 shows corresponding tests for the 7 layer profile, where the situation is different: The green curve shows the results from the 'ignorant' profile (i.e. without knowledge of the atmosphere), the red curve after sufficiently many (in this case 2) optimization steps - a slight

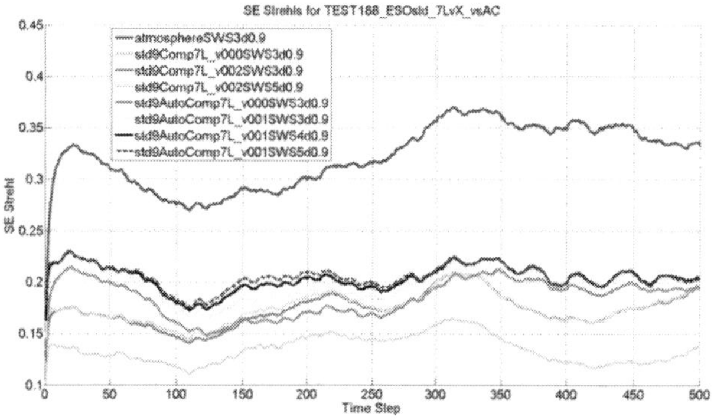

Figure 8. Comparison of different optimization iterates vs. automatic compression for 7 layers

improvement could only be reached by increasing the number of Kaczmarz iterations from 3 to 5 (cyan curve). In contrast to the 3 and 5 layer profiles, the compressed 7 layer profile performs immediately better (magenta curve) and can still be improved by one optimization step (yellow) and subsequently increasing the number of Kaczmarz iterations from 3 to 4 (black) and 5 (dashed blue).

These results indicate the following conclusions:

- For a very small number of reconstruction layers (in our case 3 or 5), there is no guaranty that a reconstruction profile resulting from compression of the atmosphere model is a good starting value for optimization - arbitrarily chosen profiles can perform better.
- For a larger number of reconstruction layers (in our case 7), it definitely pays off to use compression, i.e., to exploit as much information about the atmosphere as possible.
- Although 7 is not much less than 9, performance of the 'best possible' 7 layer profile is significantly worse than usage of the full 9 layer atmosphere model, as is evident from Figure 8 (dashed blue vs. solid blue curve).

This definitely shows that it seems recommendable to use as much reconstruction layers as possible and spend enough time in optimization, exploiting as much information about the atmosphere as possible. Note that this clearly indicates a preference to tomographic reconstruction algorithms which can be run in parallel with respect to the reconstruction layers, as is the case e.g. for the Kaczmarz algorithm.

References

[1] G. Auzinger. *New Reconstruction Approaches in Adaptive Optics for Extremely Large Telescopes*. PhD thesis, industrial mathematics institute of JKU Linz, to appear 2015.

[2] R. Ramlau and M. Rosensteiner. An efficient solution to the atmospheric turbulence tomography problem using Kaczmarz iteration. *Inverse Problems*, 28(9):095004, 2012.

Simulation and laboratory demonstration of measurement and mitigation of dome seeing

Alastair Basden, Scott Wells, Richard Myers

Department of Physics, South Road, Durham, DH1 3LE, UK

E-mail: `a.g.basden@durham.ac.uk`

Abstract. Ground layer turbulence is seen to be dominant at most telescope sites. Some fraction of this is caused by the interaction of the telescope dome and structure with the surrounding environment, leading to mixing of air at different temperatures, and hence, optical perturbations. Here we investigate the use of laminar air flow in the form of air curtains to reduce this mixing. We provide laboratory measurements of the effect of mixing and its mitigation, and detail the steps that can be taken to reduce this effect. Application for telescope dome structures is developed.

1. Introduction

Atmospherically induced perturbations form the major performance limitation for ground-based optical telescopes. Mixing of air of different temperatures in the atmosphere between telescope and target lead to changes in refractive index, and hence differential delays in the wavefronts of light incident along different lines of sight. These variations in wavefront phase limit effective telescope resolution, and mean that, if unmitigated, the largest optical telescopes have only the effective resolution of a large hand-held telescope, albeit with a larger light collecting area.

At most astronomical sites, the large majority of atmospheric perturbations are introduced in the atmosphere close to the ground, where the free flow of air is restricted both by the surrounding environment, and the telescope structure itself. We investigate the effect of air flow around a telescope dome using both compute simulation and laboratory bench models, and study the use of air curtains to provide a laminar air flow to isolate air within the dome from air outside, thus reducing mixing of air at different temperatures. In §2 we present our method and discuss our findings. In §3 we provide details of future work, and we conclude in §4.

2. Introduction

Air curtains, which produce a laminar flow of air are used in many application to separate different regions of air. In supermarkets, the are used to prevent cold exterior air from mixing with warm air in the shop (or in warmer climates, to prevent hot exterior air from mixing with cooler interior air). Air curtains are also used in refrigerated displays to retain chilled air without necessitating a physical barrier. Operating theatres within hospitals also use air curtains to provide a sterile operating space similarly to a clean-tent, but without a physical barrier. Warehouses sometimes use air curtains across wide entrances to allow free access for vehicles, whilst retaining and maintaining interior air temperature.

2.1. Dome seeing

Mixing of air within and around a telescope dome structure can lead to significant optical aberrations caused by the differing refractive indices when air of different temperatures is mixed. This effect, called dome seeing, is situation dependent, and unsensed by external atmospheric profilers, such as DIMMS, Slodar and Scidar systems. A reduction of this dome seeing, by reducing the mixing of interior and exterior air, can have a positive impact on astronomical image quality.

Placing a large glass screen across the telescope dome opening is impractical. The surface quality would have to be maintained at the ten-nm-level, and transmissive losses would impact telescope sensitivity. The alternative approach is to use an air curtain.

2.2. Air curtain design

The design of an air curtain involves the optimisation of several different parameters, in addition to a consideration of the purpose of the curtain. In the case considered here, we are aiming to reduce turbulence, and so a curtain with a laminar air flow is highly desirable (turbulent air flows are sometimes used for shop entrances to reduce the number of insects passing through the curtain).

The air curtain velocity is an important consideration, as the air within the curtain need to have ample momentum to traverse the gap. The volume of air within the curtain (i.e. its thickness) also requires consideration, as this affects the momentum of the curtain, the dispersion of air leaving the source, and the length over which laminar flow can be achieved.

The laminarity of air leaving the source should also be considered: if air flow is not linear at this point, quality of the air flow will be reduced. The introduction of flow shaping within the source can aid this, e.g. smaller tubes of air prior to the source exit.

The use of an air sink should also be considered, to provide a negative pressure across the aperture aiding the laminar flow by maintaining momentum within the curtain. This will aid the laminarity of the airflow, channelling the air from source to sink.

2.3. Optical perturbations due to airflow

It is important to note that moving air itself does not cause optical perturbations. Rather, it is the mixing of air with different temperatures that cause this, leading to refractive index differences which then move across the field of view. Turbulent eddies can facilitate this air mixing, thus increasing optical perturbations.

The key concept under investigation here is whether laminar air flow can be used to reduce this mixing, by separating air within a telescope dome from external air, which will be at a different temperature.

Mixing of air above a telescope can be caused by incident wind, a chimney effect within the telescope (air being sucked up from ground level and out of the dome), and heating effects such as mirror heating (the primary mirror of most telescopes having a large thermal mass) which will cause convection above the mirror. Telescope domes are not usually conducive for smooth airflow: obstructions, sharp corners and sudden surface gradient changes will cause eddies to form as air flows past the telescope. Introduction of an air curtain will help to smooth these eddies.

3. Air curtain modelling and testing

We have used a commercial computational fluid dynamics software package to model a telescope dome, internal and external air mass, and an air curtain across the dome aperture. The model is based on the dimensions for the dome of the 4.2 m William Herschel Telescope on La Palma, and as such has a 6 m open aperture.

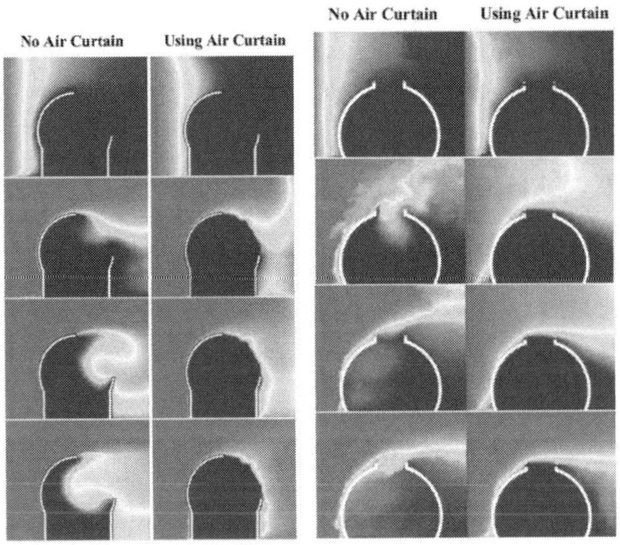

Figure 1. A figure showing mixing of interior and exterior air with wind propagating past a telescope dome, both with and without an air curtain, and with wind directions coming from behind the telescope, and from the side relative to the dome opening. The first two columns show air blowing from behind the dome (from left to right), without and with an air curtain respectively. The last two columns show air blowing across the dome aperture (again from left to right), without and with an air curtain respectively. The rows represent increasing time in 2.5 s intervals, from 2.5 to 10 s.

We assume a 5 m/s wind speed, and a 10 m/s air curtain speed in our models, and include an initial condition with a region of warmer air away from the telescope, but travelling towards it, as shown in Fig. 1. We then propagate the model for 10 s to show how this warmer air will mix with cooler air, both with and without air curtains. The initial temperature difference is 1 °C, and is necessary so that we can clearly identify the mixing of interior and exterior air. We assume warmer exterior air, though this is immaterial: it is simply so that we can identify the mixing.

Fig. 1 shows that the air curtain is able to prevent mixing of interior and exterior air within the dome, and that eddies are significantly reduced.

3.1. Qualitative laboratory measurements

We have set up a laboratory air curtain across free space, and use Schlieren imaging to observe hot air flow from a heater both with and without the air curtain, as shown in Fig. 2. It is evident that the air curtain is able to reduce turbulent eddies.

3.2. Quantitative analysis

Schlieren imaging is generally a qualitative observation tool, and so to perform a quantitative analysis of the reduction in optical perturbation using air curtains, we use a Shack-Hartmann sensor, passing a collimated beam of light through the region of turbulence under investigation. Initial tests were performed using a cuboid dome geometry, i.e. the dome aperture was a slit within a planar surface, across which the air curtain was set up, and turbulent air blown.

The air curtain source used a honeycomb arrangement of tubes and piping to linearise the

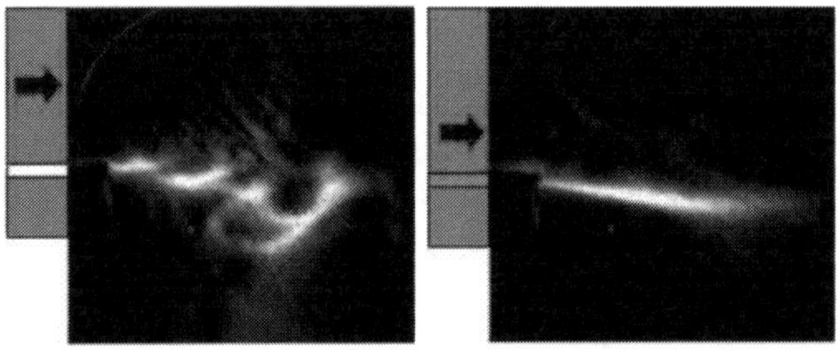

Figure 2. A figure showing air turbulence from a heater without (left), and with (right) an air curtain present. Hot air blows from left to right as does the air curtain.

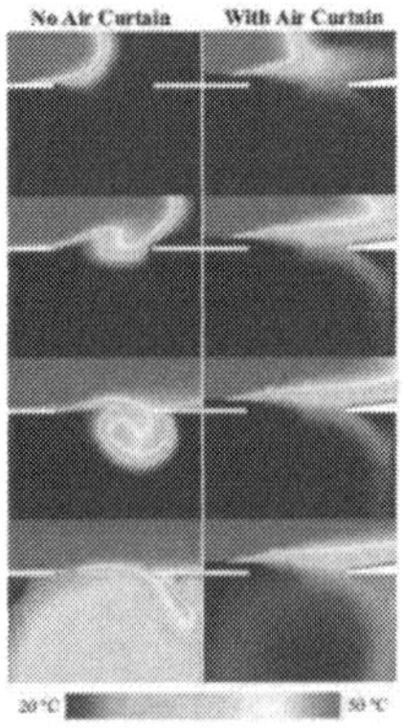

Figure 3. A figure showing modelling of the effect of an air curtain across a planar surface with an open aperture.

airflow upon exit of the source. Pressure for the air curtain was generated using a fan, and the velocity, source width and length were chosen to give a low Reynolds number.

Initial modelling of this configuration (Fig. 3 suggested that a reduction in turbulent mixing would be achieved. Shack-Hartmann slopes were reconstructed into an incident wavefront using the CuReD algorithm ([1]), and a Zernike decomposition of the time-varying incident phase obtained (Fig 4(a)). This was found to follow a generalised turbulence model power spectrum:

$$\phi_n(\kappa) = 0.033C_n^2(z)|\kappa|^{-\beta} \tag{1}$$

with $\beta = 2.7$ rather than the $\frac{11}{3}$ expected for pure Kolmogorov turbulence.

The RMS phase perturbation was found to be significantly reduced by the air curtain. As shown in Fig. 4(a), the variance of turbulence with the air curtain is significantly reduced from that without. It is also interesting to note that free-space turbulence, i.e. without a dome aperture to contend with, results in even lower variance. This may have application for dome-free telescopes, though even the presence of the telescope itself is likely to increase phase variance, and so we do not discuss this further here.

In our measurements, we find that the air curtain reduces variance by close to a factor of three compared to the case without an air curtain, across a planar aperture. In free-space, the variance is found to be a factor of two lower. Ambient air, and the air curtain alone (at ambient temperature) result in very low, almost identical phase variance, i.e. the air curtain alone is not introducing significant phase perturbations.

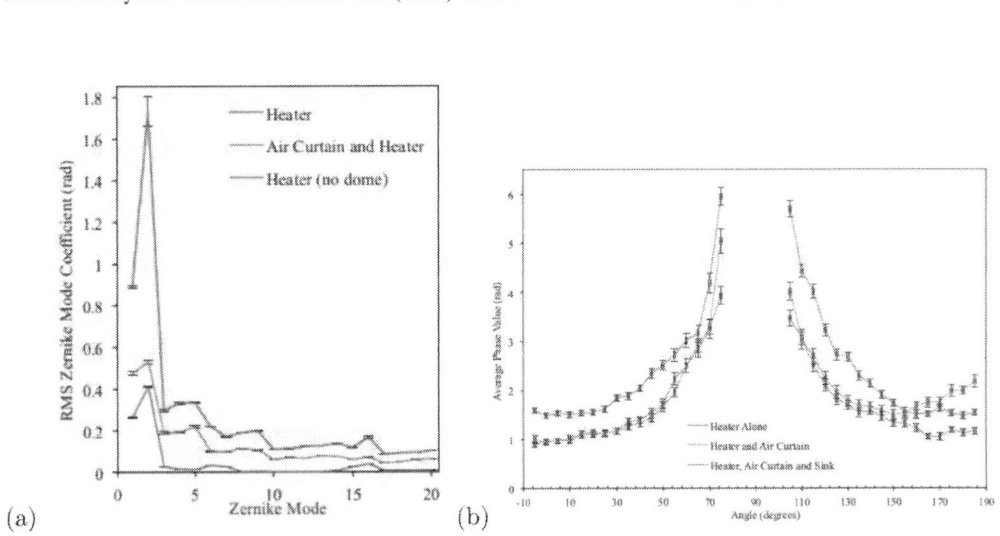

Figure 4. (a) A figure showing the normalised RMS of different Zernike modes measured by a Shack-Hartmann wavefront sensor, viewing free-space turbulence (blue), turbulence across a planar dome opening (red), and with the air curtain (green). (b) A figure showing perturbation strength as a function of angle of incidence of incoming wind. Performance is given without an air curtain (Heater Along), with an air curtain, and also including an air curtain sink. An angle of zero represents the air curtain and moving air translating in the same direction.

3.3. Laboratory investigation of a conventional dome

We developed our laboratory air curtain model to use a conventional shaped dome, with an aperture cut into a hemispherical shell. The outlet width was adjustable, and a honeycomb mesh was used to linearise the airflow before exit. An optional air curtain sink was also installed, with the ability to provide a negative pressure on the opposite side of the dome aperture from the source. The use of this sink was investigated.

The angle of arrival of the incident turbulent air was varied, from zero (being parallel to the air curtain), through to 180° (opposing the air curtain).

Fig. 4(b) shows turbulence strength (measured as phase variance) as a function of angle of arrival of the perturbing wind. It can be seen that the air curtain significantly reduces the phase variance compared to the case without an air curtain. It can also be seen that the use of a sink improves performance significantly when the wind direction is opposing the air curtain direction, though for other incidence angles, there is little difference in performance regardless of whether a sink is used or not.

The effectiveness of a sink when used with opposing incident air can be explained by the sink giving additional momentum to the curtain allowing it to propagate the aperture even when significantly perturbed by the wind.

3.4. Baffles for air curtain initialisation

When air curtains are used within the mining industry to direct airflow within mines, baffles are used to initially direct the air. Once the airflow has been set up, these baffles can then be removed, and the airflow pattern is sustained. Without baffles for initialisation, airflow is undirected, and follows all paths throughout the mine. In a similar way, we have investigated the use of baffles across the telescope dome aperture during air curtain initialisation. Here, the dome aperture is initially blocked when the air curtain is switched on. Then, after a few minutes of operation, once a steady state flow has been initiated, the baffle is slid up over the dome, revealing the aperture, whilst maintaining the steady state flow. Our investigations show that a 10% performance improvement is achieved using baffles, reducing the phase variance. We therefore recommend that baffles should be designed into any operational telescope system.

4. Future work

We have clearly demonstrated, both through laboratory testing, and computer modelling that the use of an air curtain across a telescope dome aperture can significantly reduce dome turbulence, improving the quality of astronomical images. We are now seeking to test this concept on 0.5 m class telescopes both on the roof of the Physics department at Durham University, and on a nearby observatory. This will involve the development and deployment of a portable air curtain system, complete with fan, source and sink, and optical components for testing the reduction in atmospheric perturbation, based on a Shack-Hartmann sensor to be installed on the telescopes in question.

The use of an air curtain reduces ground layer seeing, and hence can reduce the demands placed on an adaptive optics system (or, depending on scientific goals, remove the need entirely). Reduced adaptive optics system demands also feeds through into reduced requirements for real-time control systems ([2]), thus reducing instrument costs.

5. Conclusions

Dome seeing is known to have a significant turbulence contribution, something which we have demonstrated using modelling and laboratory testing. In particular, the interaction of a telescope dome with flowing air causes mixing of interior and exterior air, and hence optical aberrations. This mixing of air can be reduced using air curtains across the telescope dome aperture, leading to significantly improved astronomical seeing. The use of an air curtain sink is beneficial, further strengthening the air curtain effect, particularly when incident wind is not aligned to the air curtain.

6. References

[1] Rosensteiner M 2011 *J. Opt. Soc. AmA* **28** 2132
[2] Basden A, Geng D, Myers R and Younger E 2010 *Appl. Optics* **49** 6354–6363

Frozen flow or not? Investigating the predictability of the atmosphere.

Nazim Ali Bharmal
Department of Physics, Science Laboratories, South Road, Durham DH1 3LE, UK
E-mail: n.a.bharmal@durham.ac.uk

Abstract. The spatial description of aberrations produced by atmospheric turbulence for astronomical applications is agreed to be well modeled by the Kolmogorov theory of energy cascades. Instances of these aberrations can be easily generated using closed-form analytic models based on power-spectra following the von Kármán model. However, in astronomical applications it is also the temporal evolution of these aberrations which is of interest. The description of this aspect is less well established, and is generally assumed to follow the Taylor Hypothesis over large time-scales as an *approximation*. This can be described as follows: if the aberrations are considered part of a flow that, for example, moves over a telescope aperture, then from the Lagrangian perspective the aberrations are static and as described by an instance of the spatial description described earlier. However, from the Euclidean perspective, the aberrations seem to just translate. Hence the alternative description of the Taylor Hypothesis as "frozen flow". This work summarizes the relevant findings of the applicability of frozen flow for the astronomical context, and demonstrates the potential benefits as well as shortcomings of using the frozen flow model for temporal evolution of these aberrations.

1. Introduction

The temporal evolution of turbulence is a difficult topic, since the mechanisms are complex and can be a combination of stochastic and predictable phenomena. In the context of aberrations produced by the atmosphere and as viewed by astronomical telescopes, these can be produced by turbulence due to mixing of variable temperature air parcels. The description of these aberrations in space and time is then important for understanding the effects on the quality of light received by the telescope: the total intensity (scintillation) and distribution of light in the PSF (seeing) are two examples of aberration-caused phenomena.

The spatial description of aberrations is still not well understood, but generally considered to be adequately modeled in the free atmosphere by theories based on the Kolmogorov principle of energy cascades that describe energy transfer on various spatial scales. By modifying the models of Tatarski and other workers, the representation of spatial correlations in the aberrations can be described by isotropic power-spectra using a small number of parameters: typically r_0 and L_0 are considered sufficient with a power law form $\simeq \nu^{-11/3}$ between scales of $\sim 1 - 100\,\mathrm{m}$ to $\sim 1\,\mathrm{mm}$. The assumptions generally made are that the three-dimensional mixing can be represented in terms of optical path length change from a mean by a two-dimensional "phase screen". Such a phase screen can exist at any altitude, and there may be several. Collectively, these are called "layers" and each layer has its own independent parameter values.

However, the temporal description of how aberrations evolve is more complex. The description of layers and the consideration of how turbulence evolves, immediately suggests the concept of layers translating horizontally above a fixed point on the surface. This point can be within a telescope aperture, for example. Then the cumulative aberrations received by the telescope will alter based on the translation of the layers and the changes in mixing associated with each layer. These are called flow and boiling, respectively. A lack of wind, for example at the surface, can still have boiling associated with that layer. Correspondingly, for very high wind speeds, the flow[1] that translates dominates: from the Eulerian perspective, the layer with high wind speed associated with it is effectively frozen and simply moves over the aperture, while from the Lagrangian perspective the turbulent mixing is static. This latter description is called the Taylor Hypothesis [1] and allows for the optical path length at any one point, $\mathbf{x}$, in the telescope aperture to be defined per layer j by $\mathrm{OPD}_j(\mathbf{x}, t + \delta t) = \mathrm{OPD}_j(\mathbf{x} + \mathbf{v}_j\delta t, t)$. The AMS glossary[2] states the hypothesis is,

> An assumption that advection contributed by turbulent circulations themselves is small and that therefore the advection of a field of turbulence past a fixed point can be taken to be entirely due to the mean flow.

Since this neglects boiling, and presumes an isotropic and homogeneous turbulence process that leads to equivalent mixing and therefore refractive index change, it is also insufficient. Using published data, here a review is carried out to examines the quality of the frozen flow assumption from the perspective of astronomical adaptive optics.

2. Laboratory-based results on mixing processes

In a series of balloon-borne experiments, Coulman *et al.*[3] investigated the mixing processes that are associated with small-scale temperature fluctuations, as these lead to variations in atmospheric refractive index. The general conclusion was that there was evidence of strong mixing at the edges of well developed turbulent regions. At these edges, which were relatively confined in altitude, the mixing was undeveloped and so aberrations would be imparted on light traveling through them. Within the regions, the mixing was often fully developed and homogeneous, and the potential temperature gradient approximately equal to zero. This can be translated to a stable situation which has little or no refractive index variation. Thus the concept of layers is preserved at the edges of these regions.

A series of laboratory experiments by He *et al.*[4] on mixing shows that a modification of Taylor's hypothesis to include a velocity RMS term can permit frozen flow-like behaviour to be used as an approximation for situations where $v << \delta t$ and boiling is not a dominant term. (It can be large, however, in terms of σ_v.) Then this is evidence, albeit limited, that the general description given in the introduction is not an unreasonable state for the atmosphere: layers of refractive index variation that are associated with sufficient wind velocity for the resulting aberrations to be described by a flow.

3. Telescope-based results

The main astronomical application to date regarding the assumption of the frozen flow hypothesis has been in adaptive optics. Accordingly, here the results from some key papers are summarized and brought together for an understanding of the current state-of-the-art in frozen flow applicability over small space- and time-scales. Key uses for frozen flow include,

- Predictive (forecasting) behaviour, in time and/or space,
- Sub-sampling wavefronts for increased resolution,

[1] The Eulerian perspective is associated with a fixed point relative to the surface i.e. $\mathbf{x}(t) \equiv \mathbf{x}(0)$, while the Lagrangian is associated with a point moving with the flow i.e. $\mathbf{x}(t) = \mathbf{x}(0) + \mathbf{v}.t$.

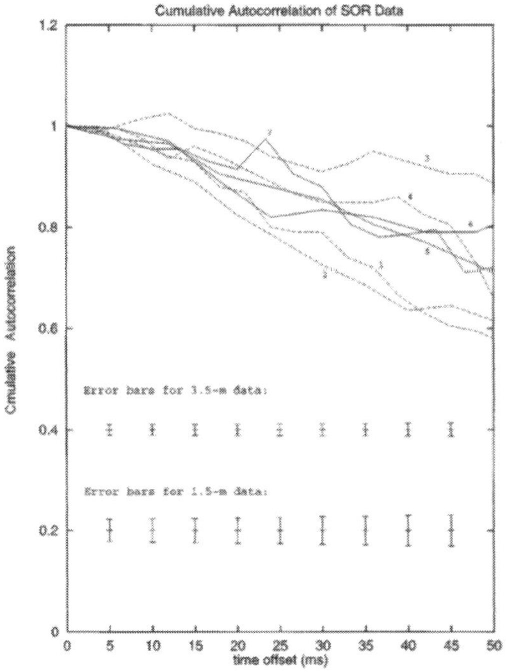

Figure 1. (From Schöck and Spillar[5]:) Demonstrating an approximate linear decay in the cross-correlation with time, whereas pure frozen flow would retain a value equal to one. The horizontal time-scale corresponds to (given typical parameters) a scale of $0.2 - -1$ m.

- Characterization of 'optical turbulence',
- Improved accuracy of AO simulations.

Generally, there are two approaches to using the data from one or more AO wavefront sensors to analyze the existence of frozen flow in layer motion: using cross correlation of slope measurements between different sub-apertures (on one WFS or between multiple, identical WFS), or Fourier analysis of all slope measurements from a WFS. The three publications considered here are Schöck and Spillar[5], Poyneer *et al.*[6], and Guesalaga *et al.*[7]. A précis of the conclusions is shown in table 1.

Table 1. Précis of the conclusions from the three works considered[5][6][7].

Telescope/site	Period	Method	Temporal correlation
Starfire, 1.5 & 3.5 m	7 nights	XC	100 ms, XC→0.5
Keck & Gemini-N	125 nights	Fourier	20–40% decay over 0.5 s
Gemini-S	3 campaigns	XC	100s of ms, XC→0.5

The results from Schöck and Spillar are summarized in figure 1 which demonstrates that unlike with frozen flow, the cross-correlation of slopes displays a linear decay. In figure 2, it is shown that there is a relationship between derived wind speed in a layer and the decay of cross-correlation from the expected frozen flow value with time. The work of Poyneer *et al.* is consistent with these figures, and suggests frozen flow has time-limited application in the context of a 8 m aperture.

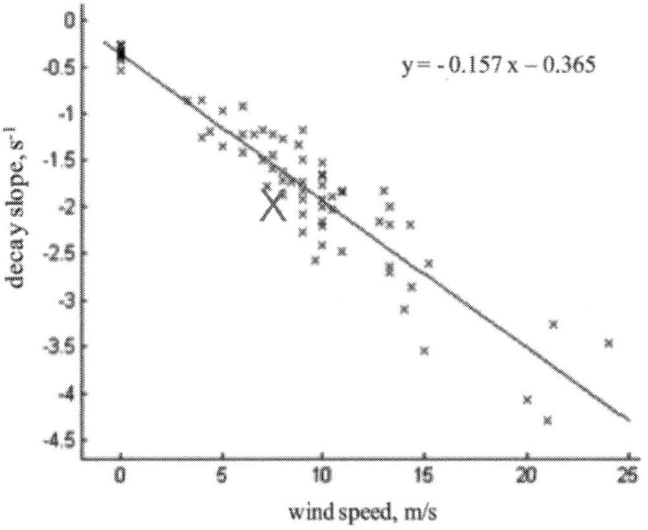

Figure 2. (From Guesalaga *et al.*[7]:) Demonstrating that cross-correlation decay is correlated with derived wind speed. The data from figure 1 was estimated to produce a cross (red in electronic figure), to show that the published work[5] is consistent with the analyses of Guesalaga *et al.*

From these works, Guesalaga *et al.*[7] state that for AO feedback loop latency ($\simeq 5$ ms), the decay of the cross-correlation is $\leq 3\,\%$. Schöck & Spillar[5] find that for cross-correlation to $\geq 90\,\%$, time-scales of 25 ms are typical. Finally, Poyneer *et al.* state frozen flow is detectable for $\geq 94\,\%$ of data analyzed and such flow contributed up to 70% of the 'high-frequency' power.

4. Summary

This short review of frozen flow has briefly investigated why the translation of layer-based phase screens in the atmosphere is a valid proposition for modelling aberrations in the context of the apertures of astronomical telescopes. The uses of frozen flow are varied, and do include the improved deconvolution of the point-spread-function[8] and predictive control[9]. The evidence suggests that frozen flow is partially valid, over small time scales (≤ 50 ms) over diverse nights at diverse observatories and sites. These then lead to the following questions as regards future research on this topic:

- The importance of accurate representation in adaptive optics simulations,
- Are decay rates always linear? Is this a sufficient temporal description at small time-scales?
- The variation over different spatial scales: are low- and high-spatial frequencies similar in behaviour?
- The lack of discovered cross-comparisons: no comparisons known by this author between techniques and/or instruments using the same and diverse datasets.

The Taylor Hypothesis of frozen flow is a convenient description for temporal evolution of phase screens that is valid over small time-scales and is partially valid over significant time-scales (≤ 100 ms). Given its ubiquitous application, establishing the limitations and where to stop using this model is important to understand its limitations.

Acknowledgments
The author wishes to acknowledge the other workers upon whose results this summary has been based.

References
[1] Taylor G I The spectrum of turbulence 1938 *Proc.R.Soc.A* **164** 476
[2] American Meteorological Society Glossary of Meteorology http://glossary.ametsoc.org/ (accessed 1/Nov/2014)
[3] Coulman C E, Vernin J, and Fuchs A. Optical seeing-mechanism of formation of thin turbulent laminae in the atmosphere 1995 *Appl. Opt.* **34** 5461
[4] He X, He G, and Tong P Small-scale turbulent fluctuations beyond Taylors frozen-flow hypothesis 2010 *Phys.Rev.* E **81** 065303(R)
[5] Schöck M and Spillar E J 2000 *J.Opt.Soc.Am.A* **17** 9 1650
[6] Poyneer L, van Dam M, Véran J-P 2009 *J.Opt.Soc.Am.A* **26** 4 833
[7] Guesalaga A, Neichel B, Cortés A, Béchet C, and Guzmán D 2014 *MNRAS* **440** 1925-1933
[8] Jeffries S and Hart M 2011 *Opt.Express* **19** 3 1975–1984
[9] Sivo G, Kulcsár C, Conan J M, Raynaud H F, Gendron É, Basden A, and Longmore A 2014 *Optics Express* **22** 19 23565-23591.

Study on the detecting ability of the adaptive astronomical telescopes

Tan BiTao[1,2,3,4], RM Myers[4] and Chen HongBin[1]

[1]Institute of Optics and Electronics, Chinese Academy of Sciences, Chengdu, China

[2] University of Chinese Academy of Sciences, Beijing, China

[3] Northwest Institute of Nuclear Technology, Xi'an, China

[4]University of Durham, Durham, United Kingdom

Email: tanbitao@aliyun.com

Abstract. Adaptive imaging systems have been developed to compensate for distortion introduced by atmospheric turbulence. The performance of its imaging quality can be evaluated by the Strehl ratio, but this does not directly quantify the detecting ability of an adaptively correcting telescope. Combining the normally detecting SNR and the telescope Strehl ratio, a new method evaluating the detecting ability of an adaptive astronomical telescope is put forward, which can give a quantified value. The new method is simulated on a computer, the simulation results indicated that the detecting ability of an adaptively correcting a 2-m telescopes improves 2 visual magnitudes. The effects of different atmospheric coherence length on detecting ability also can be quantitatively studied through the new method, which can give a scientific basis for the optimization of the design of the system and the development of implementations.

1. Introduction

Nowadays, due to the development of the space technology, there are more and more satellite objects in our space. The observation, the tracking and the recognizing of these objects is an interesting and very important task. As we known, the object itself cannot radiate, and depends on the reflecting of the sun rays illuminating on its surface, so the main observation method is to use the optical method with a telescope. Although the larger-aperture telescopes are efficient to collect the light, they are not able to resolve objects any better than can a telescope of diameter ~r_0, the characteristic size of wave-front phase variations, and the detecting ability is also reduced by the phase variations causing the Strehl ratio to be reduced. However, telescopes with adaptive optical systems, which partially correct for wavefront distortion introduced by the atmosphere, make it possible to operate large ground based telescopes at or near the diffraction limit[1-3]. At present, the study of the detecting ability of such telescopes is only based on the energy passed through the system[4-10], using the detected signal to noise ratio for analysis, but in the practical process of detecting, the detecting ability is also affected by the quality of the space resolution performance, so the detecting signal to noise ratio method can not be used well to analyze telescopes with adaptive optical systems, and it can not give the difference between telescopes with adaptive optical systems and the ones without it. The Strehl ratio is usually used to evaluate the performance of the telescope with adaptive optical systems. This is the ratio of the on-axis irradiance in the focal plane to the on-axis irradiance that would be produced with a diffraction limited telescope. It is evident that the detecting ability is also related to the Strehl ratio: a low Strehl ratio make the details of the image margin blur, which results in detecting difficulty, and reduces the detecting ability lastly.

After analysis of the characteristic of the detecting process of the telescope with adaptive optical systems, a new signal to noise model based on the Strehl ratio is put forward (SNRS), which synthesizes the SNR and Strehl ratio. Based on the SNRS, a new evaluating method for the detecting ability of the telescope is presented. The new evaluating method can quantify the effects of adaptive optical system and different atmospheric coherence lengths on detecting ability. The simulation result indicates that the detecting ability of an adaptively correcting 2-m telescopes improves 2 visual magnitudes, and the detecting ability improves nearly 1 visual magnitude as r0 increased from 5cm to 20cm.

2. Telescopes with adaptive optical systems

2.1. Schematic of the adaptive optical system

An arrangement for an adaptive optical system is shown in Fig.1.

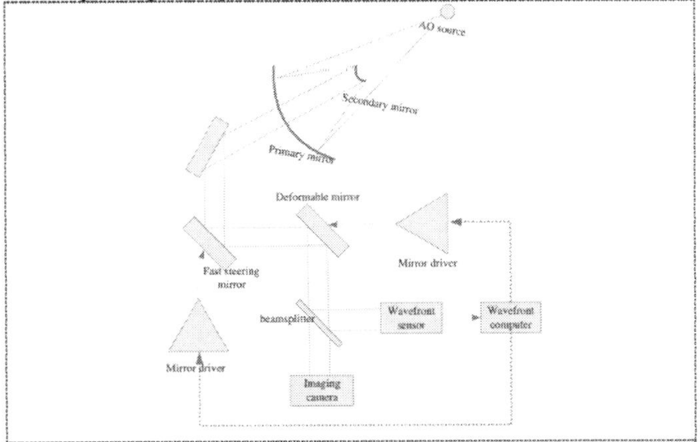

Figure 1. Schematic of an adaptive optical system

The telescope is imaging a distant point like object above the atmosphere. Although the light arriving at the telescope should be nearly a plane wave, it is instead highly distorted by passage through many localized air masses of varying temperatures (and varying indices of refraction). This aberrated wavefront is divided using a beamsplitter and sent to the main imaging camera (or other detection instrument such as an imaging spectrometer) and a wavefront sensor that controls some device to correct the wavefront phase.

2.2. Performance criteria Strehl ratio

The optical transfer function or its Fourier transform, the point-spread function, provides most of the information that is needed to describe the performance of a telescope with adaptive optical system quantitatively. For reasons of simplicity, designers of adaptive optical systems typically try to maximize a single number, the normalized Strehl ratio, which is the ratio of the on-axis irradiance in the focal plane to the on-axis irradiance that would be produced with a diffraction limited system. This is under the assumption that a high Strehl ratio, combined with an unaberrated optical train, will most likely result in a sharp point-spread function.

For telescopes without adaptive optical systems, and only affected by a turbulent atmosphere, the Strehl ratio is[2]

$$Strehl = \frac{4}{\pi D^2}\int_0^D A(\alpha)\exp[-1/2D(\alpha)]d\alpha \tag{1}$$

where $A(\alpha)$ is the convolution of the aperture function, which for a circular aperture is given by

$$A(\alpha) = \frac{2}{\pi}\left\{\cos^{-1}\left(\frac{|\alpha|}{D}\right) - \frac{|\alpha|}{D}\left[1-\left(\frac{|\alpha|}{D}\right)^2\right]^{1/2}\right\} \tag{2}$$

for $|\alpha| < D$ and $A(\alpha)=0$ otherwise. $D(\alpha)$ is the structure function of the atmospherically induced phase fluctuations, defined by

$$D(r_1 - r_2) = \left\langle [\Phi(r_1) - \Phi(r_2)]^2 \right\rangle \tag{3}$$

where $\Phi(r)$ is the accumulated random phase error, which is due to the turbulent atmosphere on the aperture plane of the telescope at coordinate **r**. The phase structure function that is due to atmospheric turbulence has been well studied and has been determined to be

$$D(\alpha) = 6.88(|\alpha|/r_0)^{5/3} \tag{4}$$

where r_0 is Fried's coherence diameter:

$$r_0 = \left[0.422(2\pi/\lambda)^2 \sec(\psi)\int_0^\infty C_N^2(h)dh\right]^{-3/5} \tag{5}$$

where λ is the wavelength of the light, ψ is the angle from zenith, and $C_N^2(h)$ is the height dependent refractive-index structure parameter.

For a partially corrected telescope, the distorted PSF consists of a sharp central peak surrounded by a halo. The width of this central peak is approximately the same as for the unaberrated system, but its amplitude is decreased. For small wavefront distortions the Strehl ratio can be approximated as[11]

$$Strehl = \exp(-\sigma_{wf}^2) \tag{6}$$

where σ^2_{wf} is the total phase variance. which is

$$\sigma_{wf}^2 = \sigma_{servo}^2 + \sigma_{dm}^2 + \sigma_{meas}^2 + \sigma_{sep}^2 + \sigma_{cone}^2 + \sigma_{es}^2 \tag{7}$$

where σ^2_{servo} is the RMS residual phase error in the closed loop, σ^2_{dm} is the residual wavefront error due to deformable mirror, σ^2_{meas} is the total wavefront measurement error of the sensor, σ^2_{sep} is the RMS error due to angular separation between the beacon source and the astronomical light of interest, σ^2_{es} is the variance due to the finite size of the guide star.

3. Theory of SNRS

3.1. Detecting SNR

The detecting SNR of the telescope is a RMS values, and can be calculated as follow[12]:

$$SNR = St[(S + Bn_{pix} + I_d n_{pix})t + I_r^2 n_{pix}]^{-\frac{1}{2}} \tag{8}$$

where S is the signal photons produced in a unit time, B is the background photons produced in a unit time, t is the integration time, I_d is the photons caused by the dark current, Ir is the photons caused by the readout noise, n_{pix} is the numbers of pixels of the target image.

Here we discuss the detection in the daytime, so it is a detection limiting by the background noise, so

$$SNR = St^{\frac{1}{2}}(Bn_{pix})^{-\frac{1}{2}} \tag{9}$$

The signal photons produced in a unit time can be calculated as follow:

$$S = \frac{\pi}{4}D^2 q_s \lambda_a h^{-1} c^{-1} \tau_o \tau_a (1-\varepsilon^2) 3.9\times 10^{-9} 2.512^{-m} = S_0 \cdot 2.512^{-m} \tag{10}$$

where D is the diameter of the system, q_s is the quanta efficiency, τ_0 is the optical permeation efficiency of the system, τ_a is the atmosphere permeation efficiency, λ_a is the average wavelength, ch is Planck's constant, c is the velocity of light, ε is the obstruct ratio of the system.

The background photons produced in a unit time can be calculated as follow:

$$B = \frac{\pi^2}{16} d^2 q_b \lambda_a h^{-1} c^{-1} B_b (1-\varepsilon^2) D^2 f^{-2} \tau_o = B_0 \cdot \frac{\pi}{4} d^2 f^{-2} \tag{11}$$

where D is the diameter of the system, d is the size of the pixel, f is the focus of the system, q_b is the quanta efficiency, τ_0 is the optical permeation cefficiency of the system, λ_a is the average wavelength, h is Planck's constant, c is the velocity of light, ε is the obstruct ratio of the system.

3.2. Model of SNRS

In the unaberrated case, most of the image flux is focused into the central peak. For the partially corrected system, some of the image flux is scattered from the central peak to the surrounding halo. We know that the energy cannot be lost, and the imaging area becomes bigger than the unaberrated case. Then the total signal in the main lobe of the image is the same as the unaberrated case, which can be defined as

$$S_c = S = S_0 \cdot 2.512^{-m} \tag{12}$$

And the total background photons produced in a unit time can be defined as

$$B_c = B_0 \cdot \omega_c \tag{13}$$

where ω_c is the apparent solid angle subtended by the imaged object as observed by the corrected telescope, which can be regarded as the solid angle related to the imaging area. For objects of small spatial extent, ω_c is approximately given by the corrected telescope resolution:

$$\omega_c = \left(2 \times 1.22 \frac{\lambda}{D}\right)^2 Strehl^{-\frac{1}{2}} \tag{14}$$

Finally, the model SNRS can be defined as

$$SNRS = S_c B_c^{-\frac{1}{2}} = Strehl^{1/2} \cdot D \cdot S_0 \cdot 2.512^{-m} (1.22\lambda)^{-1} B_c^{-\frac{1}{2}} \tag{15}$$

The limit on how faint an object can be detected is found by setting the expression for the SNRS Eq.(15) equal to a certain value, According to empirical evidence, it is usually defined as 6 when the telescope can detect the object stably. Finally, the limited detecting ability of the telescope can be used to express the result in terms of the visual magnitude as follows.

$$m = -\log_{10}(3.66\lambda B_0^{1/2} Strehl^{-\frac{1}{2}} \cdot D^{-1} \cdot S^{-1}{}_0)[\log_{10}(2.512)]^{-1} \tag{16}$$

4. Simulation

The new method of evaluating on the detecting ability based on the SNRS is simulated through computer modelling. Before the simulation, the background brightness is 14 $W/m^2 \cdot Sr$, the system permeation ratio is 85%, the atmosphere permeation ratio is 75%, the quantum efficiency is 85%, the integration time is 0.02s, the diameter of the system is 0.6m, the focus is 3m, and the pixel size is 16um.

In figure 2, the detecting ability of different diameters at some given atmospheric coherence length is illustrated. The given atmospheric coherence length is 5cm, 10cm, 20cm and no atmospheric effect. From Fig.2, we can find that the detecting ability is affected seriously by the atmospheric conditions, it improves as the increase of the atmospheric coherence length at the same diameter, and it also improves as increasing diameter at the same atmospheric coherence length. The detecting ability improves nearly 1 visual magnitude as the r_0 improved from 5cm to 20cm. The atmosphere affects the telescope seriously, the difference in the detecting ability between atmospheric effects and no atmosphere is nearly 3 visual magnitudes, and as the coherence length increased to a certain value, the detecting ability improves a little.

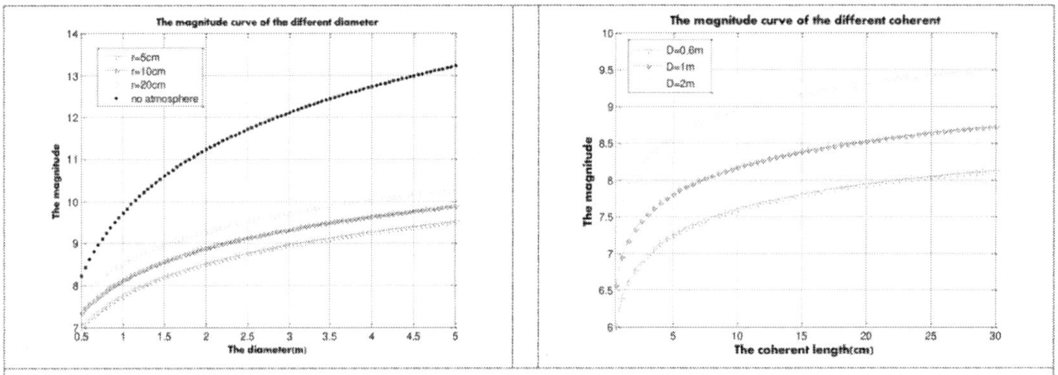

Figure 2. The magnitude curve of different diameters and coherence lengths.

In figure 3, the detecting ability of different background brightness at some given diameter is illustrated: the given diameter is 0.6m, 1m and 2m. From Fig.3, we can find that the detecting ability is reducing by the increasing background brightness.

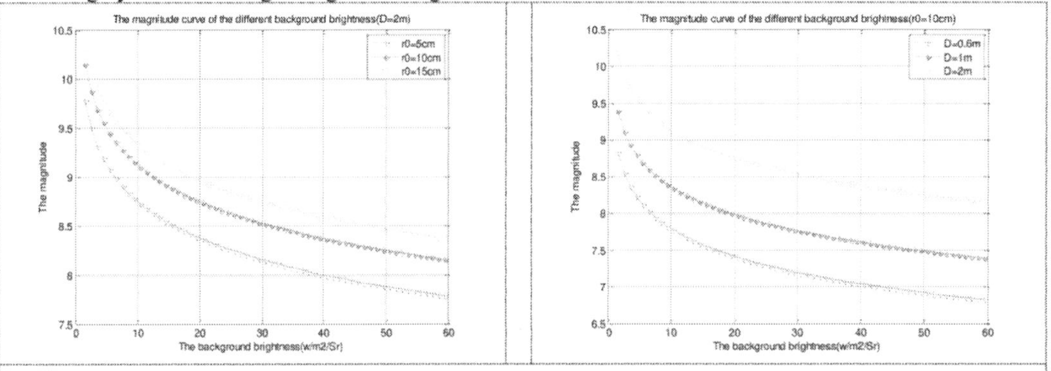

Figure 3. The magnitude curve of different background brightness

In figure 4, the detecting ability of the adaptive astronomical telescopes of a given phase error is illustrated, and this is compared to the diffraction limited telescope and the uncorrected telescope. From Fig.4, it can be seen that the detecting ability of the adaptive astronomical telescope is improved significantly compared to the uncorrected telescope, such that it improves about 2 visual magnitudes at aperture diameter 2m, and that 3 visual magnitudes at aperture diameter 4m, and the detecting ability of an adaptive astronomical telescope is near that of a the diffraction limited telescope when the phase error is below *λ/10.*

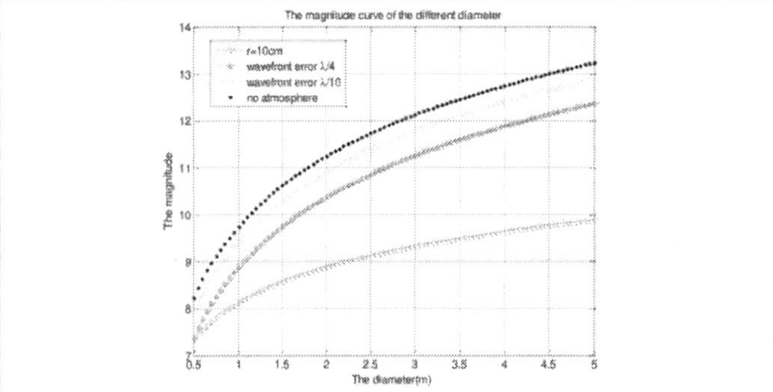

Figure 4. The magnitude curve for different conditions.

5. Conclusion

A new evaluating method for the detecting ability of the telescope, the SNRS, is defined in this paper which can give a quantified result for an adaptive astronomical telescope. The new method is used to simulate the detecting ability of the diffraction limited telescope, the atmosphereic telescope and the adaptive astronomical telescope. From the simulation results, we can find that the detecting ability of the system is affected seriously by the atmosphere and residual phase error. The detecting ability improves slowly along with the increasing of the diameter under bad atmosphere condtions.

The detecting ability of astronomical telescopes improves about 2 visual magnitudes at aperture diameter 2m, and 3 visual magnitude at aperture diameter 4m, and it will be near the diffraction limited telescope when the phase error is below $\lambda/10$. The effects on detecting ability can quantitatively studied through the new method, which can give a scientific basis for the optimization of the design of the system and the development of implementations.

References

[1] R.K. Tyson M 2010 *Principles of adaptive optics*(Florida: CRC press)
[2] Gavel D T, Morris J R and Vernon R G 1994 *J. Opt. Soc. Am. A* **11** 914
[3] Michael Hart 2010 *Appl.Opt* **49** 11
[4] Cognition R L 2013 *Proc. Int. Conf. on Sensors and systems for space applications (Baltimore)* vol 4: (Washington: SPIE) p 8739
[5] Jim K T C, Wolfshagen R, Gibson B, Pier E, Hodapp K and Onaka P 2011 *Proc. Int. Conf. on Advanced Maui Optical and Space Surveillance Technologies (Hawaii)* vol 1: (Maui: Maui Economic Development Board) p 183
[6] Seitzer, P 2011 *Proc. Int. Conf. on Advanced Maui Optical and Space Surveillance Technologies (Hawaii)* vol1: (Maui: Maui Economic Development Board) p 200
[7] Rork E W, Lin S S and Yakutjs A J 1982 *Air Force Research Laboratory Report* (USA: MASSACHUSETTS) p 117413
[8] Bitao T, Chun-yuan J, Bao-guo W and Xin Z 2009 *High Power Laser and Particle Beams* **21** 187
[9] Bitao T, Hongbin C, Qunshu W. 2013 *Acta Photonica Sinica* **42** 1340
[10] Bitao T, Hongbin C, Qunshu W. 2014 *High Power Laser and Particle Beams* **26** 0110131
[11] Lelièvre M, Lévesque M P and Buteau S 2007 *Defence R&D Report*(Canada: Valcartier) p 34
[12] Gerald C.Holst M 2010 *Electro-optical Imaging System Performance*(Washington:SPIE press)

Astronomy Developments and Site Testing in East Africa

D A H Buckley[1]
SALT Science Director, South African Astronomical Observatory,
PO Box 9, Observatory 7925, Cape Town, South Africa.

E-mail: dibnob@salt.ac.za

Abstract. Two astronomical site testing initiatives are beginning in both Kenya and Ethiopia, with the aim of selecting suitable locations in those countries for initially modest sized (1-2m) optical telescopes. The first project, in Kenya, has to date involved a study of existing meteorological satellite data, from which potential sites have been selected for further studies (see paper by Graham et al. in these proceedings). A similar program of astronomical site testing, including using DIMMs, will begin later in 2014 in the Lalibela region of northern Ethiopia, at three potential sites ranging in altitude from ~3600 to 4500 m. In parallel to this, the Entoto Observatory and Research Centre (EO), comprising twin 1-m alt-az telescopes, has been established at Mt. Entoto, on the outskirts of Addis Ababa, and science operations are due to begin in early 2015.

1. Introduction

South Africa has been supporting astronomy developments in Africa for a number of years through a variety of initiatives. Students from other African countries have been trained through the South African National Astrophysics and Space Science Program (NASSP), a 4th year Honours undergraduate program designed to prepare students for postgraduate studies. Although priority has been given to students from countries which will host the final extended of the Square Kilometre Array (SKA Phase 3) and/or the proposed African VLBI Network (AVN), students from other African countries (e.g. Ethiopia) continue to participate in the NASSP.

A number of astronomy capacity building workshops in Africa have also been supported by South African (and other) astronomers, some with IAU sponsorship. Over the past ~5 years, workshops have been held in Kenya (2010), Ethiopia (2011), Uganda (2012) and Rwanda (2014). In addition, the East African Astronomical Society (EAAS) has been established, where there has been significant developments in establishing university courses, and most recently the Mt. Entoto Observatory and Research Centre, which will support teaching and research initiatives in the region.

2. Entoto Observatory (EO) and Research Centre

This centre was established by the Ethiopian Space Science Society (ESSS) in 2012 and is supported by both public and private universities in Ethiopia. Currently EO is financed by the member universities, ESSS and the Ethiopian government.

[1] To whom any correspondence should be addressed.

Situated on the ~3100m Mt. Entoto, on the outskirts of Addis Ababa, the observatory was established as a multi-purpose facility to support teaching, research training, science research in astronomy and space science, and for public outreach. It consists of a facility including a visitor's centre with accommodation for students, lecturers and researchers, lecture rooms, IT and technical services and two recently installed twin 1.0m telescopes.

The ESSS vision for EO is to make Ethiopia an effective and extensive user of space science and technology, especially satellite science and technology applications, supporting all aspects of development within the country and to become a global contributor to the development of astronomy, space science and related sciences within ten years. This is to be achieved through developing the necessary capacity and research training and to build regional, continental and international networks to foster research and development in space science, astronomy and related technologies. In addition the promotion of space science, astronomy and related technologies will be achieved through establishing outreach programs.

The current status of EO is the all building construction has been completed, including the two domes for the 1.0-m telescopes, which were installed by early-2014 and first light subsequently obtained on both. Fully internet connectivity has been established to EO. In addition, the science instruments have been delivered, including two sets of site testing equipment. Final commissioning is in progress and is expected to be completed by early-2015.

Figure 1. The two domes for the twin 1-m telescopes at Entoto Observatory following completion of the construction in 2013. The trees have subsequently been trimmed!

3. The EO Telescopes and Instruments

The two EO telescopes are 1-m alt-az Ritchey-Chretian telescope, supplied by the German company, *ASTELCO*. The telescopes are compact, fast slewing f/8 systems, with two Nasmyth focii and two-element correctors giving a 45 arcmin field and a platescale of 25.8 arcsec/mm. Image de-rotators are installed on one foci, to support imaging CCD cameras, while the second focus can support high speed cameras, a fibre feed or visual eyepieces (for public viewing). The telescopes feature active control of the thin primary mirrors using a Shack-Hartman wavefront camera, to provide optimal alignment of the mirrors.

The science instruments for the EO telescopes consist of the following:

- A Finger Lakes FLI-PL4301E-1 camera (2048 × 2048 × 24μm) with a 21 arcmin square field and 0.62 arcsec/pixel.
- A Finger Lakes FLI-PL230-42 camera (2048 × 2048 × 15μm) with a 13 arcmin square field and 0.39 arcsec/pixel, although this camera serves as the detector for the fibre-fed FLECHAS echelle spectrograph
- An SBIG STXL-11002 active tip/tilt correcting camera (4008 × 2762 × 9μm) with a 15 × 10 arcmin field and 0.23 arcsec/pixel.
- An SBIG STL-6303E camera (3060 × 2040 × 9μm) with a 12 × 8 arcmin field and 0.23 arcsec/pixel.
- An Andor iXon EM+ 888 high speed camera with an Electron Multiplication CCD (E2V CCD97, 1024 × 1024 × 13μm) with a 5.7 arcmin field and 0.34 arcsec/pixel.
- A fibre-fed R=40,000 échelle spectrograph, FLECHAS, covering the range 390 – 670nm in one exposure on the FLI-PL230-42 camera. A 50μm optical fibre with image slicer is utilized to provide two slices.

These instruments are designed to support both training of student in observational astronomy, but will also provide significant research capability. Although the EO is close to a city, with consequent issues of light pollution, the skies are not so bright as for a similar city in other parts of the world. No seeing measurements have yet been made for the site, although anecdotal evidence indicates only modest seeing (1-2 arcsec). Despite these constraints, it is anticipated that EO's facilities will allow significant science research to be undertaken.

Figure 2. The author with the first of the EO 1-m *ASTELCO* telescopes, during factory acceptance testing in Feb 2013.

4. Potential Science at Entoto Observatory
In 2011 the author was asked by ESSS to develop a draft list of science drivers to support the establishment of the EO.

While the trend in astronomy is to build *bigger* telescope to study fainter objects, there is still compelling forefront research that can – and indeed is being – conducted on modest sized (1-m class) telescopes. Current examples of science drivers associated with such telescopes include:

- Rapid follow-up observations of Novae, Supernovae and Gamma Ray Bursters (GRB)
- Observations of extra-solar planet transits
- Detection of extra-solar planets by precision radial velocimetry
- Detection of extra-solar planets by gravitational microlensing observations
- Target of Opportunity multi-wavelength observations (e.g. X-ray binaries)
- Synoptic monitoring of eruptive variable stars (e.g. recurrent novae, novae, symbiotic stars)
- Time series photometry of pulsating variable stars (e.g. Cepheids, white dwarfs, roAp stars)
- High time speed photometry of binary stars (e.g. cataclysmic variables, LMXBs, contact binaries)
- Search for Kuiper belt objects in the Solar System
- Observations of Near Earth Objects
- Active Galactic Nuclei variability studies

The list is not meant to be exhaustive, but rather illustrative of the various current topics for which astronomers with frequent access to a 1-m class telescope could make a substantial research contribution. They are focussed on stellar astronomy, since the 1-m aperture is not really sufficient to do competitive extra-galactic research, although there are some areas that could be considered (e.g. AGN).

Worldwide, there are plenty of examples of observatories with modest sized telescopes whose research output is significant and important. Indeed new observatories with 1-m class telescopes continue to be built. The tendency for modern telescopes design is to automate them as much as possible, something which is readily achievable with modern computing technology. This can lead to a better operational efficiency and has led to "roboticizing" telescopes: turning them into remote – or robotically – operated telescopes. A network of such telescopes can be used in "campaign mode" where several are used to undertake observations of the same objects, but at different times. This can maximize scientific productivity, sometimes giving results which are more than simply the "sum of the whole". Examples include multi-longitude observations of variable stars, where the daily aliasing problems of single sites can be reduced. The Mt Entoto Observatory could be an important node in such multi-longitude observing campaigns.

5. The Future

With the establishment of the Entoto Observatory and Research Centre in Ethiopia, currently consisting of twin, well-instrumented, 1m telescopes, observational optical astronomy has been enabled. This will clearly benefit Ethiopian astronomers and those in the East African region. Continuing support will be needed to assist in the development of research programs and training of student, and this is already happening though the appointment of astronomers from outside the country with the required experience and background. It is hoped that visiting astronomers (e.g. for ~ a month or longer timescales) will also play a role in assisting in the development of competitive research programs and new collaborations.

Finally, a site testing initiative to select a potential new observatory site in the north of Ethiopia, in the Lalibela region, is soon to begin. Two sites (at 3600 and 4500m) have been set up with the requisite equipment (DIMMs, automatic weather stations, all sky cameras, sky brightness monitors) and infrastructure support (roads, housing) has been established. Over the period of several years, this site testing will be undertaken by students from EO member universities. Eventually a new telescope is planned to be established at this site.

Journal of Physics: Conference Series **595** (2015) 012006 doi:10.1088/1742-6596/595/1/012006

Error sources in SLODAR turbulence profile fitting

Tim Butterley, James Osborn and Richard Wilson

Centre for Advanced Instrumentation, Department of Physics, Durham University, South Road, Durham, DH1 3LE

E-mail: timothy.butterley@durham.ac.uk

Abstract. This paper discusses some of the main error sources that affect use of the Slope Detection and Ranging (SLODAR) technique to recover atmospheric turbulence profiles from Shack-Hartmann wavefront sensor data. The most significant factor we address here is that of temporal sampling – we examine the effect of poorly-averaged slope correlations due to low wind speeds (or data sets that are insufficiently long) on spatial covariances of wavefront slopes. We discuss how this can affect the observed outer scale of the turbulence. A dual SCIDAR-SLODAR instrument is described that is ideally suited to exploring the impact of these effects and, potentially, correcting them. Some preliminary results from the instrument are presented.

1. Introduction

Slope Detection and Ranging (SLODAR) [1] is an optical turbulence profiling technique that can be implemented using a dedicated instrument or using wavefront slope measurements from a facility adaptive optics system. However, the technique is easier to apply on a small telescope (e.g. 0.5 m [2]) than on a large telescope (e.g. 8 m [3]). This paper describes why, and presents some approaches to ameliorating some of the problems on larger telescopes.

Section 2 gives a brief overview of the main similarities and differences between the Scintillation Detection and Ranging [4] (SCIDAR) and SLODAR techniques. Section 3 explains the factors that contribute to the response of a SLODAR instrument to a layer of turbulence. Section 4 describes a combined instrument that provides simultaneous Scintillation Detection and Ranging (SCIDAR) and SLODAR profiling and section 5 describes an approach to correcting temporal errors in SLODAR profiles using velocity profiles measured using SCIDAR. Section 6 contains the conclusions.

2. Comparison between the SCIDAR and SLODAR techniques

SCIDAR and SLODAR are both 2-star triangulation turbulence profiling techniques. SLODAR uses a Shack-Hartmann wavefront sensor (WFS) to measure local wavefront slopes across the telescope pupil; SCIDAR measures the flux. In both cases the strength and height of turbulence is recovered from peaks in a spatial cross-covariance map. In general, SCIDAR has more resolution elements than SLODAR so yields higher resolution turbulence profiles.

In both techniques, the turbulence profile reconstruction process involves fitting response functions to the cross-covariance peaks. SCIDAR response functions are narrow and localised. SLODAR response functions have wings that extend across much of the covariance map. Examples of both are shown in figure 1.

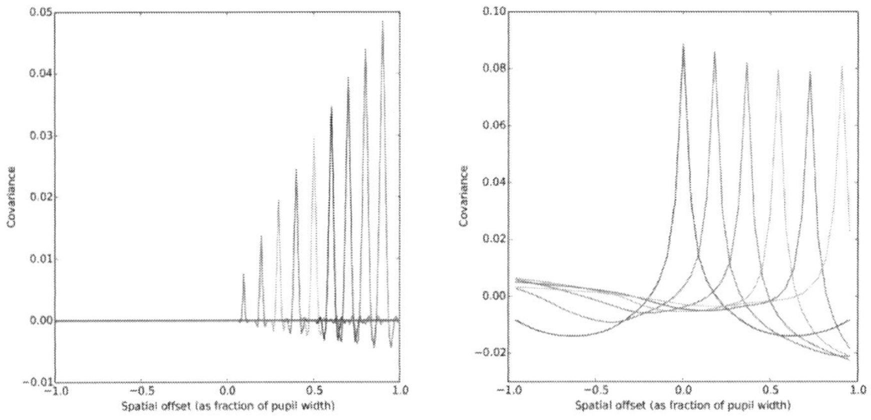

Figure 1. Response functions for (left) SCIDAR and (right) SLODAR. The SLODAR response functions are for wavefront slopes in the longitudinal direction i.e. parallel to the separation direction of the two stars. SLODAR response functions for the transverse direction (not shown) have broader peaks and more extended wings than longitudinal functions.

SLODAR can provide an estimate of the total seeing via a model-fit to the autocovariance (i.e. data from a single star). This is similar to the differential image motion monitor (DIMM) [5], and is independent of the turbulence profile fit. A total seeing estimate for SCIDAR has to come from summing all layers in the turbulence profile, so errors in the profile fit can propagate into the seeing estimate.

3. SLODAR response functions

A SLODAR response function is a model of the cross-covariance measured in response to a single layer of turbulence at a given altitude. Response functions can be generated theoretically [6] or using a Monte Carlo simulation. A response function depends on the following parameters:

(i) Subaperture size.
(ii) Δ, the spatial offset of the peak in the covariance map.
(iii) The height of the turbulence (if scintillation can not be neglected) – this is related to Δ, but also depends on separation of the target stars.
(iv) The shape of the pupil function (assuming we are subtracting tip/tilt to remove telescope shake and tracking errors).
(v) The temporal baseline of the measurement (i.e. the averaging time for the cross-covariance).
(vi) The outer scale of the turbulence.
(vii) The translational velocity of the turbulent layer.

The first five parameters relate to the design of the instrument (or the observing method) and they are generally known. The last two are both properties of the turbulence being measured, are generally unknown and can vary with altitude. Our main concern here is these two parameters and how they interact with parameter (v), the temporal baseline.

In the regime where the covariance is well temporally-averaged (i.e. high wind speed and long temporal baseline) the shape of the response functions is dominated by the outer scale. A larger outer scale leads to broader correlation peaks, as can be seen in Figure 2.

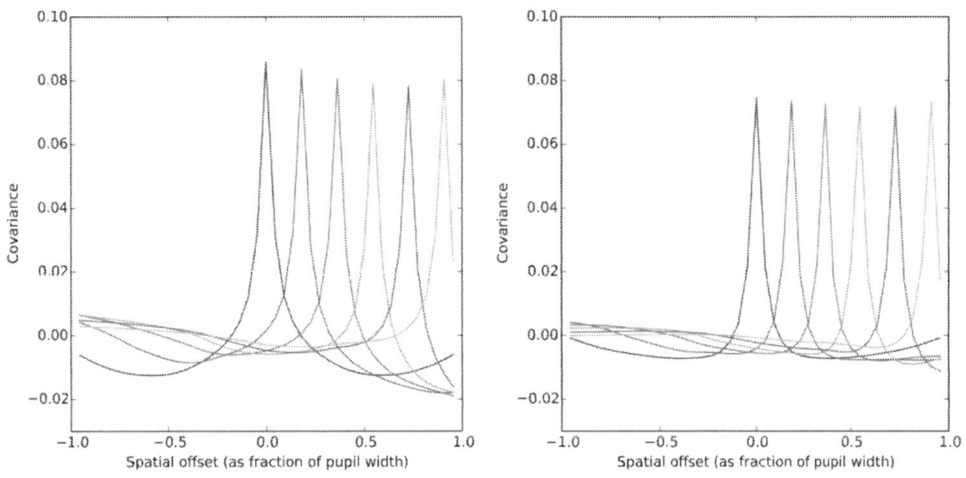

Figure 2. Theoretical SLODAR response functions (longitudinal direction only) for Von Karman turbulence seen by a 2.54 m aperture. Left: $L_0 = 20$ m, right: $L_0 = 5$ m.

The assumption that the covariance response to a layer is well-averaged is not always correct. Covariance measurements of very slowly moving turbulent layers can take a long time to converge, and this effect is exacerbated on larger telescopes (where the pupil crossing time is longer). Larger scale aberrations converge more slowly than smaller scale aberrations so a poorly-converged covariance measurement tends to have a smaller apparent outer scale. This effect can be demonstrated by making outer scale measurements from the same wavefront sensor data using different temporal baselines. An example is shown in figure 3, where the outer scale has been fitted to the same autocovariance data multiple times but with the datasets truncated to different durations.

A simplistic view is that the temporal baseline can be increased until the apparent outer scale is no longer observed to change but this may be unfeasible in practise, especially when the wind speed is particularly low. Firstly, it may be impractical to take sufficiently long datasets e.g. due to other constraints in how the instrument is operated. Secondly, the turbulence profile may vary over shorter timescales than the time required to make a well-averaged covariance measurement.

The example presented here is for the "overall" outer scale i.e. averaged over the whole atmosphere. The outer scale and the turbulence velocity are not necessarily the same at every altitude. The vertical profiles of the outer scale and velocity must be taken into account when generating response functions if they are to fit the measured cross-covariance correctly. Failure to include this effect will lead to errors in the strength and/or height of turbulent layers in the fitted profile.

4. The dual SCIDAR-SLODAR instrument

One approach to investigating temporal effects in SLODAR is to compare with simultaneous SCIDAR data; SCIDAR is not very sensitive to the outer scale and does not require such long datasets to provide well-averaged cross-covariance measurements. The ideal experimental arrangement is to make measurements using both techniques along the same line of sight – for this purpose a dual SCIDAR-SLODAR instrument was built.

The dual SCIDAR-SLODAR instrument was installed by Durham University as a guest

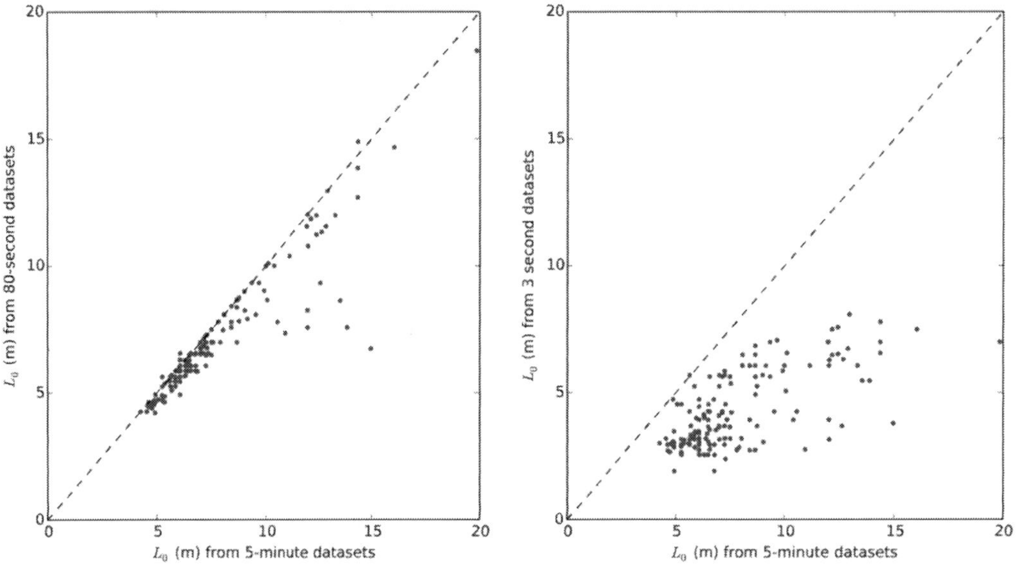

Figure 3. Demonstration of the effect of insufficient temporal averaging on the measured outer scale. Plots show the outer scale measured from a 5-minute data set against the outer scale from the same data cut down to (left) 80 seconds and (right) 3 seconds.

instrument on the 2.54 m Isaac Newton Telescope (INT) on La Palma in July 2014. It is a modification of an existing generalised stereo SCIDAR [7] that has been used to support the CANARY AO demostrator [8]. A dichroic with a cutoff wavelength of 615 nm is used to divert red light into a SLODAR channel. The blue light continues to the SCIDAR optics. The SLODAR channel consists of a single Shack-Hartmann wavefront sensor (WFS) with the two stars interleaved on one Andor "Zyla" SCMOS camera. The WFS has 22×22 subapertures and is suitable for observing pairs of stars with separations of approximately 13 arcseconds or 30 arcseconds.

In addition to measuring the turbulence profile, the SCIDAR instrument measures the wind velocity profile i.e. wind velocity as a function of height. Velocities can only be measured at heights where turbulence is present, but an interpolation routine can be used to estimate the profile at other heights. An example interpolated wind speed profile is shown in figure 4.

5. Wind correction of SLODAR response functions

The wind velocity profile measured from SCIDAR can be used to improve the SLODAR response functions to account for the combination of translation velocity and finite temporal baseline. The response function at each height can be generated (via Monte Carlo simulation) with the correct wind velocity. Two examples of such response functions (for different temporal baselines) are shown in figure 5. For the purpose of this demonstration all layers are assumed to be moving in the same direction but, in order to achieve the best fit possible, the wind direction information for each layer from SCIDAR should also be incorporated. An outer scale of 30 m is assumed here.

Fitting response functions that have been corrected for the wind profile in this manner implies a relatively short temporal baseline could be chosen deliberately in order to allow a turbulence

Journal of Physics: Conference Series **595** (2015) 012006 doi:10.1088/1742-6596/595/1/012006

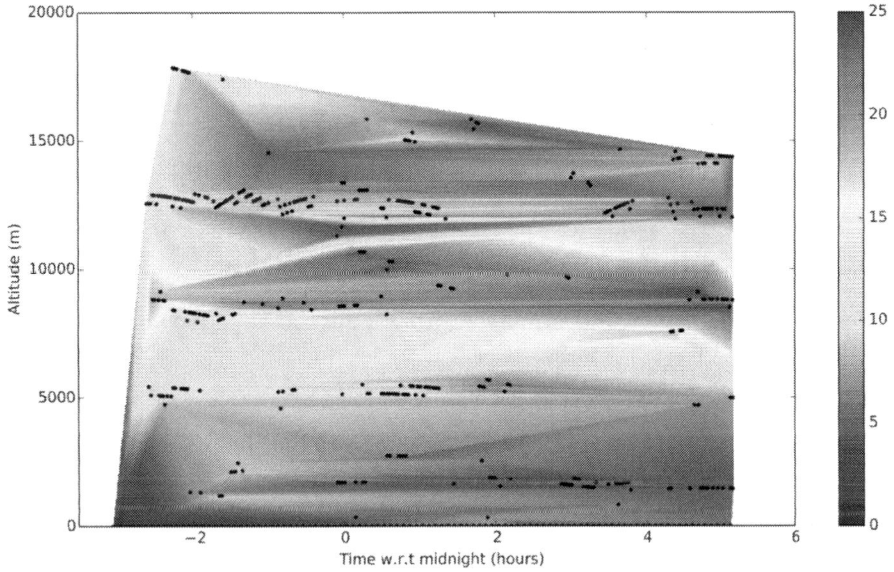

Figure 4. Interpolated wind speed profile (with colour scale in m/s) for the night starting 16 July 2014.

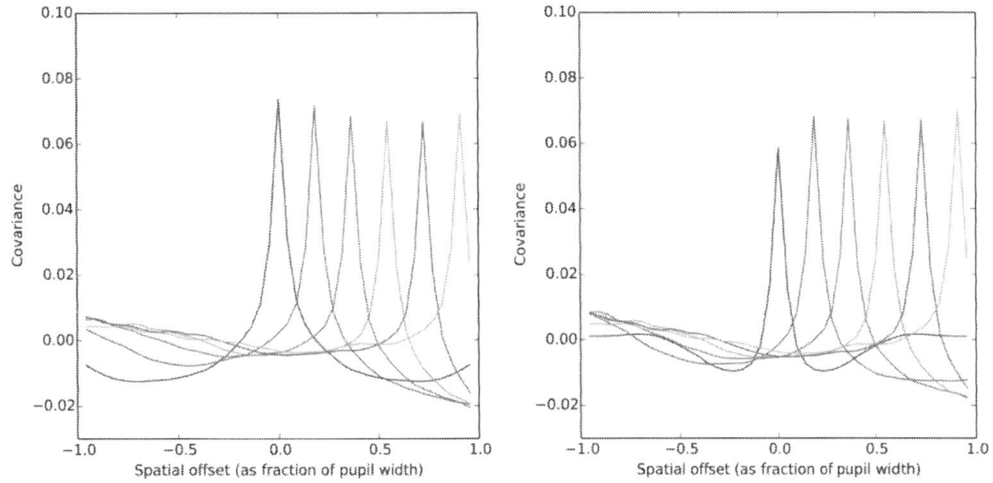

Figure 5. Simulated SLODAR response functions (longitudinal direction only) including correction for the wind speed profile. Left: temporal baseline 10 seconds; right: temporal baseline 1 second. All layers are assumed to be moving in the same direction.

profile to be measured in a shorter amount of time. This correction also suggests more confidence could be placed in outer scale estimates that are obtained via a slope covariance fit.

6. Conclusion

SLODAR turbulence profiles are fitted using response functions that depend on numerous parameters including the outer scale, wind speed and temporal baseline for measurements. Outer scale and temporal effects can vary with altitude and can be difficult to distinguish from one another. This problem tends to be more pronounced on larger telescopes. One approach to solving this problem is to use wind velocity profiles (from SCIDAR or other sources) to compensate for these effects but it remains to be seen whether this allows the outer scale profile to be measured. Further data from the dual SCIDAR-SLODAR instrument will be used to continue this investigation.

Acknowledgments

The Isaac Newton Telescope is operated on the island of La Palma by the Isaac Newton Group in the Spanish Observatorio del Roque de los Muchachos of the Instituto de Astrofisica de Canarias. FP7/2013–2016: The research leading to these results has received funding from the European Communitys Seventh Framework Programme (FP7/2013–2016) under grant agreement number 312430 (OPTICON).

References

[1] Wilson R W 2002 *Mon. Not. R. Astron. Soc.* **337** 103
[2] Wilson R W, Butterley T, Sarazin M 2009 *Mon. Not. R. Astron. Soc.* **399** 2129
[3] Cortés A, Neichel B, Guesalaga A, Osborn J, Rigaut F, Guzman D 2012 *Mon. Not. R. Astron. Soc.* **427** 2089
[4] Vernin J and Roddier F 1973 *J. Opt. Soc. Am.* **63** 270
[5] Tokovinin A 2002 *PASP* **114** 1156
[6] Butterley T, Wilson R W and Sarazin M 2006 *Mon. Not. R. Astron. Soc.* **369** 835
[7] Shepherd H, Osborn J, Wilson R W, Butterley T, Avila R., Dhillon V and Morris T J 2014 *Mon. Not. R. Astron. Soc.* **437** 3568
[8] Gendron E, Vidal F, Brangier M, Morris T, Hubert Z, Basden A, Rousset G, Myers R, Chemla F, Longmore A, Butterley T, Dipper N, Dunlop C, Geng D, Gratadour D, Henry D, Laporte P, Looker N, Perret D, Sevin A, Talbot G and Younger E 2011 *A&A* **529** L2

Natrium-Teide experiment. Sodium layer observations at Teide Observatory (Canary Is.)

J.A. Castro-Almazán,[1,2], J.J. Fuensalida,[1,2], A. Alonso[1,2], and I. Montilla[1,2]

[1]Instituto de Astrofísica de Canarias, E-38200, La Laguna, Spain
[2]Dept. Astrofísica, Universidad de La Laguna, E-38200, La Laguna, Spain

E-mail: jcastro@iac.es (JACA); jjf@iac.es (JJF); aas@iac.es (AA); imontilla@iac.es (IM)

Abstract. With the aim of characterizing the atmospheric Na layer, the Teide Observatory facilities have been used in the last years to carry out a long term monitoring experiment on LGS generation by low power laser excitation. The LGSs are generated by a Rodamine 6G Dye tunable laser launched from the Optical Ground Station with a typical output power of 400 – 500 mW. The LGS images are obtained in a bistatic configuration with a baseline of 165 m. The experiment has been improved in several aspects in order to boost the Na photon flux return. The `natrium`-Teide experiment has generated a dataset of more than 400 images of Na LGS traces obtained in 37 nights. The main results are going to be been published soon.

1. Introduction

Laser Guide Stars -LGS- has been widely proposed to extend the domain of stable operation for Adaptive Optics (AO) systems in large telescopes. The overall Strehl ratio achieved by an AO system is depending on all errors involved in the wavefront sensor function. LGSs contribution to this error directly depends upon two main aspects, the LGS spot profile and the Signal to Noise Ratio (SNR). Therefore, an accurate characterization of the Na layer is imposed. Three main aspects must be dealt with such a characterization: the abundance of atoms, the layer altitude and thickness (that is directly related with the elongation projected due to the perspective and the finite thickness of the layer) and the stability of the layer.

With the aim of characterizing the atmospheric Na layer, the Teide Observatory facilities ($\approx$ 2400m, 16°30.5′ W, 28°18.1′ N) have been used in the last years to carry out a long term monitoring experiment on LGS generation by low power laser excitation: the `natrium`-Teide experiment. Details on the experiment have been best explained by Chueca et al. [1]. We employ the 1 m Optical Ground Station[1] (OGS) as launching telescope (see figure 1), while the backscattered photons are received in the 0.8 m IAC80 telescope[2], 165 m away, in a bistatic configuration.

[1] The Optical Ground Station was built as part of European Space Agency research in the field of inter-satellite optical communications (`http://www.iac.es/telescopes/pages/en/home/telescopes/ogs.php?lang=ES`)
[2] `http://www.iac.es/telescopes/pages/en/home/telescopes/iac80.php?lang=ES`

Figure 1. The Optical Ground Station while launching the Rodamine 6G dye tunable laser onto the atmosphere for the generation of a Na LGS spot.

2. Methodology

The OGS is equipped with a Coudé room which facilitates laser operations. The LGS are generated by a Rodamine 6G dye tunable laser (Coherent 899 − 21) pumped by a Coherent Verdi Solid-State Green laser (15 W, 532 nm). The tuning to the D_2 sodium spectroscopic line is achieved using a birrefringent filter inside the resonant cavity with a bandwidth of ≈ 10 MHz, allowing an unique oscillation mode. The fine tuning is carried out with a pair of ethalons and the wavelength control is made splitting $\approx 2\%$ of the beam energy to a Michelson interferometer.

The beam is projected onto the mesosphere with the telescope and an off-axis optical arrangement. The typical output power is 400 − 500 mW. After a resonant scattering process with the Na atoms in the Mesosphere (≈ 90 km), the LGS is imaged with the CCD of the 0.8 m IAC80 telescope, in a bistatic configuration with a baseline of 165 m (see figure 2). We use a 2 nm narrow BW filter and exposure times of 300 sec. For the basic astronomical reduction we have developed a pipeline based on IRAF. The height determination is obtained by triangulation. As the telescope pointing information has not enough resolution, natural star trails are used to obtain the zenith angle. The detailed equations of the geometry of the system for absolute heigh determination are in Castro-Almazán et al. [2]. The Na abundance is obtained with relative photometry. We obtain images of the natural star trails in the R Johnson band for calibration and we use Sextractor.[3]

[3] `www.astromatic.net/software/sextractor`

Journal of Physics: Conference Series **595** (2015) 012007 doi:10.1088/1742-6596/595/1/012007

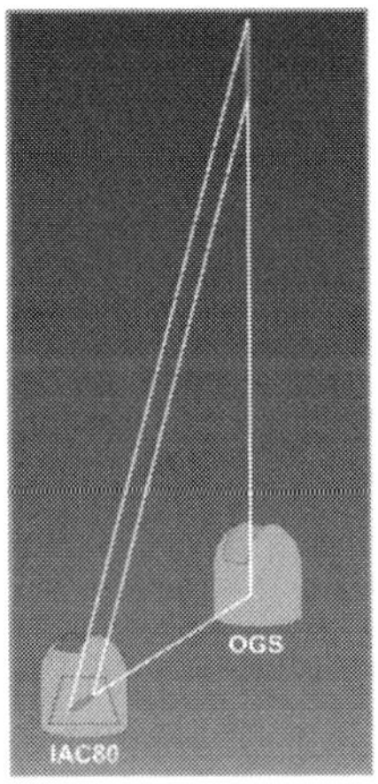

Figure 2. The bistatic configuration used for the experiment, launching to the zenith.

2.1. Experiment improvements

The experiment has been improved in several aspects in order to boost the Na photon flux return, for example:

- Increasing the pumping efficiency from Ar laser (25 W, 514 nm) to High-Power Optically Pumped Solid State (Verdi, 15 W, 532 nm).
- Reducing the optical elements in the launching path.
- Changing the launching optical configuration from 'on axis' to 'off axis' to avoid the secondary mirror obscuration (see figure 3).

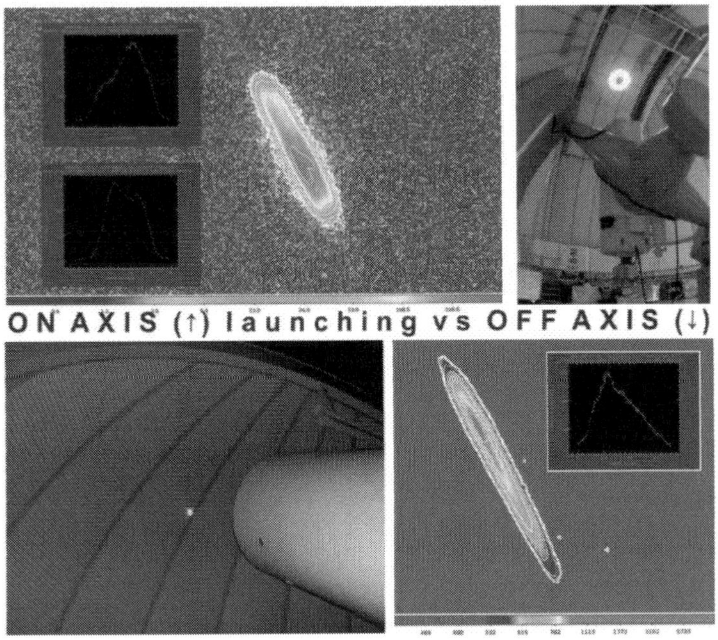

Figure 3. The effect of changing the launching from 'on-axis' configuration to 'off-axis'. The secondary obscuration is avoided, improving the photon flux return. An optical arrangment was designed to collimate the beam axis on the Mesosphere level.

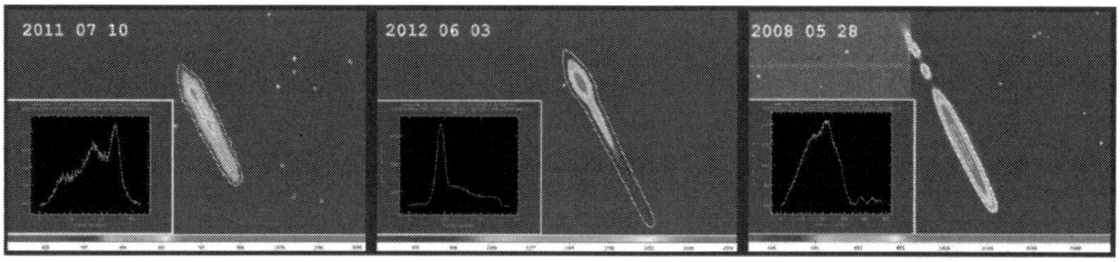

Figure 4. Three typical images obtained with the `natrium` experiment with a doubled sporadic layers [3], an enhanced layer [4] and a triple layer.

3. Results

The `natrium`-Teide experiment has generated a dataset of more than 400 images of Na LGS traces obtained in 37 nights. The main results are going to be been published soon, and will be auxiliary data for the incoming Laser Guide Star for AO at Gran Telescopio Canarias (GTC) in the neighbouring Roque de los Muchachos Observatory ($\approx$ 140 km away). The figure 4 show three typical images obtained with the experiment with remarkable phenomena, as sporadic layers [3] and enhanced layers [4].

Acknowledgments

We would like to acknowledge the Teide Observatory and the IAC Headquarters staff for permanently showing an excellent will in the achievement of the tasks associated to this work. This work has been supported by IAC and the Spanish "Plan Nacional de Astrofísica" of the Ministry of Science under grants AYA2003-07728 and AYA2006-13682.

References

[1] Chueca S, Fuensalida J, Alonso A and Reyes M 2004 *Optics in Atmospheric Propagation and Adaptive Systems VII* (*Proc. SPIE* vol 5572) ed Gonglewski J and Stein K pp 392–399
[2] Castro-Almazán J, Fuensalida J, Alonso A and Chueca S 2008 *Ground-based and Airborne Telescopes II* (*Proc. SPIE* vol 7012) ed Stepp L and Gilmozzi R pp 701240–1–12
[3] Simonich D, Clemesha B and Batista P 2005 *Adv. Space Res.* **35** 1976–1980
[4] Clemesha B 1990 *Adv. Space Res.* **10(10)** 10(59–70)

Application of climatological monitoring for the candidate CTA site at Izaña (Tenerife)

J.A. Castro-Almazán,[1,2] C. Muñoz-Tuñón,[1,2] R. Sanz[3], R. García-López[1,2], I. Puerto[1,2], A.M. Varela[1,2] and M. Serra[1,2]

[1]Instituto de Astrofísica de Canarias, E-38200, La Laguna, Spain
[2]Dept. Astrofísica, Universidad de La Laguna, E-38200, La Laguna, Spain
[3]Western Canary Is. Regional Meteorological Center, Agencia Estatal de Meteorología, c/San Sebastian 77, Santa Cruz de Tenerife, Spain

E-mail: `jcastro@iac.es (JACA); cmt@iac.es (CMT)`

Abstract. The location of Izaña (Tenerife, Canary Islands) has been proposed for hosting the Cherenkov Telescope Array (CTA). To support the site-testing campaign, we obtained climate normals from the 10 year long series recorded at Izaña by the atmospheric Observatory (CIAI), we have also compared the wind regime with a simultaneous 10 year database at the neighbouring astronomical site (OT) and with the available simultaneous preliminary wind data for the proposed CTA location. The results show a smooth temperature profile with sporadic but moderate extremes and a low daily (24 h) temperature range. Rainfall is extremely sporadic (287 mm/y) with low intensity. Wind speed and gusts show high stability and moderate values and dispersions. The lowest winds are in the summer. Speed episodes greater than 100 km/h are very rare and 11 data points have been recorded with gusts above 150 km/h over the 10 years. The dominant wind direction is WNW. There is an excellent correlation for wind speed between the CIAI and OT, with a slight difference in the dominant wind direction (N versus WNW). The wind data recorded directly at the CTA location also show very good correlation in trends, the wind speed at the CTA site being $\approx$ 24% lower. The dominant wind direction at the CTA site is NW.

1. Introduction

The Izaña summit ($\approx$2400 M.S.L.), on Tenerife (Canary Islands, Spain) hosts both the Teide astronomical Observatory[1] (OT) and a meteorological observatory, the Izaña Atmospheric Research Centre[2] (CIAI). The area has been proposed as a candidate site for the Cherenkov Telescope Array in the northern hemisphere[3] – CTA-North (see Fig. 1 for the location details). To support the CTA site-testing campaign, we have obtained climate normals over a ten-year period. Ten year studies are recommended for providing an accurate picture of a location's climatic conditions as they minimize distortion in climate series resulting from climate change and other trends [1]. The results presented here have been extracted from an extensive report that was provided to the CTA site selection team.

[1] `www.iac.es/eno.php?op1=3&lang=en`
[2] `www.izana.org`
[3] `www.cta-observatory.org`

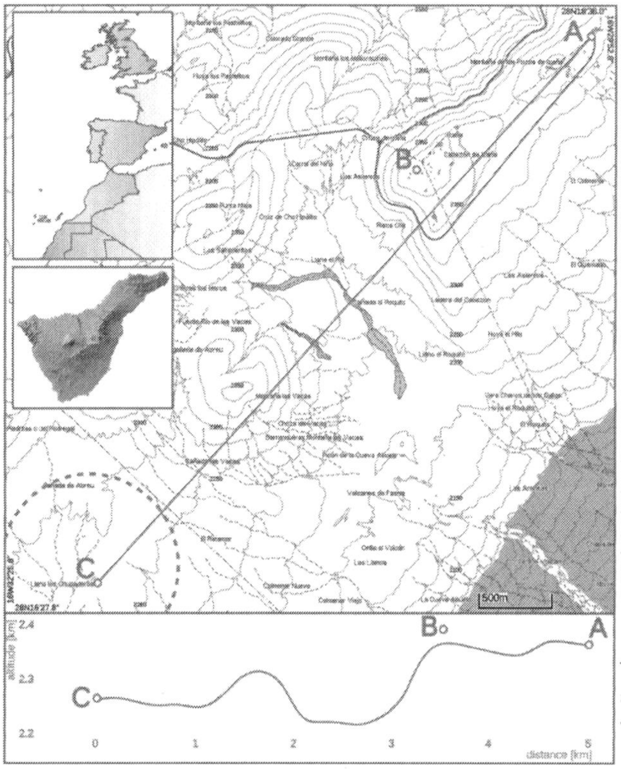

Figure 1. Izaña topographic map. "A" is the Atmospheric Observatory (CIAI) station point (28°18.49′ N, 16°29.98′ W; 2367 M.S.L.); "B" is the Astronomical Observatory (GONG) station point (28°18.04′ N, 16°30.73′ W; 2395 M.S.L.); "C" is the CTA proposed area (≈28°16.5′ N, ≈16°32.0′W; ≈2270 M.S.L.). Below is the topographic profile linking "A", "B" and "C".

2. The stations

The meteo station at CIAI was selected on the basis of its excellent data quality. The CIAI is part of the Spanish Agencia Estatal de Meteorología[4] (AEMet). The selected station is part of the WMO[5] with index 60010. The sensors are installed at the top of a 10 m mast.

Furthermore, to analyse possible local inhomogeneities, wind regime comparisons have been carried out using a station at the astronomical site (OT), ≈1.5 km away. The station selected supports the operation of the GONG[6] astronomical facility. The sensors are located on a 7 m mast.

While this work was being carried out, the CTA team set up their own weather station on a 10 m mast at the exact location of the candidate site, some 5 km from the CIAI. In order to extend the parameters obtained for Izaña to the CTA site, a preliminary wind comparison with the available CTA data was carried out.

3. Data at CIAI

The data, except for precipitation, were recorded with an automatic weather station (AWS) at a frequency of 1/10 min, suitable for the separation of night- and day-time analysis. Precipitation was obtained from the non-automated series as hourly cumulative by observers. All central tendency estimates have been calculated as median values in order to look for statistical

[4] `www.aemet.es/en/`

[5] World Meteorological Organization.

[6] Global Oscillation Network Group, `http://gong.nso.edu/sites/elteide.shtml`

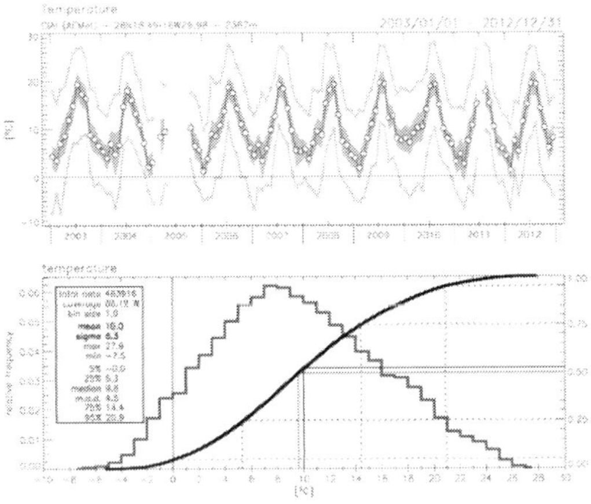

Figure 2. Temperature [°C].
Top: Time series. Monthly medians (white circles), dispersion (±m.a.d.; shadow) and absolute monthly max and min (thin lines).
Bottom: Histogram (blue; left axis) and cumulative distribution (black; right axis).

robustness in the results. The median absolute deviation (m.a.d.)[7] has been used for dispersions. Additionally, the mean and standard deviation are also given in the figures. Special care has been taken to avoid including periods where we estimated sampling errors above 10%. Statistical homogeneity was assumed for the series. No inconsistencies of note were found when the data were explored for the presence of outliers. Wind data were rejected when wind speed equalled 0 simultaneously with temperatures below 2°C (frozen anemometer effect).

The database includes temperature, precipitation, relative humidity, wind speed (10 min averages) and gusts.

4. Temperature

Temperature is dominated by a smooth profile, with very sporadic extreme events, and a low temperature range (the difference between daily max and min). Figure 2 summarizes the temperature statistics at the CIAI. The median temperature is 9.6°C and the absolute minimum recorded is -7.5°C. Temperature is over 0°C for 95% of the time. The temperature Range is small, with a median value of $\approx$7°C and is highly stable throughout the year.

5. Precipitation and Relative Humidity

Table 1 summarizes the precipitation as yearly cumulative. Median rainfall at the CIAI is 287 ± 158 mm/y. This value corresponds to an environment classified as arid and semi-arid[8]. The precipitation is even less frequent and intense during night time. Figure 3 shows the precipitation intensity in mm/h. The precipitation episodes correspond to singular events, typically less than 1 hour, that rarely reach a 'heavy' level. The *no precipitation* stage corresponds is applicable to 96% of the data. The precipitation is concentrated between October and March with noticeable minima in December and January. Snow and hail occur 11.7 and 1.1 days/year respectively, mainly in February and March, with some episodes in autumn and spring.

[7] The m.a.d. is a robust estimator for dispersion. For normally distributed samples, 50% of the sample falls in median±m.a.d. – m.a.d.$\approx 1/2(P75 - P25)$ – and $1\sigma \approx 1.4826\cdot$m.a.d.

[8] Extract from the Arid Zone classification based on rainfall of the FAO (Food and Agriculture Organization of the United Nations)[2]:

Hyper-arid zones	< 100 mm/y
Arid zones	100 – 300 mm/y
Semi-arid zones	300 – 800 mm/y

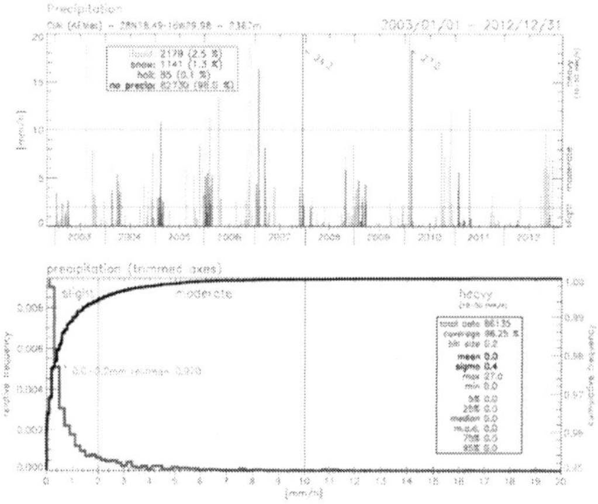

Figure 3. Precipitation Intensity [mm/h]. *Top:* Time series. Liquid phase (sky blue) includes rain and drizzle. Frozen phase is shown separately as snow (deep blue) and hail (green). The intensity references follow the UK Met Office synoptic scale for rain showers [3]: slight (0–2 mm/h), moderate (2–10 mm/h), heavy (10–50 mm/h) and violent (> 50 mm/h). *Bottom:* Histogram (blue; left axis) and cumulative distribution (black; right axis). Y-axes have been trimmed for detailing structures other than 0 mm/h.

The relative humidity is low and follows the precipitation cycle. The median value is 29% and the 25th percentile is 15%.

Table 1. Yearly precipitation [mm/year] (total –PR, liquid –PR_{liq}– and frozen –PR_{froz}).

	central tendency		dispersion	
[mm/year]	Median	Mean	m.a.d.	Sigma
PR	287	354	93	158
PR_{liq}	209	217	70	125
PR_{froz}	81	137	40	120

6. Wind

Figure 4 presents the main results for wind (speed and direction) at the CIAI. Wind speed shows moderate values. The median (22.3 km/h) and the upper percentiles (P95= 52.9 km/h) perfectly allow a safety operation and ensure a low percentage of downtime due to wind for most astronomical installations. Wind speed is ≈12% higher at night. The lowest wind speed values are in summer.

Both wind speed and gust show stability with moderate dispersion and high correlation. Wind gust is ≈ 30% greater than wind speed and ≈ 20% more spread. Wind gust does not show significant differences between night and day. In the 10 year period there are only 11 points, concentrated in 4 days, with gusts over 150 km/h.

The dominant direction has a clear component from the WNW, both for speed and gust, with punctual 'winter south component' episodes. The most intense wind direction roughly coincides with the predominant wind direction.

7. Wind comparison between CIAI, OT and the CTA

Once we established the climate normals for the CIAI, we compared the wind regime with a simultaneous 10 year database at the OT (the GONG station) and with the available simultaneous data from the CTA site.

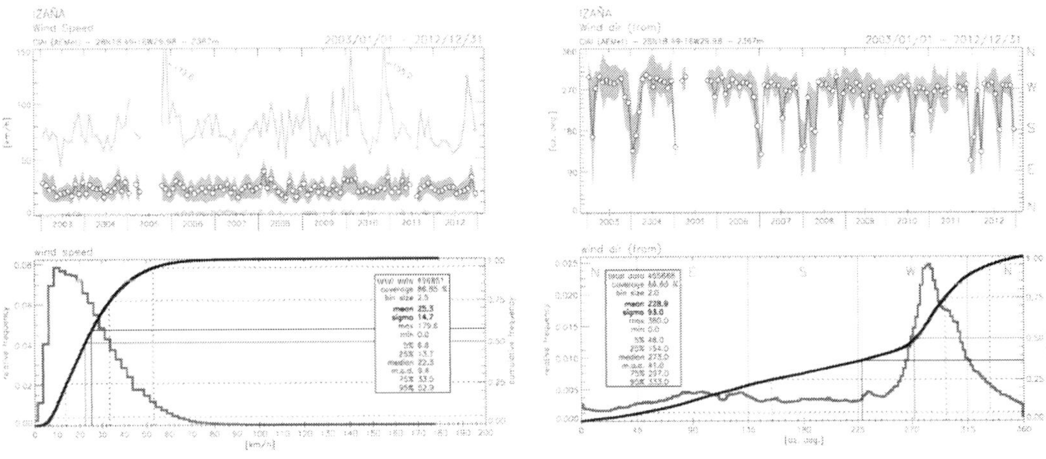

Figure 4. Wind speed [km/h] (left) and direction (right).
Top: Time series. Monthly medians (white circles), dispersion (±m.a.d.; shadow) and absolute monthly max and min (thin lines). Peaks outside the Y top range are specifically labelled.
Bottom: Histogram (blue; left axis) and cumulative distribution (black; right axis).

7.1. GONG station

Meteo data at GONG are recorded with an AWS at a frequency of 1 min^{-1}. The extent covered by GONG ensures a direct comparison with the CIAI. Unfortunately, no wind gusts are available. The data were also examined and corrected for the presence of outliers and inconsistencies. The GONG 1 min^{-1} series had to be processed to match the CIAI standard (1/10 min). The smoothing was carried out by grouping the data in 10 min steps and computing the median.

Figure 5 shows the main statistics for GONG wind in comparison with the CIAI series (magenta). Wind speed behaviour is very similar in the two locations with a correlation of 0.85. The GONG wind direction also show a clear dominant direction but $\approx 80°$ northward from that of the CIAI (N versus WNW).

7.2. CTA station

In order to extend the normal values of the CIAI to the CTA site, a wind comparison was carried out with the first year of available data (December 2012 to January 2014). We have selected only simultaneous data that guarantee a correct comparison. The sampling rate for CTA is 1 min^{-1}, so the data were processed to match the CIAI standard (1/10 min). Figure 6 summarizes the main results at the CTA in conjunction with the CIAI (magenta). There is excellent agreement between the wind speed trends and also for gusts, the wind being generally lower at the CTA. The median wind speed is 15.8 km/h at the CTA and 20.9 km/h at the CIAI, which implies a relative difference of $\approx 24\%$ (the median gust is 21.8 and 27.8 km/h, for the CTA and CIAI, respectively).

The wind speed and wind gust direction coincide at the CTA. The predominant direction at the CTA is NW. The strongest wind direction at CTA is also NW.

8. Final summary

In this article we have presented a brief overview of the analysis carried out in order to study the climate at Izaña (Tenerife). A full text, with the complete information, will be published soon [4]. It is important to emphasize the mildness of the climate as measured for the temperature and precipitation at this height of 2400 m. Also important is the relative homogeneity of the

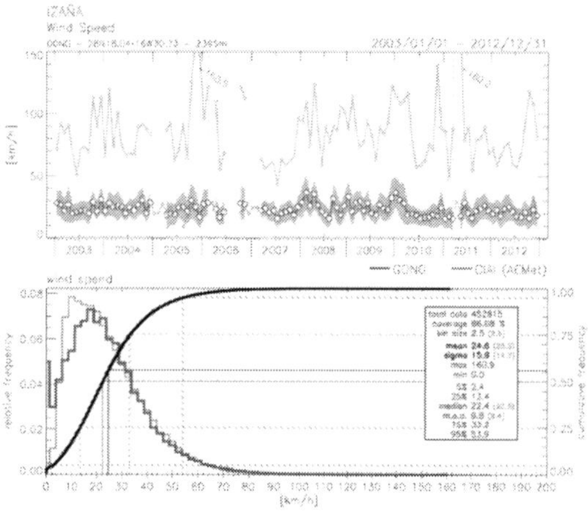

Figure 5. Wind speed [km/h] at GONG, ≈ 1.5 km from CIAI (magenta).
Top: Time series. Monthly medians (white circles), dispersion (±m.a.d.; shadow) and absolute monthly max and min (thin lines).
Bottom: Histogram (blue; left axis) and cumulative distribution (black; right axis).
Note 1: 160.9 km/h is the sensor saturation. Note 2: the peak around 0 km/h in the GONG histogram have been identified as an artefact from the original series.

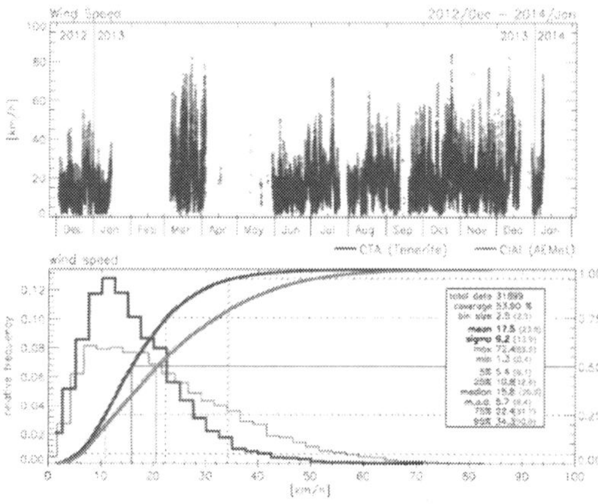

Figure 6. Wind speed [km/h] at CTA), ≈ 5 km from CIAI (magenta).
Top: Simultaneous time series.
Bottom: Histogram (left axis) and cumulative distribution (right axis).

wind pattern when comparing values at the CIAI with the astronomical observatory (≈1.5 km away) and the CTA candidate site (≈5 km apart).

Acknowledgments
This report made use of 10 years of data recorded by the Spanish Agencia Estatal de Meteorología (AEMet) at the Izaña Atmospheric Research Centre (CIAI) and by the Global Oscillation Network Group (GONG) at Teide Observatory (OT). Additionally, some wind data from the Cherenkov Telescope Array (CTA) *atmoscope* station at Tenerife were also used for comparison. We are grateful to all these sources for the kind distribution of their data. We particulary thank Pere Lluís Pallé from GONG and Diego Sierra from the IAC Network department. We are also grateful to Terry Mahoney for the language correction of the manuscript. This work was funded by the IAC.

References

[1] Huang J, van den Dool H M and Barnston A G 1996, *Journal of Climate*, **9**, 809–817
[2] Food and Agriculture Organization of the United Nations (FAO) 1989 Arid zone forestry: A guide for field technicians, book, FAO Conservation Guide T0122, `http://www.fao.org/docrep/t0122e/t0122e00.htm`
[3] UK Met Office 2011 Water in the atmosphere, Fact sheet No. 3, `http://www.metoffice.gov.uk/media/pdf/4/1/No._03_-_Water_in_the_Atmosphere.pdf`
[4] Castro-Almazán J, Muñoz-Tuñón C and Sanz R 2015, *Journal of Climate*, **(to be submitted)**

Journal of Physics: Conference Series **595** (2015) 012009 doi:10.1088/1742-6596/595/1/012009

The Atmospheric Monitoring Strategy for the Cherenkov Telescope Array

MK Daniel[1] **for the CTA Consortium**[2]

[1] Department of Physics, University of Liverpool, Liverpool, L69 7ZE. UK.

[2] http://www.cta-observatory.org/

E-mail: `michael.daniel@liverpool.ac.uk`

Abstract. The Imaging Atmospheric Cherenkov Technique (IACT) is unusual in astronomy as the atmosphere actually forms an intrinsic part of the detector system, with telescopes indirectly detecting very high energy particles by the generation and transport of Cherenkov photons deep within the atmosphere. This means that accurate measurement, characterisation and monitoring of the atmosphere is at the very heart of successfully operating an IACT system. The Cherenkov Telescope Array (CTA) will be the next generation IACT observatory with an ambitious aim to improve the sensitivity of an order of magnitude over current facilities, along with corresponding improvements in angular and energy resolution and extended energy coverage, through an array of Large (23 m), Medium (12 m) and Small (4 m) sized telescopes spread over an area of order $\sim$km^2. Whole sky coverage will be achieved by operating at two sites: one in the northern hemisphere and one in the southern hemisphere. This proceedings will cover the characterisation of the candidate sites and the atmospheric calibration strategy. CTA will utilise a suite of instrumentation and analysis techniques for atmospheric modelling and monitoring regarding pointing forecasts, intelligent pointing selection for the observatory operations and for offline data correction.

1. Introduction

Ground based gamma-ray astronomy is not able to directly measure the high energy photons, for the atmosphere is thankfully[1] opaque to them. Instead this branch of astronomy exploits an indirect method, by measuring optical photons of Cherenkov light (concentrated toward the blue end of the spectrum) produced from the faster than light-in-air passage of the particles produced in an extensive air shower (EAS) that results from the multiple interactions that redistribute the energy from a single very high energy primary particle to many lower energy ones, see figure 1 top. The Cherenkov light is generated high in the atmosphere (typically the first interaction is at 20 km altitude and the maximum of emission between 8-12 km depending on the primary energy and the atmospheric density profile) and spreads out at a characteristic angle that depends on the density of the medium (for air this is about a degree) (see fig 1 bottom for how this varies depending on latitude and season). When Cherenkov photons that are not attenuated reach ground level they are thus fairly uniformly spread in a pool of order 100 m radius and drop off rapidly after that. Placing a telescope anywhere within that pool of Cherenkov light means the EAS can be detected and so this provides a very large effective area for the detection of very low flux sources (a few photons per minute is a bright source). The extent of the Cherenkov

[1] for the general purpose of daily life

image means it is immune to many of the atmosphere's characteristics that affect traditional ground based optical astronomy, such as seeing/scintillation effects, but with the atmosphere comprising such a fundamental component of the detection system, and acting as a calorimeter, it is still very much subject to changes that are beyond our control and must be monitored for and measured instead.

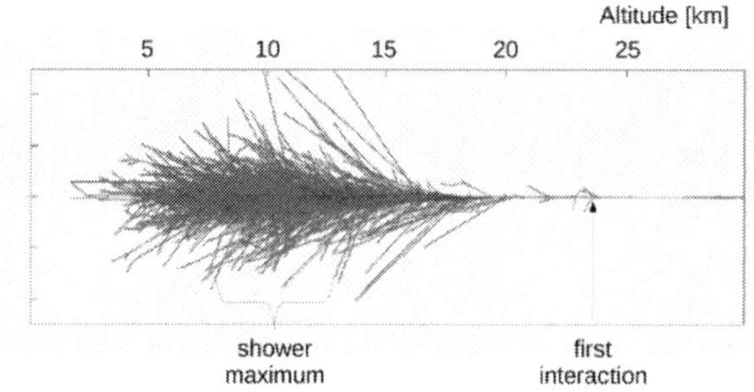

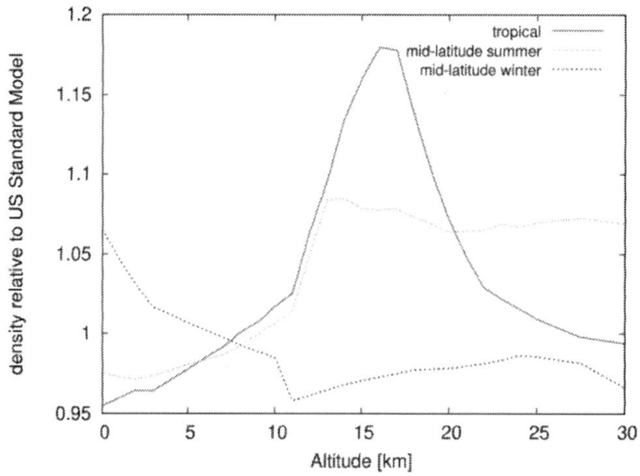

Figure 1. Top: A representation of an Extensive Air Shower initiated by a 1 TeV gamma-ray photon. Most of the Cherenkov light is generated below 15 km altitude, with the shower maximum at 10 km. **Bottom:** The density profile of the tropical (solid red line), mid-latitude summer (dashed green line) and mid-latitude winter (dotted blue line) relative to the US Standard model showing the variation due to location and season. The variation is greatest in the development region between the first interaction and where shower maximum occurs.

The Imaging Atmospheric Cherenkov Technique (IACT) is the most sensitive method of detecting astrophysical sources of very high energy (VHE, E> 10 GeV) gamma rays. The Cherenkov Telescope Array (CTA)[2] [1] will be the next generation IACT facility, providing the community with an open-access observatory for the observation of gamma rays with energies from a few tens of GeV to hundreds of TeV with unprecedented sensitivity and angular and energy resolution. Whole sky coverage will be achieved by operating at sites in both hemispheres: the southern array, covering an area of about $10\,\mathrm{km}^2$ with O(100) telescopes of all sizes, is optimised for the highest energy observations of the galactic plane; and the northern array, covering an area of about $1\,\mathrm{km}^2$, is optimised for the low energy thresholds best suited to observing extragalactic sources.

The CTA has set some very stringent requirements, including: the uncertainty on the effective collection area to be $< 12\%$; the systematic error on the absolute intensity of Cherenkov photons at each telescope (post calibration) must be $< 8\%$ such that the systematic error on the energy scale must not exceed 10%; a final systematic error on the localisation of a point-like source of 3 arcseconds (under favourable observing conditions). Furthermore, the observatory will need to maximise the duty cycle and keep the maintenance budget low. As the atmosphere is such an integral part of the detection system for IACT, in order to achieve these ambitions the CTA requires a suitably ambitious atmospheric monitoring strategy compared to what has come before, which will be detailed in the following sections.

2. Atmospheric Monitoring Strategy

The goal of the CTA atmospheric monitoring strategy is to both characterise the systematic uncertainties in gamma-ray event reconstruction and to minimise any loss of data due to non-optimal atmospheric conditions. With an appropriate atmospheric monitoring strategy, not only can the energy and flux of an astrophysical source be correctly retrieved but also the overall duty cycle of the observatory can be increased. The CTA strategy is split into 5 tasks

Climatology Characterisation of the site conditions and simulation input. This task will begin as soon as the site decision is made, even before telescopes are installed. The use of weather stations and radiosondes will build up the database of atmospheric conditions that can be input to and tested against data models like GDAS[3], WRF[4] and SENES[5] to ensure they are reliable enough descriptions of the site conditions, such as has been successfully achieved for the Auger site [2]. Also a profound knowledge of aerosol size distributions, their occurence and variability shall be obtained.

Weather Nowcast, Forecast, Alerts & Protection Protect the instrumentation, inform observing schedule. Instruments providing readily available information of the weather and short term forecasts are not only useful for data-taking, but for planning maintenace trips, etc to the extended array. Many of these will be commercial instruments to provide standardised reliable 24/7 remote operations and automated alerts.

Online Smart Scheduling Find the optimal target in non-ideal observing conditions.

Offline Data Selection Certify good quality data. If the conditions do not allow for unaffected/unbiased observations then the recoverable data must be flagged against those which must be rejected.

Offline Data Correction Retrieve correct flux and energy estimates. Data taken during some non-optimal atmospheric conditions can be corrected by the use of specially adapted Monte Carlo simulations or clever algorithms making use of the atmospheric data.

[2] www.cta-observatory.org
[3] http://www.ncdc.noaa.gov/data-access/model-data/model-datasets/global-data-assimilation-system-gdas
[4] http://www.wrf-model.org/index.php
[5] http://www.senes.ca/

Equipment is primarily designed to serve one of the tasks, but may be used for others as well, e.g. for cross-checks and calibration. Where possible passive monitoring devices that do not interfere with Cherenkov observations or take away from source observation time are preferred, but not always possible.

3. Atmospheric Monitoring Equipment

A mixture of custom built and commercial instruments will comprise the bulk of the atmospheric monitoring equipment.

3.1. ATMOSCOPE

The ATMOSCOPE (Autonomous Tool for Measuring Site COndition PrEcisely) [3] is a sensor station comprised of

- a commercial weather station for measuring temperature, humidity, pressure and wind;
- a Light of Night Sky (LoNS) instrument for measuring the sky brightness through B and V band filters;
- an All Sky Camera for monitoring night sky quality.

designed to be deployed and operated in a remote site without power supply and ethernet connection. The ATMOSCOPEs were cross-calibrated and then deployed to each of the CTA site candidates to provide an independent and unbiased comparison for the purpose of site selection. In addition the units can be compared to existing devices at sites and satellite data to provide a better understanding of long term historical datasets.

3.2. All Sky Camera

The All Sky Camera (ASC) is an astronomical CCD camera with a fish-eye lens to determine the night sky quality from the cloud fraction (percentage of the sky covered by clouds) and the night sky brightness (in mag/arcsec^2) [4]. An upgraded system to the one used in the site selection study will be used to estimate sky quality and determine areas of the sky free from cloud for the CTA central scheduler.

3.3. FRAM

The F/(Ph)otometric Robotic Atmospheric Monitor (FRAM) [5] is a wide field astronomical imaging camera on a robotic mount to monitor stars in the whole field of view of the Cherenkov telescopes (up to $10^\circ \times 10^\circ$). From measuring the light flux of the stars in the same direction we can determine the atmospheric extinction with high temporal and spatial resolution without interfering with the Cherenkov observations.

3.4. UVscope

The UVscope [6] is a portable, NIST-calibrated, multi-pixel photon detector developed at INAF IASF-Palermo (Italy) to support experimental activities in the fields of high-energy astrophysics and cosmic-rays, but thanks to its features and operational flexibility the instrument can be used in a wide range of applications. The instrument works in single photon counting mode to directly measure the light flux in the 300-650 nm wavelength range. The atmospheric transparency can be evaluated by tracking a star measuring the absorption of the flux as a function of the air mass travelled. The UVscope can also measure the diffuse emission of the night-sky-background (NSB) in the same field of view of a given telescope [7]. In the CTA framework, UVscope would follow the same pointing of a given CTA telescope or group of telescopes with a quantum efficiency well matched in the wavelength region of interest for Cherenkov light detection [8].

3.5. Cherenkov Transparency Coefficient
A Cherenkov telescope is always going to be the most sensitive instrument to the weather conditions that affect Cherenkov light yield. The Cherenkov Transparency Coefficient (CTC) method [9] combines the information of the average pixel gain and optical throughput of the telescope to the expected telescope trigger rates for cosmic-ray protons to determine if the trigger energy threshold of the telescope has changed due to atmospheric effects. This enables the CTC to be used as a direct and free data quality selection parameter, but will naturally suffer from the fact it can only do so post data acquisition. A high correlation factor of the CTC with independent aerosol optical depth measurements by the MISR satellite and onsite lidar and IR radiometer systems at the H.E.S.S. site in Namibia means such independent systems could be used to automatically interact with the central autoscheduler to determine an optimum pointing position. For full offline data correction, though, a system providing height information is required.

3.6. Lidar
Because the Cherenkov light is generated and transported within the atmosphere just treating it like a simple filter can lead to a biased estimate of the amount of attenuation [10]. Instead a differential transmission profile is needed, meaning instrumentation with height resolution, for which a lidar is best suited. Also, something capable of distinguishing between different types of attenuation, from the Rayleigh scattering of the molecular component of the atmosphere and the Mie scattering of aerosols, is highly desirable for CTA to reach its goal energy resolution and energy scale bias requirements. A Raman lidar should enable us to decouple the molecular and aerosol components to $\sim 5\%$ uncertainty. Several institutes in CTA are involved with developing Raman lidar systems. Generically these lidars are characterised by having large reflectors of ≥ 1.5 m diameter, powerful Nd:YAG lasers operated at 2-3 wavelengths (1054 nm, 532 nm, 355 nm) with filters at 355 nm and 532 nm for the elastic backscatter component and 387 nm and 607 nm for the N_2 inelastic scatter Raman lines. The different groups are prototyping independent mechanical, optical and steering solutions. The Laboratoire Univers et Particules de Montpellier, Institut de Fisica d'Altes Energies (IFAE) and the Universitat Autònoma de Barcelona (UAB) are re-purposing some CLUE Cherenkov telescopes for this purpose; the Centro de Investiciones Laser y sus Aplicaciones (CEILAP) group have a custom designed solution based on a multiple mirror system. More details of these devices can be found in [11]. Lidar laser shots will affect the acquisition of Cherenkov events, but will occur at low frequency (~ 20 Hz) and can be easily flagged and removed in the post-observation data analysis.

3.6.1. Ceilometer Primarily used by the aviation industry and designed to be eye safe and remotely operated, a ceilometer is a low power version of a lidar utilising an infra-red (905 nm) laser diode and summing many thousand return traces to detect the height of clouds. A Vaisala CL51 model ceilometer has the capability to report up to three cloud layers in a range up to 13 km and gives backscatter profiling for vertical visibility up to 15 km. In favourable conditions the backscatter profile can also be used to monitor boundary layer structures [12]. Use of an infra-red device means that the heights of clouds anywhere in the sky can be determined without interference to Cherenkov observations. If the height of the cloud is above shower maximum then data can still be collected with minimal bias, for lower cloud then appropriate data recovery techniques could be applied [13, 14].

3.7. Global Networks, Satellite Data and Atmospheric Models
Satellites such as MODIS, GOES and METEOSAT are available to offer accurate measurements in the regions not (yet) covered by ground-based instrumentation and can fill in the long term climatology history for a region, albeit at coarse scale and, unless geostationary, often low

cadence. Ground-based atmospheric monitoring networks provide input to global atmospheric models such as GDAS, WRF or SENES. Networks are also important because they impose standardised algorithms, thus reducing the potential for bias. AERONET is a network of Cimel sun photometers installed at some 550 stations worldwide that provides the most accurate measurement of aerosol optical depth, to $\delta\tau \sim 1\%$. CTA is investigating the mutual benefits both of profiting from the wealth of data provided by these networks and from providing access to the data of the atmospheric monitoring equipment of the CTA observatory, given these will be located at remote sites normally not covered by ground-based installations.

4. Summary

The requirements for the next generation ground based gamma-ray observatory to have unprecedented sensitivity in flux estimation and minimal uncertainties in energy reconstruction also demands a comprehensive atmospheric monitoring strategy. The suite of instruments planned for the CTA sites will perform pointed observations and provide all-sky characterisations that will provide an accurate site climatology, enable smart online scheduling in addition to offline data selection and correction where necessary. This will serve to not only minimise systematic errors, but also boost the useful operational duty cycle of CTA.

Acknowledgments

We gratefully acknowledge support from the agencies and organizations listed under Funding Agencies at this website: http://www.cta-observatory.org/.

References

[1] Actis M, Agnetta G, Aharonian F, Akhperjanian A, Aleksić J, Aliu E, Allan D, Allekotte I, Antico F, Antonelli L A, Antoranz P, Aravantinos A, Arlen T, Arnaldi H, Artmann S, Asano K, Asorey H, Bähr J, Bais A, Baixeras C and et al 2011 *Experimental Astronomy* **32** 193–316 (*Preprint* 1008.3703)
[2] Keilhauer B and Will M 2012 *European Physical Journal Plus* **127** 96 (*Preprint* 1208.5417)
[3] Vincent S and CTA Consortium f t 2014 *ArXiv e-prints* (*Preprint* 1403.5075)
[4] Mandat D, Pech M, Hrabovsky M, Schovanek P, Palatka M, Travnicek P, Prouza M and Ebr J 2014 *ArXiv e-prints* (*Preprint* 1402.4762)
[5] Prouza M, Jelínek M, Kubánek P, Ebr J, Trávníček P and Šmída R 2010 *Advances in Astronomy* **2010** 849382
[6] Maccarone M C, Catalano O, Giarrusso S, La Rosa G, Segreto A, Agnetta G, Biondo B, Mangano A, Russo F and Billotta S 2011 *Nuclear Instruments and Methods in Physics Research A* **659** 569–578
[7] Segreto A 2011 *International Cosmic Ray Conference* **3** 129
[8] Maccarone M C, Segreto A, Catalano O, La Rosa G, Russo F, Sottile G, Gargano C, Biondo B, Fiorini M, Incorvaia S and Toso G 2014 *Society of Photo-Optical Instrumentation Engineers (SPIE) Conference Series* (*Society of Photo-Optical Instrumentation Engineers (SPIE) Conference Series* vol 9149)
[9] Hahn J, de los Reyes R, Bernlöhr K, Krüger P, Lo Y T E, Chadwick P M, Daniel M K, Deil C, Gast H, Kosack K and Marandon V 2014 *Astroparticle Physics* **54** 25–32 (*Preprint* 1310.1639)
[10] Bernlöhr K 2000 *Astroparticle Physics* **12** 255–268 (*Preprint* astro-ph/9908093)
[11] Doro M, Gaug M, Pallotta J, Vasileiadis G, Blanch O, Chouza F, D'Elia R, Etchegoyen A, Font L, Garrido D, Gonzales F, López-Oramas A, Martínez M, Otero L, Quel E, Ristori P and for the CTA consortium 2014 *ArXiv e-prints* (*Preprint* 1402.0638)
[12] Münkel C 2006 *Society of Photo-Optical Instrumentation Engineers (SPIE) Conference Series* (*Society of Photo-Optical Instrumentation Engineers (SPIE) Conference Series* vol 6367)
[13] Nolan S J, Pühlhofer G and Rulten C B 2010 *Astroparticle Physics* **34** 304–313 (*Preprint* 1009.0517)
[14] Garrido D, Gaug M, Doro M, Font L, López-Oramas A and Moralejo A 2013 *Proceedings of the 33rd ICRC, Rio de Janeiro* arXiv:1308.0473

Scintillation Noise in Exoplanet Transit Photometry

Dóra Föhring[1*], Richard Wilson[1], James Osborn[1], Vik Dhillon[2]
[1]Department of Physics, Centre for Advanced Instrumentation, University of Durham, South Road, Durham, DH1 3LE, UK
[2]Department of Physics and Astronomy, University of Sheffield, Sheffield S3 7RH, UK
[*] E-mail: dora.fohring@durham.ac.uk

Abstract. Transit photometry is a powerful technique for studying exoplanets. Transit observations from the ground of targets of magnitude V= 10 or brighter, however, are limited by scintillation noise due to Earth's atmosphere. Through turbulence profiling using instruments such as the stereo-SCIDAR, we have shown to able to accurately model scintillation noise, which is essential in order to fully account for the error budget of the observation. Through numerical modelling we find that employing scintillation reducing techniques enables an improvement of a factor between 1.36 – 1.6 on the astrophysical parameters.

1. Introduction

Time series photometry is a powerful technique widely used in all areas of astronomy, including in the study of X-ray binaries (e.g. [1]), pulsars (e.g.[2]), white dwarfs (e.g. [3]), asteroseismology (e.g. [4]), brown dwarfs (e.g. [5]), exoplanets (e.g.[6, 7, 8]), as well as Kuiper Belt and Oort Cloud objects via occultations (e.g. [9]). In particular for exoplanet science, transit photometry provides a vast wealth of information not obtainable through other methods. Quantities obtained through transit photometry include the absolute dimensions of planetary and stellar radii, inclination and semi-major axis, when combined with an extra constraint. Photometry of the secondary eclipse allows the inference of unseen planets from timing, enables the measure of planets dayside flux emission and gives constraints on the planet's orbital eccentricity.

Ground-based photometric observations play an important role in the field of exoplanet science as they are three orders of magnitudes less costly than space-based observations. For this reason, the availability of observing time from the ground is far higher than that in space. However, observations from the ground have the major drawback of suffering from the degrading effects of the Earth's atmosphere. For time-series photometry, scintillation arising from the Earth's atmosphere is a major source of noise on observations which causes the quality of data from the ground to be significantly worse compared to that obtained from space.

Scintillation arises due to high altitude optical turbulence in the Earth's atmosphere. As opposed to *seeing*, which arises from perturbations in phase, scintillation is a result of interference during propagation, and cannot be corrected by most adaptive optics techniques.

2. When is Scintillation a Problem?

The effects of scintillation have been well studied in the past [10] [11][12][13]. For exposure times t_{exp} of typical for exoplanet photometry (of a few seconds) and typical telescope diameters D, the noise due to scintillation can be calculated via [14],

$$\sigma_I^2 = 10.7\cos(Z)^{-3}\int_0^\infty \frac{C_n^2(h)h^2}{V_\perp(h)}\mathrm{d}hD^{-4/3}t_{\mathrm{exp}}^{-1}, \tag{1}$$

which is a function of the integrated sum of the turbulence strength, $C_n^2(h)$, the horizontal windspeed, $V_\perp(h)$ and the height of each layer h. Because of this, in order to accurately model scintillation effects in long exposures, we require measurements of the wind speed $V_\perp(h)$ in addition to the turbulence profile. We note that since transit observations are performed during the course of several hours, the effect of changing airmass, arising from the change zenith angle Z, also has an observable effect on the quality of the transit data (see [15]).

For the average turbulence profile for La Palma, this equates to a magnitude of $V = 10.36$, below which scintillation noise is the dominant source of noise on photometry. While this was calculated for a small, $D = 0.5$m telescope, this value only has a weak, $D^{-1/3}$ dependence and is independent of the exposure time used.

3. Measurements of Scintillation

Between 2014-03-13 and 2014-03-17 we ran a campaign on the Isaac Newton Telescope on La Palma using the using the stereo-SCIDAR instrument for turbulence profiling [16]. From data we were able to produce estimates of the expected amount of scintillation over the course of the night. Figure 1 shows an example of the calculation of the atmospheric parameter $\sum C_n^2(h)h^2/V_\perp$ obtained from the stereo-SCIDAR profile for the night of 2014-03-15. The perpendicular wind velocity $V_\perp$ was obtained by interpolating the velocity vectors obtained from stereo-SCIDAR to the whole atmosphere.

The scintillation measurements were compared to the actual variance on simultaneous transit observations made using ULTRACAM on the 4.2m William Herschel Telescope and the 0.5m *pt6m*. Figure 2 shows an example of a comparison for two stars of brightness around 13.2 Vmag (see [17]). As the night progressed, the scatter on the ULTRACAM measured data was found to improve, coinciding with the scintillation prediction from stereo-SCIDAR, indicating that the noise additional to the known photon and instrument noise is caused by scintillation. The 3 nights of comparisons yielded a total of 390 data points which were found to have a Pearson correlation coefficient of 0.9, indicating a good agreement.

4. Correcting for Scintillation and Improvements on Astrophysical Parameters

Previous results for scintillation correction using conjugate-plane photometry (see [18]) have shown that, depending on the atmospheric conditions, we can obtain an improvement of a factor of 2 on transit photometry. Using Markov-chain Monte-Carlo methods, we found that this would enable an improvement in the range of 1.36 - 1.6 on the scatter on all astrophysical parameters.

5. Conclusions

Scintillation is a limiting source of noise for ground-based observations of bright targets. We have shown that we can obtain an accurate measure of photometric scintillation noise through the measurement of the atmospheric turbulence profile. We are currently working on techniques for scintillation reduction, one of which is conjugate plane photometry, which has been shown to be capable of reducing scintillation noise to the level of photon noise. Modelling has shown that such an improvement would result in a measurement improvement of a factor of ~ 1.5 on all astrophysical parameters.

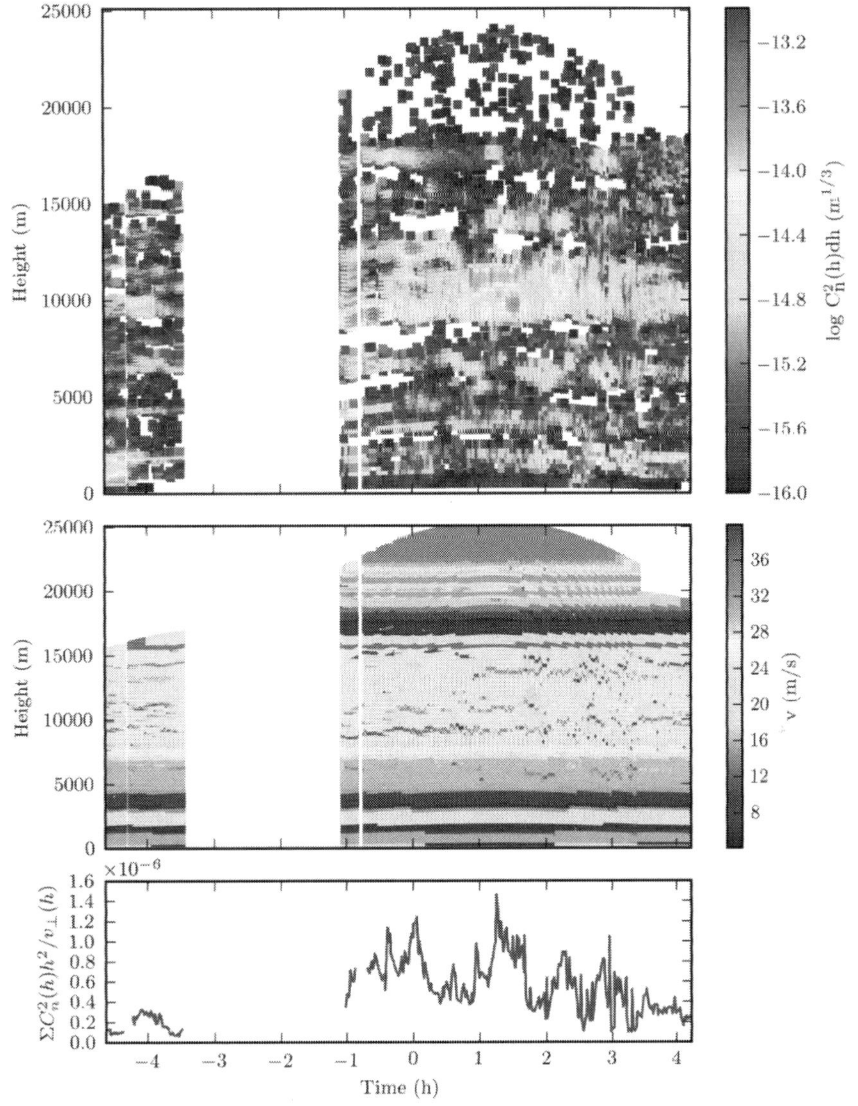

Figure 1. Atmospheric data for the night of 2014-03-15 obtained using Stereo-SCIDAR on the 2m Isaac Newton Telescope Telescope. The top panel shows the measured C_n^2 profile with height. Data points with C_n^2 weaker than $10^{-16}\mathrm{m}^{2/3}$ have been removed for clarity. The middle panel shows the determined average wind velocity for the layers for each night. The bottom panel shows the calculated sum of $C_n^2(h)h^2/V_\perp$ from the information in the top two panels.

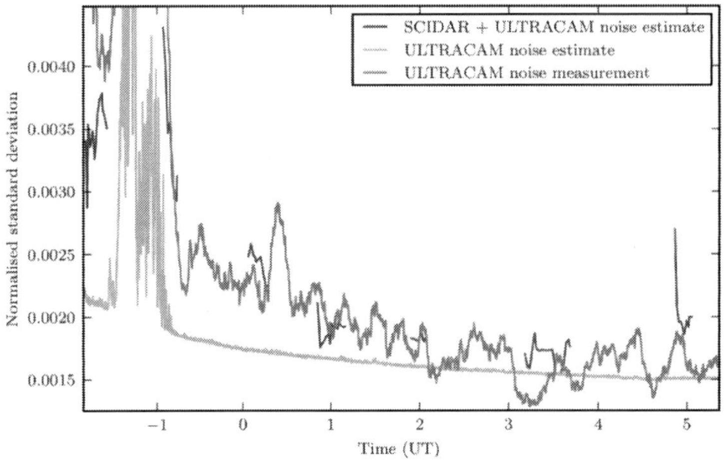

Figure 2. Comparison between predicted and measured scintillation from ULTRACAM on 2014-03-16 in the g' band. The red is line shows the normalised standard deviation of the measured flux from ULTRACAM. The prediction from SCIDAR (blue) takes into account scintillation, which is scaled according exposure time, telescope diameter and airmass of the observation, and the predicted noise from ULTRACAM. The predicted ULTRACAM noise includes photon noise and the small amount of system noise in ULTRACAM, shown in yellow. The sharp spikes in the yellow line are caused by the transparency variations and the gradual curve in the line is due to the changing extinction due to changing airmass.

References

[1] D. Chakrabarty. High-Speed Optical Photometry of the Ultracompact X-Ray Binary 4U 1626-67. *ApJ* , 492:342–351, January 1998.

[2] V. S. Dhillon, T. R. Marsh, F. Hulleman, M. H. van Kerkwijk, A. Shearer, S. P. Littlefair, F. P. Gavriil, and V. M. Kaspi. High-speed, multicolour optical photometry of the anomalous X-ray pulsar 4U 0142+61 with ULTRACAM. *MNRAS* , 363:609–614, October 2005.

[3] P. Bergeron, S. K. Leggett, and M. T. Ruiz. Photometric and Spectroscopic Analysis of Cool White Dwarfs with Trigonometric Parallax Measurements. *ApJS* , 133:413–449, April 2001.

[4] G. Handler, R. R. Shobbrook, M. Jerzykiewicz, K. Krisciunas, T. Tshenye, E. Rodríguez, V. Costa, A.-Y. Zhou, R. Medupe, W. M. Phorah, R. Garrido, P. J. Amado, M. Paparó, D. Zsuffa, L. Ramokgali, R. Crowe, N. Purves, R. Avila, R. Knight, E. Brassfield, P. M. Kilmartin, and P. L. Cottrell. Asteroseismology of the β Cephei star ν Eridani - I. Photometric observations and pulsational frequency analysis. *MNRAS* , 347:454–462, January 2004.

[5] É. Artigau, S. Bouchard, R. Doyon, and D. Lafrenière. Photometric Variability of the T2.5 Brown Dwarf SIMP J013656.5+093347: Evidence for Evolving Weather Patterns. *ApJ* , 701:1534–1539, August 2009.

[6] David Charbonneau, Timothy M. Brown, Robert W. Noyes, and Ronald L. Gilliland. Detection of an extrasolar planet atmosphere. *The Astrophysical Journal*, 568(1):377, 2002.

[7] D. Pollacco, I. Skillen, A. Collier Cameron, B. Loeillet, H. C. Stempels, F. Bouchy, N. P. Gibson, L. Hebb, G. Hébrard, Y. C. Joshi, I. McDonald, B. Smalley, A. M. S. Smith, R. A. Street, S. Udry, R. G. West, D. M. Wilson, P. J. Wheatley, S. Aigrain, K. Alsubai, C. R. Benn, V. A. Bruce, D. J. Christian, W. I. Clarkson, B. Enoch, A. Evans, A. Fitzsimmons, C. A. Haswell, C. Hellier, S. Hickey, S. T. Hodgkin, K. Horne, M. Hrudková, J. Irwin, S. R. Kane, F. P. Keenan, T. A. Lister, P. Maxted, M. Mayor, C. Moutou, A. J. Norton, J. P. Osborne, N. Parley, F. Pont, D. Queloz, R. Ryans, and E. Simpson. WASP-3b: a strongly irradiated transiting gas-giant planet. *MNRAS* , 385:1576–1584, April 2008.

[8] D. Föhring, V. S. Dhillon, Nikku Madhusudhan, T. R. Marsh, C. M. Copperwheat, S. P. Littlefair, and Richard W. Wilson. Ultracam z-band detection of the secondary eclipse of wasp-12b. *MNRAS* , 435(3):2268–2273, 2013.

[9] S. J. Bickerton, J. J. Kavelaars, and D. L. Welch. A Search For Sub-km Kuiper Belt Objects with the Method of Serendipitous Stellar Occultations. *AJ* , 135:1039–1049, March 2008.

[10] F. Roddier. The effects of atmospheric turbulence in optical astronomy. *Progress in optics. Volume 19. Amsterdam, North-Holland Publishing Co., 1981, p. 281-376.*, 19:281–376, 1981.

[11] D. Dravins, L. Lindegren, E. Mezey, and A. T. Young. Atmospheric Intensity Scintillation of Stars, I. Statistical Distributions and Temporal Properties. *PASP* , 109:173–207, February 1997.

[12] D. Dravins, L. Lindegren, E. Mezey, and A. T. Young. Atmospheric Intensity Scintillation of Stars. II. Dependence on Optical Wavelength. *PASP* , 109:725–737, June 1997.

[13] D. Dravins, L. Lindegren, E. Mezey, and A. T. Young. Atmospheric Intensity Scintillation of Stars. III. Effects for Different Telescope Apertures. *PASP* , 110:610–633, May 1998.

[14] S. Kenyon, J. Lawrence, M. C. B. Ashley, J. W. V. Storey, A. Tokovinin, and E. Fossat. Atmospheric scintillation at Dome C, Antarctica: implications for photometry and astrometry. *IAU Special Session*, 7, August 2006.

[15] J. Osborn. Scintillation correction for astronomical photometry on large and extremely large telescopes with tomographic atmospheric reconstruction. *MNRAS* , 446:1305–1311, January 2015.

[16] J. Osborn, R. W. Wilson, T. Butterley, T. Morris, D. Föhring, and R. Avila. Characterising atmospheric optical turbulence using stereo-scidar. *Proc. of IOP: conference series, Adapting to the Atmosphere*, 2015.

[17] D. Föhring. *The Effect of Scintillation on Ground-Based Exoplanet Transit Photometry*. PhD thesis, Durham University, 2014.

[18] J. Osborn, R. W. Wilson, V. S. Dhillon, R. Avila, and G. D. Love. Conjugate-plane photometry: reducing scintillation in ground-based photometry. *MNRAS* , 411:1223–1230, February 2011.

Adaptable adaptive optics and image processing at Fraunhofer IOSB

S Gladysz, P Marin Palomo, A Zepp and K Stein
Fraunhofer Institute of Optronics, System Technologies and Image Exploitation IOSB
Gutleuthausstraße 1, 76275 Ettlingen, Germany

Abstract. Research activities in the Adaptive Optics Group at the Fraunhofer Institute of Optronics, System Technologies and Image Exploitation (IOSB) in Ettlingen, Germany, revolve around imaging and laser propagation through strong turbulence, especially along horizontal paths. We are developing simulations, theoretical models, image processing software and adaptive optics systems. This paper gives an overview of two application areas bound by the same deficiency: both image processing and laser correction systems often require information about average turbulence strength at the time of their operation in order to function properly or to maximize their effectiveness.

1. Introduction
Performance of electro-optical systems is always affected by environmental factors. One of such factors is atmospheric turbulence which limits the attainable resolution delivered by ground-based [1] or space-based [2] telescopes. Similarly affected are the directed-energy [3], or horizontal-path imaging and laser communications systems [4].

Measurements of optical turbulence at Fraunhofer IOSB in Ettlingen have a long tradition [5-7]. More specifically, the research group Optics of the Atmosphere carries out vertical and horizontal turbulence measurements in various climates around the World in order to predict the performance of imaging and laser systems when deployed in less-than-perfect conditions. On the other hand, the Adaptive Optics group focuses on turbulence correction, in software and in hardware. The overlap between the research themes of both groups is therefore significant. This overlap forms the topic of this paper. Specifically: how information about average turbulence strength at the time of the observations can be used to maximize the performance of adaptive optics (AO) and image processing systems. This premise will be demonstrated by two examples: digital holographic wavefront sensor (DHWS) and estimation of turbulence strength directly from target images.

2. Digital holographic wavefront sensor
The most widely applied wavefront sensor for the measurement of turbulence-induced phase errors is the Shack-Hartmann sensor. This sensor has several advantages, the most significant being its clear principle of operation. On the other hand it has some limitations, namely the sensor relies on a time consuming matrix-vector multiply algorithm to reconstruct the wavefront, and secondly the error rate of the wavefront reconstruction algorithm becomes significant with the onset of strong scintillation which causes obscuration or saturation of sections of the detector [8].

An interesting alternative is the holographic wavefront sensor [9-11] which directly measures the strengths of the Zernike modes, hence avoiding extra processing time. Several approaches have been proposed with regard to the core of this sensor, i.e. the diffractive optical element (DOE) which contains the holograms of one or several Zernike (or actuator) modes. The two main categories of these approaches are concerned with the type and generation method of the hologram, i.e. analogue, using a holographic plate, or digital, using a multiplexed computer generated hologram (CGH). The former approach has the advantage of low cost and high spatial resolution, while the latter offers more flexibility. It is this flexibility (or adaptability) which is the focus of the current communication. Other properties of the DHWS which we have tested, e.g. its insensitivity to scintillation, are described in detail elsewhere [11].

The implementation of holographic wavefront sensor for one wavefront aberration is as follows: with one DOE, two holograms are recorded one after another. The object beams are symmetrically arranged converging beams – they form two foci (called here $A+$ and A-) behind the hologram plate where the detector(s) will subsequently be positioned (figure 1). The reference beams are collimated and have conjugated wavefronts (with phases $2\pi aZ_i(x,y)$ and $-2\pi aZ_i(x,y)$) corresponding to a specific Zernike mode Z_i. For the first hologram, the amplitude of the chosen aberration is $+a$, where a is the maximum amplitude of the chosen mode that the sensor will be able to measure. For the second hologram, the amplitude is $-a$. After recording, this setup consisting of DOE and two detectors per mode can be used as wavefront sensor. The incoming light will be diffracted into positions $A+$ and A-. The normalized difference of intensities $(I_{A+} - I_{A-}) / (I_{A+} + I_{A-})$, integrated over a small area on the detector, is proportional, within a certain range, to the amount of the aberration mode contained in the input wavefront.

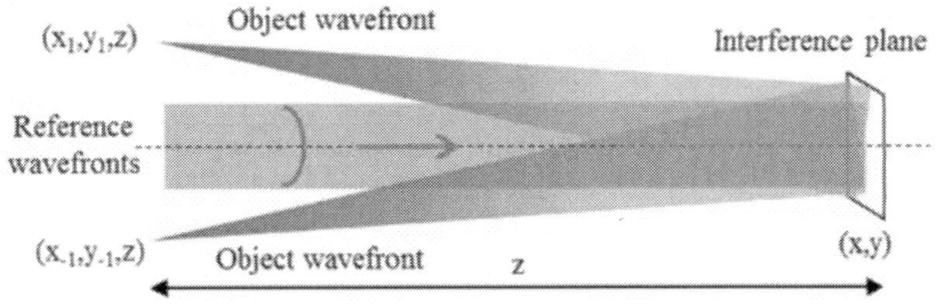

Figure 1. Recording scheme of one aberration mode in the holographic wavefront sensor, from [11].

There are two properties of holographic wavefront sensor which were just alluded to in the above paragraph: (1) its operation is bound to a range $\pm a$, unless one can change a after hologram recording, and (2) the operational characteristics of the sensor such as sensitivity or linearity are connected to integration area on the detector, in case of a CCD or CMOS array, or to pinhole/light-sensitive area in case of avalanche photodiodes. Both of these properties have been extensively tested in [11] and we only give a summary here.

The advantage of the DHWS is that its properties can be changed on the fly ("adaptable adaptive optics"). These include: the number of Zernike modes sent to the spatial light modulator (SLM), and the range of operation (a). One may want to decrease the former if turbulence gets stronger, while the former could be made smaller after the AO loop is closed in order to increase both linearity and sensitivity of the sensor to AO residual aberrations.

Defocus-only DHWS is shown in figure 2. It consists of a beam expander, a defocus generation system, CGH displayed on the SLM and the CCD. The two images on the camera corresponding to the defocus channel are separated by 16 mm and the plane of the camera is placed 800 mm from the SLM. In this experiment, 1.4λ has been chosen for a to be encoded in the CGH. Previous measurements of the atmospheric conditions in Ettlingen, Germany, showed that this amplitude range is sufficient to cover the defocus variability over our paths of interest but the range can be very easily changed by sending new holograms to SLM when turbulence gets stronger.

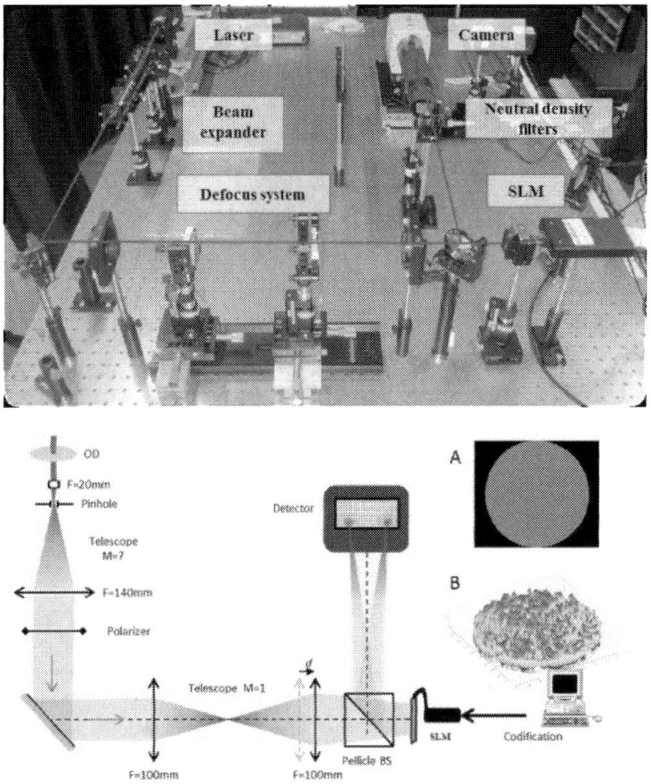

Figure 2. Top: Setup used for DHWS calibration and defocus measurement. Bottom: diagram of the same setup. Insets: **(A)** Multiplexed CGH after the mod 2π operation is applied; **(B)** Three-dimensional representation of the CGH and therefore of the optical wavefront after being diffracted by the SLM. From [11].

3. Self-calibrating turbulence mitigation in software

Image restoration methods such as speckle imaging [12,13], Richardson-Lucy deconvolution [14,15], Wiener-filter deconvolution [16] and others rely on the availability of a PSF, or in general, a transfer function. This transfer function combines the effects of: diffraction by the sensor's aperture, turbulence, and static aberrations of the imaging system.

The turbulent PSF is completely specified by known optical parameters, such as observing wavelength and the aperture of the sensor, and one unknown describing the integrated effect of the atmospheric turbulence between the source and the observer. This unknown is usually the Fried's parameter r_0.

In night-time astronomy usually an unresolved reference star is observed simultaneously with, or after the target to naturally provide an estimate of the PSF. Solar observations unfortunately do not provide reference point sources. Same is true for surveillance applications. In ground-to-ground scenarios, methods based on image motion are most commonly used [17] but their accuracy is limited by non-atmospheric sources of image jitter (such as platform vibration, movement).

Object-cancelling transformation, as a way of removing object being observed from the image formation equation, was first proposed by von der Lühe [18]. We proposed an original transformation, which we call "Fourier contrast" method [19]. This transformation is more accurate than the original spectral ratio [18] and additionally it is insensitive to image motion.

Image formation equation, expressed in the Fourier domain, is:

$$I(\vec{u}) = O(\vec{u})H(\vec{u}) \tag{1}$$

where $\vec{u}$ is a spatial frequency vector in the Fourier plane and $I(\vec{u})$, $O(\vec{u})$ and $H(\vec{u})$ stand for Fourier transforms of the (instantaneous) image, object, and speckle PSF, respectively. For each frequency we calculate the mean values and standard deviations of the power spectra:

$$C_I(\vec{u}) \equiv \frac{\mathrm{var}\left(|I(\vec{u})|^2\right)^{1/2}}{\left\langle |I(\vec{u})|^2 \right\rangle} = \frac{|O(\vec{u})|^2}{|O(\vec{u})|^2}\frac{\mathrm{var}\left(|H(\vec{u})|^2\right)^{1/2}}{\left\langle |H(\vec{u})|^2 \right\rangle} = \frac{\mathrm{var}\left(|H(\vec{u})|^2\right)^{1/2}}{\left\langle |H(\vec{u})|^2 \right\rangle} \tag{2}$$

where $\langle . \rangle$ denotes average and var (.) denotes variance. We use the term "contrast", and denote it with letter C as is customary in research pertaining to speckle. The object's information gets cancelled and one only needs to have models of the quantities on the right-hand side of equation (2) which involve a set of known parameters like telescope aperture or observation wavelength and one unknown, that is r_0.

We used this approach to estimate turbulence strength from observations of quasi-point sources (figure 3) where the estimation could be done globally, on the whole image, and extended, anisoplanatic scenes, where local estimation on isoplanatic fragments of the scene was necessary (figure 4). Subsequently, we used bispectrum image restoration technique to recover sharp representations of the objects. The whole process was therefore automated, and no parameter had to be tweaked to obtain these results.

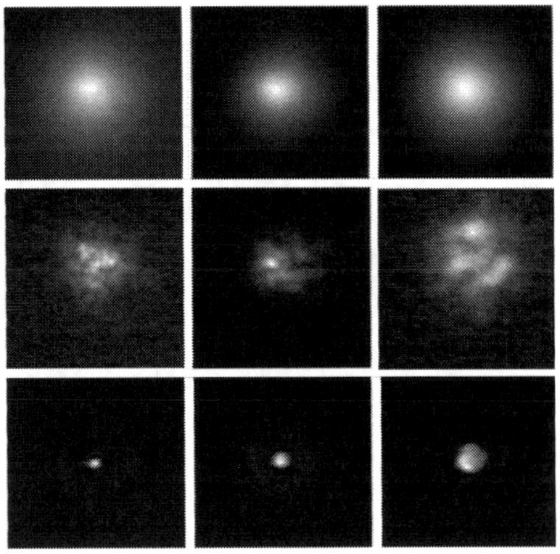

Figure 3. Left to right: 1-, 2.5- and 5-cm source. Top to bottom: long exposures, short exposures, and bispectrum reconstructions. Data taken over 2.5 km horizontal path. The sources are white-light lamps.

Figure 4. Left: single frame. Centre: shift-and-add (global-motion-removed) image composed of 500 frames. Right: bispectrum reconstruction for 20 frames based on r_0 estimated directly from the sequence. Distance to the target was approximately 7 km. Height above ground was changing between 15 and 40 m along the propagation path. The refractive index structure parameter C_n^2 obtained from r_0 was roughly 5×10^{-15} $m^{-2/3}$.

4. Conclusions and outlook

We demonstrated the utility of knowledge of average turbulence strength, even for adaptive optics systems which should be able to correct turbulence in real time. Also for turbulence correction software it is extremely useful to know r_0 or C_n^2 at the time of the observations. Work is underway to build demonstrators utilizing the ideas described in this paper.

Acknowledgments
Research sponsored by BAAINBw/WTD91.

References

[1] R. Davies and M. Kasper, "Adaptive Optics for Astronomy," Annual Review of Astronomy and Astrophysics 50, 1-47 (2012).
[2] N. Hajlaoui, C. Chaux, G. Perrin, F. Falzon, and A. Benazza-Benyahia, "Satellite image restoration in the context of a spatially varying point spread function," J. Opt. Soc. Am. A 27, 1473-1481 (2010).
[3] D. V. Murphy, "Atmospheric-turbulence compensation experiments using cooperative beacons," Lincoln Laboratory Journal 5, 25-44 (1992).
[4] M. Vorontsov, J. Riker, G. Carhart, V. S. R. Gudimetla, L. Beresnev, T. Weyrauch, and L. C. Roberts, Jr., "Deep turbulence effects compensation experiments with a cascaded adaptive optics system using a 3.63 m telescope," Appl. Opt. 48, A47-A57 (2009).
[5] K. Stein, K. Weiss-Wrana, and A. Kohnle, "IR imaging of a point target through inhomogeneous media along a slant path of 2.3 km", Proceedings of SPIE Vol. 2580, pp. 86-105 (1995).
[6] K. Weiss-Wrana and L. S. Balfour, "Statistical analysis of measurements of atmospheric turbulence in different climates", Proceedings of SPIE Vol. 4538, pp. 93-101 (2002).
[7] D. Sprung, P. Grossmann, E. Sucher, K. Weiss-Wrana, and K. Stein, "Stability and height dependant variations of the structure function parameters in the lower atmospheric boundary layer investigated from measurements of the long-term experiment VERTURM (vertical turbulence measurements)", Proceedings of SPIE Vol. 8178, 817809 (2011).

[8] L. Yatcheva, R. Barros, S. Keary, and S. Gladysz, "Laser setup for measuring ground-level turbulence," these proceedings
[9] M. A. A. Neil, M. J. Booth, and T. Wilson, "New modal wave-front sensor: a theoretical analysis," J. Opt. Soc. Am. A 17, 1098–1107 (2000).
[10] G. Andersen and R. Reibel, "Holographic wavefront sensor," Proc. SPIE 5894 (2005).
[11] P. Marin Palomo, A. Zepp, and S Gładysz, "Characterization of the digital holographic wavefront sensor," Proceedings of SPIE, 9242, 92421T, (2014).
[12] A. Labeyrie, "Attainment of diffraction-limited resolution in large telescopes by Fourier-analyzing speckle patterns in star images," Astronomy & Astrophysics 6, 85-87 (1970).
[13] A. W. Lohmann, G. Weigelt, and B. Wirnitzer, "Speckle masking in astronomy: triple correlation theory and applications," Applied Optics 22, 4028-4037 (1983).
[14] W. H. Richardson, "Bayesian-Based Iterative Method of Image Restoration," J. Opt. Soc. Am. 62, 55-59 (1972).
[15] L. B. Lucy, "An iterative technique for the rectification of observed distributions," Astronomical Journal 79, 745-754 (1974).
[16] N. Wiener, "Extrapolation, interpolation, and smoothing of stationary time series," (John Wiley, New York, 1949).
[17] S. Zamek and Y. Yitzhaky, "Turbulence strength estimation from an arbitrary set of atmospherically degraded images," J. Opt. Soc. Am. A 23, 3106-3113 (2006).
[18] O. von der Lühe, "Estimating Fried's parameter from a time series of an arbitrary resolved object imaged through atmospheric turbulence," J. Opt. Soc. Am. A 1, 510-519 (1984).
[19] S. Gladysz, R. Baena Gallé, R. Johnson, and L. Kann, "Image reconstruction of extended objects: demonstration with the Starfire Optical Range 3.5m telescope," Proceedings of SPIE, Volume 8535, 85350M-1-13 (2012).

Search for an Astronomical Site in Kenya (SASKYA) update: Installation of on-site automatic meteorological stations

Edward Graham[1*], Richard Vaughan[3], David Buckley[2], Koi Tirima[41]

[1]Lews Castle College, University of the Highlands and Islands, Stornoway, HS1 2LG, Scotland.

[2]South African Astronomical Observatory, Observatory, Cape Town 7925, South Africa.

[3]InCA Nairobi Ltd, Centenary House, 3rd Floor, P.O. Box 97 – 00606, Westlands, Nairobi, Kenya.

[4]Inoorero University, Inoorero Centre, Parklands, PO BOX 60550-00200, Nairobi, Kenya.

E-mail: edward.graham@uhi.ac.uk

Abstract. At present, there are few large telescopes (1-metre diameter mirror or larger) in sub-Saharan Africa, and the objective of the Search for an Astronomical Site in Kenya ("SASKYA") project has been find the best possible sites in Kenya for such an observatory. An initial study using ERA-interim reanalyses and high resolution meteorological model output identified three preferential sites, based on a selection of thirteen candidate mountain peaks across Kenya. These three sites are Mount Kulal (2.72°N, 36.93°E, 2293m); OlDonyoNyiro (2.09°N, 36.85°E, 2752m) and Warges (0.95°N, 37.40°E, 2688m). Recently (as of Spring 2014), an automatic meteorological station has been installed on the summit of Warges mountain, but security issues have prevented the installation of one on Mount Kulal. Furthermore, at the former site, the lack of suitable roadway through the heavily forested mountain-side will probably preclude further investigation. Hence, an additional but lower site, near to Warges (Ololokwe mountain, 0.83°N, 37.53°E, 1920m), which has a road to the summit, will be investigated for suitability in the near future. Following a period of successful meteorological observation at either or all of the above-mentioned sites, comparison and validation studies will be enacted using climate reanalyses. If further funding permits, all-sky cameras will be installed and high resolution EUMETSAT satellite imagery will also be consulted to determine local cloudiness coefficients.

[1] To whom any correspondence should be addressed (edward.graham@uhi.ac.uk)

1. Introduction

The overall project goal of the Search for an Astronomical Site in Kenya ("SASKYA") project is to identify the best possible site in Kenya for astronomical observation, using state-of-the-art meteorological, climatological and geophysical datasets and other relevant information. Long-term cloud cover, integrated water vapour, relative humidity, air temperature, wind and vertical velocity datasets of the ERA-interim climate model reanalysis data were analysed in order to determine the best possible site(s). Data from the highest resolution meteorological model available for Kenya, namely the United Kingdom Met Office (UKMO) Africa Limited Area Model (LAM), was also used in an attempt to narrow an original shortlist of 13 potential candidate sites down to the most favoured sites. Aerosol data from the FriOWL software were also consulted [1].

2. Overview of Astroclimatological Parameters

In a comprehensive study on site selection for large telescopes, Graham [1] lists the main atmospheric constraints on astronomical viewing as the following:

a) Cloudiness
b) Atmospheric absorption due to atmospheric water vapour
c) Stability and turbulence of the atmosphere above an observatory
d) Observatory-level night-time relative humidity
e) Windspeeds at observatory level
f) Dust and aerosol loadings

In relation to (a), it is obvious in saying that night-time cloud-free skies are essential for successful operation of any astronomical observatory. It is important to add, however, that due to the typical long-exposure times, even partially clouded skies are detrimental to observations [2]. In northern Chile, total sky cloudiness fractions of less than 10% are widespread; in Kenya, we may consider sites only if they have a cloudiness fraction of 30% or less.

In the case of (b), because the primary absorber of infra-red radiation in the Earth's atmosphere is attributable to water vapour, the knowledge of total integrated amount of water vapour (IWV) in any atmospheric column is important for astronomical observation. Sarazin [2] quotes an IWV value of less than 5mm as being requisite for the satisfactory operation of large telescopes at visible wavelengths, and a value of less than 3mm at infra-red wavelengths. Such values will be difficult to achieve for lengthy periods in Kenya, except at high altitude.

The stability of the atmosphere and degree of microscale turbulence due to thermal effects largely controls the magnitude of astronomical "seeing" variable ("seeing" can be understood here as the "blurriness" of a stellar image over time). Although no author has yet been able to accurately reproduce seeing statistics based on solely micro-meteorological or climatological data for lengthy time periods, broad-scale correlation between gently subsiding (and adiabatically warmed) airmasses and areas of reduced seeing have been deduced by Graham [1]. Here, we will attempt to use this knowledge to infer spatio-temporal regions of improved seeing conditions across Kenya.

Night-time relative humidity, if too high, can be detrimental for astronomical observation due to the collection of dew or frost on optics and lenses. Similarly, too low or too high windspeeds can also preclude observation, due to (i) incomplete flushing of the telescope enclosure leading to thermal "dome" seeing, or (ii) telescope and enclosure shake due to strong winds, respectively. High aerosol or dust loadings are also problematic for the successful operation of telescopic optics in desert regions.

1.2 Data Sources

The data used in this study come from the European Centre for Medium Range Weather Forecasting (ECMWF), the UK Met Office and the National Center for Environmental Prediction / National Center for Atmospheric Research (NCEP-NCAR) in the USA. The various datasets used include the ERA-interim reanalysis [3], ERA40 reanalysis [4] and the NCEP-NCAR reanalysis [5]. High resolution data from United Kingdom Met Office (UKMO) Limited Area Model (LAM) for Africa is also used for the first time in this study. Presently, the UKMO is supporting a major technical co-operation project for Numerical Weather Prediction in Africa, and has therefore made the Africa-LAM available to academic researchers [6].

Reanalyses are the most accurate representations of the Earth's atmosphere for the past 50 years that are currently available. They are based on a state-of-the-art combined numerical weather prediction model and data assimilation scheme, using a wide range of observations such as satellites, ships, buoys and weather balloons. The scientific validation of the meteorological and climatological variables described by these reanalyses systems has been shown to be satisfactory in most circumstances [4, 5, 6]. Although recognising that no single reanalysis dataset is perfect in its representation of the Earth's atmosphere through time, these reanalysis datasets represent the culmination of an enormous effort by the meteorological and climatological communities across the globe over the past 50 years to accurately depict the state of the atmosphere.

The ERA-interim reanalysis product is currently publicly available at 0.75° latitude/longitude resolution (or about 75km horizontal scale), although much higher resolution data (e.g. EUMETSAT satellite data) has already been assimilated into the model using the data assimilation procedure known as 4D-VAR [4].

The Africa-LAM is a non-hydrostatic model (meaning non-geostrophic flow may be permitted in certain cases e.g. such as local mountain-valley winds or lake-breezes), with full prognostic equations used (meaning the best possible representation of the atmospheric state is determined). There are two configurations of the model, one with 38 vertical levels (L38) and another with 70 vertical levels (L70); output from the L70 model was used only in this study. The horizontal grid resolution of the L70 configuration is 12km [6].

1.3 The Candidate Sites

Initially, thirteen candidate sites were chosen across Kenya (Table 1). Of these, two lay at relatively high altitude to the west of the Rift valley (Kapcholio and Loita hills); the remainder were to be found across a range of altitudes to the east of the Rift valley, including Warges, Mount Kulal and Mount Marsabit in the Chalbi desert.

Immediately, because of the relative close proximity (0.3° lat/lon) between the Ndoto and OlDonyoNyiro mountains (the reanalysis data is unable to differentiate within this distances), it was assumed a priori that OlDonyoNyiro was the more favourable of the two peaks due to its higher altitude (and thus lower IWV), and thus the Ndoto peak was dropped from the list.

As the research progressed, however, it soon became apparent that some of the aforementioned sites as listed in Table 1 (e.g. Loita hills, Kapcholio) were not of high enough quality, and therefore could also be dropped from any further analyses. At the same time, however, closer inspection of the Kenyan terrain and vegetation (using Google Earth, site reconnaissance and other means) revealed new mountain peaks in the vicinity of the best performing sites. It was decided to conduct analyses for these additional sites as well.

Table 1. The thirteen candidate sites analysed during SASKYA. The latitude, longitude and altitude of each site are given.

Name	Latitude (°N)	Longitude (°E)	Altitude (m)
Ol Donyo Nyiro	2.09	36.85	2752
Mount Kulal	2.72	36.93	2293
Mount Marsabit	2.28	37.95	1420
Kapcholio	2.12	35.15	2790
Ndoto	1.75	37.17	2297
Warges	0.95	37.40	2688
Loita Hills	-1.70	35.80	2642
Mtelo	1.39	35.21	3325
Cherangany hills	1.21	35.27	3369
Mt. Furroli	3.45	38.02	1869
Ol DonyoLenkyo	1.45	37.09	2550
Nr. Sololo	3.36	38.39	~1300
Mt. Kenya	-0.09	37.19	4417

3. Results and Discussion

Thirty years (1981-2010) of ERA-interim, ERA40 and NCEP-NCAR climate model reanalyses data were analysed in detail (for the Kenyan region), and two years (2010-2011) of Africa-LAM high resolution meteorological model, made available by the UK Met Office, were also consulted. The results of this part of the study have been presented in more detail by [7, 8, 9, 10, 11].

With respect to cloud cover, findings suggest the mountain peaks of Mount Kulal and OlDonyoNyiro are the most favourable, although the whole region bordering and including the Chalbi desert (from Warges north to Mount Marsabit) appears reasonably cloud-free, with annual averages of about 30% of cloud cover. Significant seasonal variation in cloud cover can be expected, however, as well as notable inter-annual and interdecadal oscillations. The prevalence of high ice cirrus cloud across Kenya may be problematic.

Annual averaged integrated water vapour values for all sites (except Mount Kenya) are uncomfortably high, ranging from approximately 10 to 14mm over the 2,000 to 2,500 metre altitude range. Lowest IWV values are found in the south-east of Kenya; highest values in the north-west, and lower IWV values are to be found east of the Rift valley. One must ascend to a level of at least 4,000 metres in order to obtain consistent IWV values of 5mm or less – and Mount Kenya is the only site in Kenya where this is achieved for large parts of the year. Because the presented IWV values are almost uniformly high, it is recommended that ground-truthing of the reanalyses IWV values be conducted on the same peaks.

The vertical velocity variable was used as proxy to help determine areas of improved "seeing" conditions, by identifying regions of gently descending air. The results confirm that the whole area lying between Mount Kulal, OlDonyoNyiro and the Chalbi desert is again the most favourable region in Kenya. There is a deep turbulent surface boundary layer, however, and any site selection would need to take this into account. Furthermore, there are important seasonal variations in the variable, so good astronomical seeing conditions cannot be expected anywhere all-year round.

Based on the results of the above studies, at a SASKYA project meeting in South Africa in December 2012, the peaks of Mount Kulal, Warges and Ol Donyo Nyiro were selected as the three most preferable sites to date. More recently (as of Spring 2014), a solar powered automatic meteorological station has been installed on the summit of Warges mountain, and it currently transmitting live data every 5 minutes to a central server in South Africa by means of the 4-G cellular phone network. Unfortunately, security issues have prevented the installation of an automatic weather station on the summit Mount Kulal, following the theft of some initial equipment. Furthermore, at the former site, the lack of suitable roadway through the heavily forested mountain-side will probably preclude further investigation. Hence, an additional but lower site, near to Warges (Ololokwe mountain, 0.83°N, 37.53°E, 1920m), which has a road to the summit, will be investigated for suitability in the near future. It is worth noting that it would be impossible to build an observatory on the very high slopes of Mount Kenya because it lies within a national park (the construction of any buildings is prohibited).

Following a period of successful meteorological observation at either or all of the above-mentioned sites, comparison and validation studies will be enacted using climate reanalyses. If further funding permits, all-sky cameras will be installed and high resolution EUMETSAT satellite imagery will also be consulted to determine local cloudiness coefficients.

Outlook

After the initial climatological studies took place (as described above), infrastructural and cultural issues are expected to begin to play a more important role in the final site selection process. Trade-offs between the best possible site in climatological terms, and the practicalities of installing remote equipment in isolated, inaccessible areas will become important considerations. Once 12-24 months of weather data have been collected and analysed (ideally after comparison with reanalyses products), and also a complementary study completed using EUMETSAT satellite cloud mask data, it should become reasonably clear which final site(s) in Kenya are most preferable for SASKYA.

References

[1] Graham, E. (2008). The FriOWL Guide: A site selection tool for extremely large telescopes using climate data. Forschungsbericht Nr. 2008-09-MW, Institut fur Angewandte Physik, Universitat Bern. Available online at: http://amao.saao.ac.za/wp-content/uploads/2012/12/E-ELT_PhD_thesis.pdf

[2] Sarazin, M. (2003). Current knowledge of the Outer Scale and other parameters. Mini Workshop on Atmosphere knowledge and Adaptive Optics for 8 to 100 m telescopes, held at the European Southern Observatory, Garching, Germany, 13 October 2003.

[3] Dee et al., (2011). The ERA-Interim reanalysis: configuration and performance of the data assimilation system. Quarterly Journal of the Royal Meteorological Society, 137: 553-597.

[4] Uppala, S.M., Kållberg, P.W., Simmons, A.J., Andrae, U., da Costa Bechtold, V., Fiorino, M., Gibson, J.K., Haseler, J., Hernandez, A., Kelly, G.A., Li, X., Onogi, K., Saarinen, S., Sokka, N., Allan, R.P., Andersson, E., Arpe, K., Balmaseda, M.A., Beljaars, A.C.M., van de Berg, L., Bidlot, J., Bormann, N., Caires, S., Chevallier, F., Dethof, A., Dragosavac, M., Fisher, M., Fuentes, M., Hagemann, S., Hólm, E., Hoskins, B.J., Isaksen, L., Janssen, P.A.E.M., Jenne, R., McNally, A.P., Mahfouf, J.-F., Morcrette, J.-J., Rayner, N.A., Saunders, R.W., Simon, P., Sterl,

A., Trenberth, K.E., Untch, A., Vasiljevic, D., Viterbo, P. and Woollen, J. (2005). The ERA-40 re-analysis. Quarterly Journal of the Royal Meteorological Society, 131, 2961-3012. doi:10.1256/qj.04.176.

[5] Kalnay, E., Kanamitsu, M., Kistler, R., Collins, W., Deaven, D., Gandin, L., Iredell, M., Saha, S., White, G., Woollen, J., Zhu, Y., Chelliah, M., Ebisuzaki, W., Higgins, W., Janowiak, J., Mo, K.C., Ropelewski, C., Wang, J., Leetmaa, A., Reynolds, R., Jenne, R. and Joseph, D. (1996). The NCEP/NCAR Reanalysis Project. Bulletin American Meteorological Society 77, 437-471.

[6] UK Meteorological Office (UKMO; 2012). Limited Area Model for Africa (Africa-LAM), NCAS British Atmospheric Data Centre. Available from http://badc.nerc.ac.uk/view/badc.nerc.ac.uk__ATOM__DE_2f0971ba-f76d-11df-8151-00e081470265

[7] Graham, E., Vaughan, R., Buckley, D., and K. Tirima (2014). Search for an Astronomical Site in Kenya (SASKYA) using Climate Reanalyses and High-Resolution Meteorological Model Data. Theoretical and Applied Climatology (re-submitted with minor corrections completed).

[8] Buckley, D. A. H., Belay, S., Biressa, T., Catala, L., Graham, E. Pickering, T., Tirima, K., Vaughan, R. and Wanyoike, C. (2014). *Astronomical Site Testing Initiatives in Africa*. International Astronomical Union MEARIM 2014 Regional Meeting, Beirut (1-6 Sep 2014).

[9] Graham, E., Vaughan, R., Buckley, D., and K. Tirima (2014). *Search for an Astronomical Site in Kenya (SASKYA): Installation of on-site automatic meteorological stations*. Presentation at the University of the Highlands and Islands Research Conference (5-7 November 2014), Elgin, Scotland.

[10] Graham, E. (2011). *SASKA: Search for an Astronomical Site in Kenya. An analysis of candidate sites using meteorological re-analyses datasets*. Final Report: Version 1 (issued: 27 March 2012).

[11] Graham, E. (2011). *SASKYA-2: Search for an Astronomical Site in Kenya Stage 2 (Examination and characterisation of five remaining candidate sites using high resolution data from the United Kingdom Met Office's L70 African Model)*. Version 1 (issued: 29 January 2013).

Acknowledgments

The ECMWF ERA-Interim data used in this study/project have been provided by ECMWF/have been obtained from the ECMWF data server. The NCEP-NCAR data has been provided by the NOAA-CIRES team and the National Centers for Environmental Prediction / Atmospheric Research. The United Kingdom Met Office Limited Area Model for Africa is kindly made available by the UK Met Office, through the British Atmospheric Data Centre (BADC).

Funding and expenses for this project was provided by the Joint Kenya-South Africa Science and Technology programme, INCA communications (Nairobi, Kenya), the South African Astronomical Observatory (Cape Town, South Africa), the South African Large Telescope (SALT), the Carnegie Trust for the Universities of Scotland and the University of the Highlands and Islands (Scotland) – thank you to all.

Lessons learned from developing turbulence profilers for telescopes' instruments

Andrés Guesalaga[1], Benoit Neichel[2], Clémentine Béchet[1], Javier Valenzuela[3,1]
[1]Universidad Católica de Chile, Vicuna Mackenna 4860, Santiago, Chile
[2]LAM, Rue Frederic Joliot Curie, 13013 Marseille, France
[3]European Southern Observatory, Alonso de Córdova 3107, Vitacura, Chile
E-mail: aguesala@ing.puc.cl

Abstract. This article presents the results obtained during three years of developing turbulence profilers for two different telescopes; namely Gemini South and the future Adaptive Optics Facility (AOF). The profilers are embedded in a facility instrument that provides the data from the Shack-Hartmann wavefront sensors which feed the SLODAR approach used to generate the profiles. The main results focused on two unsolved problems: dealing with the dome seeing and the effect of the atmosphere outer scale on the accuracy of the profilers.

1. Introduction

The performance of wide-field Adaptive Optics (AO) system is expected to be optimized with a good knowledge of the vertical distribution of the turbulence strength in the atmosphere [1]. As the first laser-based Multi-conjugate AO system offered to the astronomers community, GeMS (Gemini Multi-conjugate AO system) has already started to study two different methods for the estimation of the Cn^2 profile out of its AO telemetry [2]. These two techniques rely on the Slope Detection and Ranging (SLODAR) approach [3], using the cross-covariances of simultaneous wavefront sensor measurements each pointing at a different guide star in the field of view. In previous works, the SLODAR methods have been adapted to account for the cone effect when using laser guide stars (LGS) [2,4]. This leads to the estimation of the Cn^2 profile over a non-equally discretized representation of the vertical turbulence distribution. The two methods have already been successfully used on GeMS telemetry recorded during the 2011, 2012 and 2013 campaigns [2,5,6]. A statistical analysis of these results has also been addressed in a recent paper [6]. We study the application of previous work to a new AO configuration, defined by the operating modes of the Adaptive Optics Facility (AOF) at the Very Large Telescope (VLT, Paranal Observatory of ESO). The AOF consists of an evolution of one of the VLT unit telescopes to a laser driven adaptive telescope with a deformable secondary mirror and four LGSs [7]. Because of its wide-field and tomographic AO modes, ESO is studying how to integrate one of this profiling technique into the AOF.

2. Results at Gemini

2.1. *Profiling and dome seeing*

In the literature, two different approaches have been used to analyse the data from SLODAR optical triangulation [2]. The first approach deduces the Cn^2 and wind profiles from the deconvolved spatio-temporal cross-correlations of the. The second one does not apply such deconvolution, but rather models the spatial covariances of the slopes using a specific turbulence statistics (e.g. Kolmogorov or von Kármán): the turbulent layers heights and strengths are recovered by fitting theoretical impulse response functions to the cross-covariance of the slopes measurements.

These impulse response functions express how a theoretical turbulent layer located exactly at the centre of a bin corresponds to the measured slope covariance functions.

Essential elements in this method are the auto-correlation submaps. They not only provide an estimate of the measurement noise, but also give an idea of how well the data matches the theoretical statistics (usually Kolmogorov or von Kármán) used to generate the theoretical submaps. The central point of the auto-correlation submaps corresponds to the centroid from each subaperture correlated with it. As the measurement noise is fully correlated at the central point of this submap, this component will be the sum of the slope and the noise variances. Hence, the difference between the central points of the theoretical and measured auto-correlation submaps (Fig. 1) gives an estimate of the noise variance.

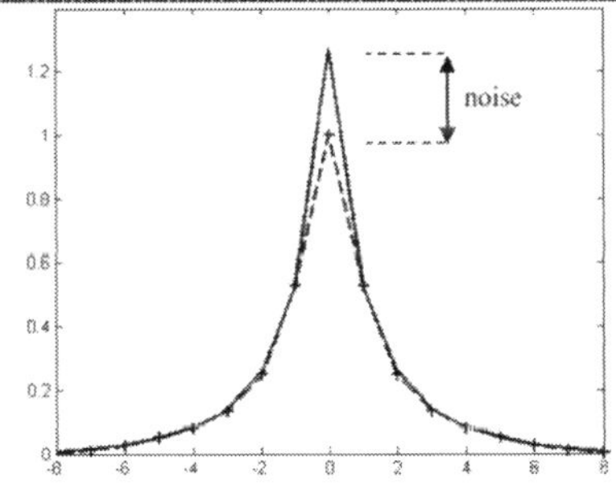

Figure 1. Central slice of the X-slopes submaps in the Y direction. The solid line is the measured covariance and the dashed line in the theoretical impulse function. The difference in the central pixel corresponds to noise (data from April 15th, 2011).

In many cases encountered with GeMS data, the profiles obtained by the model fitting and the deconvolution approaches differed similar to the ones shown in Fig. 2. The model fitting method produces a relatively strong negative Cn^2 at the bin immediately above the one at 0km. On the contrary, the estimation by deconvolution from measured autocorrelation do not exhibits such unrealistic estimate. This point is of major importance to define which estimation approach is the most adequate for our objective, and we discuss here the reasons for such discrepancy.

During the two-year measurement campaign, strong dome seeing conditions were detected in a significant number of cases (~ 30% of the recorded conditions). Fig. 2 (left) shows an example of the slope variance in the X direction for the five WFSs after static aberrations were removed. The brighter pixels correspond to the subapertures with abnormally high variance that in this case was explained by turbulence behind the secondary mirror. Such strong dome seeing typically narrows the auto-covariance impulse response as shown in Fig. 2 (right panel). A clear departure from the theoretical shape (crosses) is observed for the measured function (continuous line) with a noticeable "elbow" around the third component off the centre.

Turbulence models such as von Kármán (finite outer-scale values) or Kolmogorov only account for the atmospheric turbulence and not for the non-Kolmogorov turbulence inside the dome. Trying to fit the measured covariance submaps to these theoretical statistical models can lead to erroneous profile estimations. Figure 3 (left) provides an example of such case where a strong negative value in the turbulence intensity appears at lower altitudes. A common characteristic of such disturbances is that they simultaneously show up in the same subapertures in all five WFSs, so the effect is the same as turbulence at the ground, making the estimation of the lower bins particularly sensitive to this effect.

On the other hand, the use of the deconvolved cross-correlation implicitly accounts for the contribution of dome turbulence as the deconvolution function in equation (3) is constructed from the actual measurements. When this dome contribution is large, a narrower convolution function is obtained, leading to realistic estimates of the lower layers. However, this autocorrelation function also

includes the turbulence from the upper layers, which should behave closer to Kolmogorov statistics. This is our understanding of the degraded layer strength estimations for the upper layers.
The approach based on the deconvolved temporal cross-correlation has been extensively tested in several campaigns during 2011, 2012 and 2013.

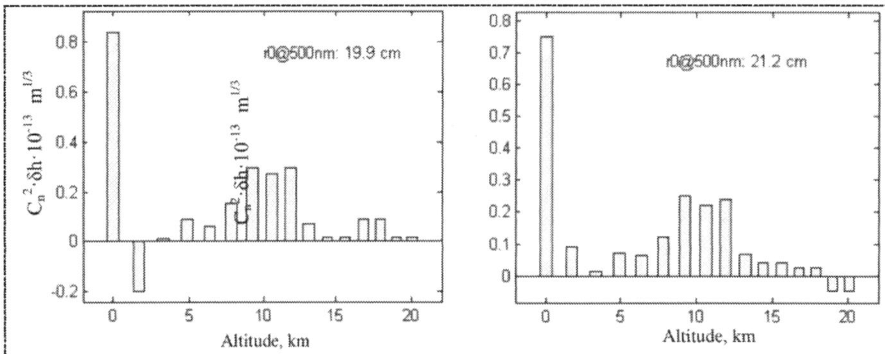

Figure 2. Estimation using the model fitting (top) and the deconvolved cross-correlation (bottom) methods. The former shows a higher value at the first bin and a negative estimation at the second, due to non-Kolmogorov turbulence (data from November 4th, 2012)

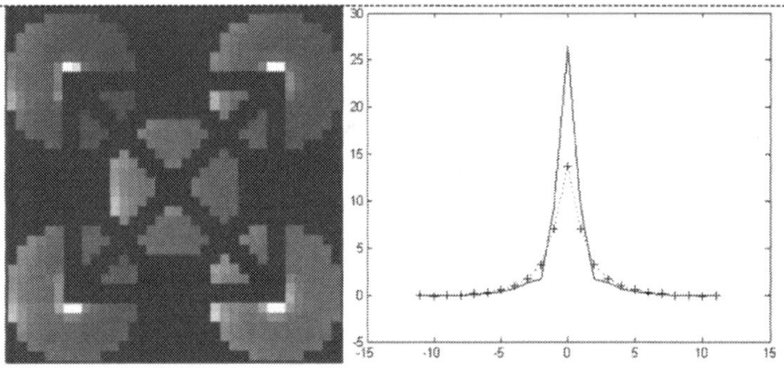

Figure 3. Left: variance of the subaperture slopes in X direction for the 5 WFSs. Right: central slices of the auto-covariance response in X direction for a theoretical Kolmogorov (dotted) and measured (continuous line) turbulences. Notice the "elbow" forming around the $\pm$ 2 components in the abscissa (data from December 16th, 2011).

2.2. Frozen flow decay ("melting")

Using temporal correlation sequences, the layers can be tracked and their speed and direction estimated. As the individual correlation peaks are being tracked, the degree of correlation can also be computed, providing an estimation of the evolution of the layer, i.e. the correlation decay or the rate of 'melting' as it crosses the telescope's field of view.
By selecting cases where the turbulence layers can be clearly isolated and tracked we focus on the analysis and interpretation of the observed dynamics of the turbulence intensity. In particular, we have estimated the rate of de-correlation of the layered turbulence as it passes over the telescope. Data collected in observation campaigns during 2011 through 2013, allowed us to gather about 74 sequences of turbulence layers where their decay rate could be determined. The set includes different conditions of altitudes, strength and velocities. In the case of the wind speed, their values ranged from 0 m/s for strong dome seeing conditions (9 cases) to speeds up to 24 m/s.

In the following analysis, we estimate the rate of de-correlation by: i) tracking the correlation peaks; ii) computing the correlation energy around the maximum. A 3x3 window is used for layers above the first bin. For the first bin (ground and dome), only the central pixel is considered; iii) normalizing the correlation values to the first correlation result in the sequence (peak at $t = 0s$).

As mentioned before and exemplified in Fig. 3, turbulences inside the dome are fairly common during operation of large telescopes. We have found that this type of turbulence remains stationary with zero translation velocity. The dome turbulence is spatially located in the same place in all WFSs, so it is seen as a layer at the ground. Hence, the profiler assigns its energy to the first bin and as this phenomenon is quasi static a distinctive stationary peak will appear during the temporal correlation sequence.

Fig. 4 (left) shows the profile for strong dome turbulence and in the right panel, the intensity of the central pixel in the sequence of time delayed correlation is shown. The plot (normalized to its value at $t = 0$) shows a rapid initial decay, that gradually stabilizes to a rate of $0.32 s^{-1}$. This fast initial decay was found to be caused by a ground layer passing over the telescope that at the start of the temporal correlation is merged with the dome seeing. As time passes, this part of the turbulence moves away from the central peak so its correlation with the fixed frame is lost, but the dome turbulence will evolve much slower, maintaining a positive match for a longer time. By measuring the slope of the asymptotic dashed line, an estimate of the rate of de-correlation can be found.

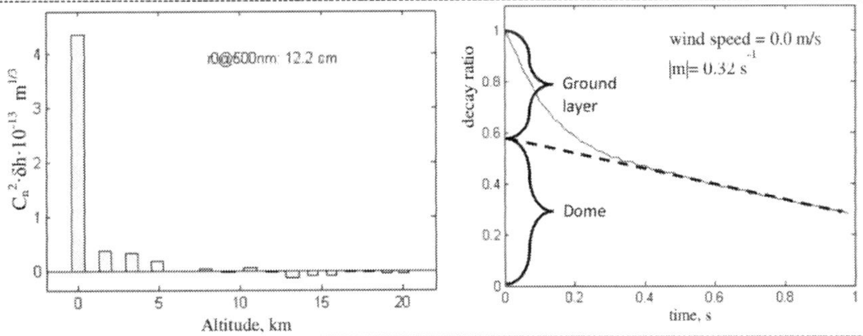

Figure 4. Left: turbulence profile in altitude, showing a strong component at the ground layer; Right: decorrelation rate of turbulence inside the dome (dashed line) and separation from the ground layer turbulence above the telescope (data from November 7th, 2012)

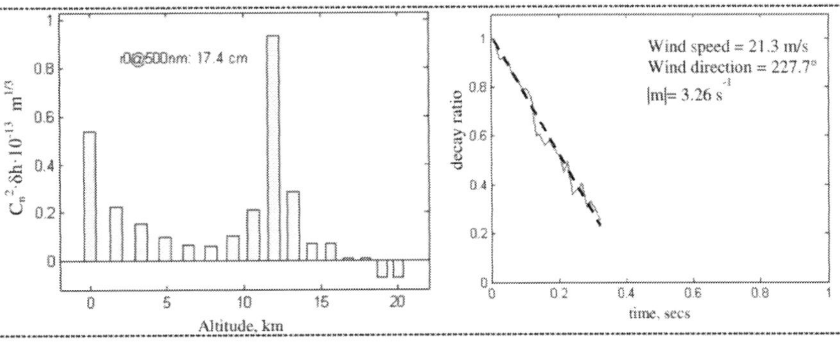

Figure 5. Left: profile for a turbulence between 11.9 and 13.2 Km from data of April 17th, 2013; Right: estimation of the turbulence melting rate (dashed line)

The profile in Fig. 5 (left panel) presents a typical case of a strong layer in the jet stream at an altitude of around 12 Km. The measured speed is 21.3 m/s with a decay ratio of $3.26 s^{-1}$. A steeper decay in the cross-correlation is observed in this case when compared to the previous cases (Fig. 5, right panel).

An interesting result of this turbulence "melting" rate comes from an integrated analysis for several cases. Fig. 6 plots the relative decay ratios against wind speed under different turbulence conditions and time. The data for this analysis were collected at Gemini South during the last two years.
A clear pattern can be observed between the dissipation rate and the wind speed with a remarkable resemblance to a linear dependence. Fitting a linear function to the set of points gives a slope of about -6.4 m^{-1}. This suggests that the 'melting' of the turbulence for short time scales, depends on the distance travelled rather than speed or altitude.
Fig. 6 highlights that many layers have been found with decay rate between -1 s^{-1} and -3 s^{-1}. Considering for instance a layer with a decay rate of -2 s^{-1}, which occurs for a wind speed around 10m/s, the correlation of this layer with the shifted one that occurred 250ms earlier (more than 125 AO cycles) still contains half of the energy in the original correlation peak. For all the cases presented in Fig. 7 the decay rate is smaller than -5 s^{-1}, meaning that the correlation decreases by a factor lower than 2 after 100ms, i.e. more than 50 cycles of the GeMS AO loop.

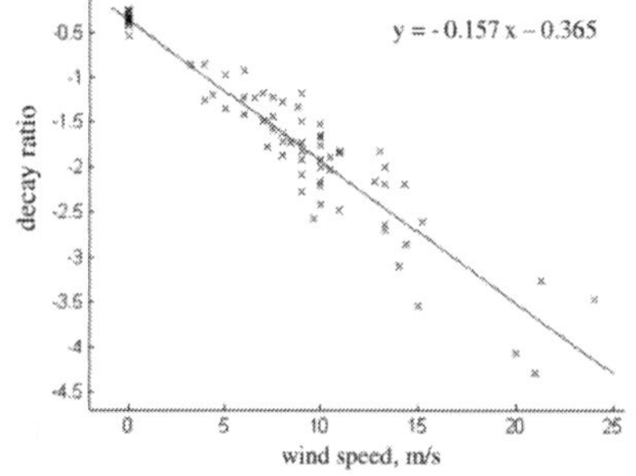

Figure 6. Rate of decay in the cross-correlation of slopes with a clear linear dependence between wind speed and turbulence correlation decay. The slope of the linear fit is in m^{-1}, suggesting that this de-correlation can be expressed in terms of the distance travelled by the layers across the pupil.

3. Results in the AOF

In this section, the profiling methods are applied on slopes sequences simulating the GALACSI wide-field mode (WFM) configuration [7].

3.1. Outer scale influence

The turbulence strength, characterized by the Fried parameter r_0, can be different for each layer, but the outer scale L_0 for a von Karman model is used as a common value. In previous studies of these profiling methods on Gemini South LGS system, GeMS, a large outer scale of 150 m was used, being almost equivalent to assuming a Kolmogorov model as seen from the 8-meter aperture of the telescope [2,5]. In the standard case of GALACSI WFM presented above, if such an outer scale L_0 = 150 m is assumed instead of the true value L_0 = 25 m, the resulting profiles are the ones shown in Fig. 7.
Fig. 7 shows that the error on the outer scale assumption leads to major error in the estimation of the r_0. The MI estimation of r_0 is wrong by 6.3%, and the one provided by the FD method is erroneous by 9.9%. The correlation criterion for the MI method drops to 88% and to 92.5% for FD. Note also that the MI method produces negative values of much greater magnitude, than in Fig. 7. Although the general profiles patterns seem quite similar as in Fig. 7, a significant difference appears for the altitude layer at 7.75 km in the case of the MI method. On its side, the FD method provides modified values compared to Fig. 7 overall for the lowest layers, but not significantly for the higher ones. While the MI really seems to detect a false layer at 7 km, the FD profile is not affected in this region.
The results of Fig. 7 highlight the importance of a correct L_0 assumption to properly estimate r0 and the relative profile. Therefore, in this section, the influence of the outer scale assumptions on the

profilers is studied more extensively, showing by how much it can impact the r0 estimate and the profile distribution.

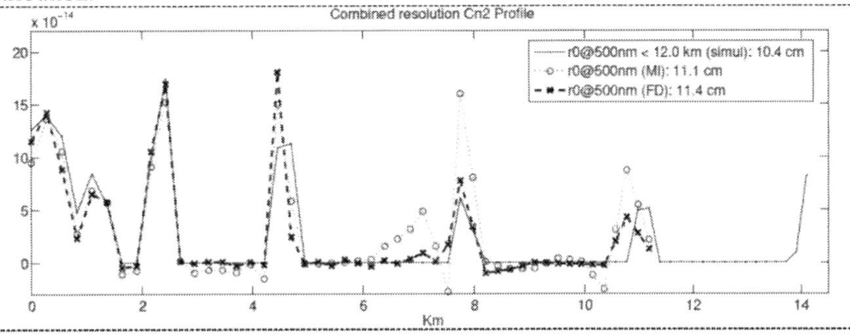

Figure 7. Same telemetry data is used as in Fig. 1, but the profiling methods are setup assuming an outer scale L_0 = 150 m instead of the true value L_0 = 25 m.

GALACSI wide field mode have been simulated with the same random seed, the same noise, but changing the value of the atmosphere outer scale $L_0^{simulation}$. For the registered data telemetry out of these simulations, the profilers have been applied assuming different values of outer scale L_0 and enhancing the influence of the latter on the result.

The estimation of r_0 is presented for MI method on the left and for FD method on the right. When the outer scale value of simulated atmosphere, $L_0^{simulation}$ matches the outer scale L_0 assumed in the structure function for a profiling method, it is observed that the error on r0 is only a few percent points when the assumed outer scale is correct. For atmospheres with short outer scale, the MI method is very sensitive to the assumed L_0 value. An overestimation of L0 can make the error jump to almost 100% of error. For the FD, the largest errors observed on r_0 are not so important, but still reach 25% error.

The quality of the profiling method cannot only be summarized by the error r0. The correlation factor is used again on the similarity of the simulated outer sale $L_0^{simulation}$ and the value of L_0assumed in the each method. This provides a complementary assessment of the profiling method for which we expect to get the maximum correlation when $L_0^{simulation} = L_0$. It is observed that for the range of outer scale considered in the simulations, the correlation stays greater than 80% for the MI method and greater than 90% for the FD one. Together with these results a superior performance of the FD method can thus already be highlighted. This last method is less sensitive to a wrong assumption on the outer scale value, but can still lead to significant error on r_0.

3.2. Finding a good L_0

The results in the previous section raise the problem of a good agreement between the real atmosphere outer scale and the assumed value in the profilers. A one-dimensional representation of these theoretical auto-covariance functions, normalized to their maximum, is illustrated on the left of Fig. 9. The various curves reveal how the outer scale L_0 affect the shape of the auto-covariance, from L_0 = 3 m to L_0 = 150 m. Actually, on this plot, the curves for L_0 = 100 m and L_0 = 150 m cannot be distinguished, meaning that the response of the system is practically the same for a Kolmogorov turbulence or for a von Karman one with outer scale greater than 100 m. For shorter outer scales, the shape of the auto-covariance varies with L_0.

The fitting is made using the 5x5 central sampled values of the autocorrelation except the central point which is the variance of the slopes in a subaperture. This variance value computed from the telemetry is affected by the noise measurement and thus is not used for the fit to the theoretical autocovariance function. The L_0 value producing the least weighted square error is kept for the L_0 assumption when applying the profiling methods. It has been checked that this simple approach is able to properly identify the outer scale value in all the diverse simulations that have been run for this study. An example is shown on the right of Fig. 10, where the measured autocovariance (line with crosses) for an AO simulation of GALACSI WFM with an average LGS flux of 40 photons per subaperture per frame and an outer scale $L_0^{simulation} = 25$ m is compared to its best fitted autocovariance (line with squares).

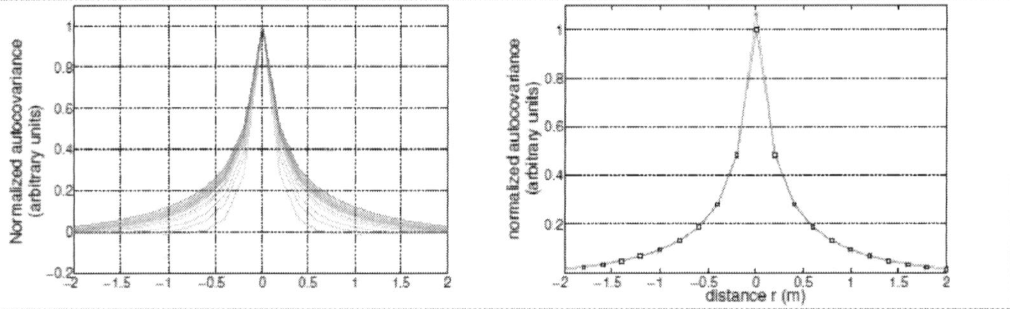

Figure 8. Left: Autocovariance functions obtained for the GALACSI WFM configuration and normalized to their maximum value. The various curves stand for varying outer scales ranging from 3 to 150 m.

4. Further work and conclusions

As highlighted in the paper, the correct assumption on the outer scale is crucial in order to estimate properly the Cn^2 distribution and the r_0 of the atmosphere. The outer scale fitting method presented above in that sense constitutes a key element to be included in the algorithms. Nevertheless, previous experience with on-sky data taken by GeMS instrument revealed the significant probability to face dome seeing or a particular turbulence contribution which would have a different outer scale than the rest of the atmosphere. In such case, the assumptions made on the turbulence model do not apply anymore and both methods are expected to provide degraded estimations of the Cn^2 profiles and r_0.
It remains for further work to find a way to take into account various outer scale at different heights. Such improvement of the profiling method is currently studied for the MI approach, for which it is expected to be easier to enlarge the matrix model including impulse response to various outer scale values.

Acknowledgments
This work was supported by the Chilean Research Council (CONICYT) grant Fondecyt 1120626

References

1. T. Fusco and A. Costille, Impact of Cn2 profile structure on wide-field AO performance, in Adaptive Optics Systems II, 7736, pp. 77360J, 2010.
2. A. Cortes, B. Neichel, A. Guesalaga, J. Osborn, F. Rigaut, and D. Guzman, Atmospheric turbulence profiling using multiple laser star wavefront sensors, MNRAS 427, pp. 2089-2099, Dec. 2012.
3. R. W. Wilson, Slodar: measuring optical turbulence altitude with a Shack-Hartmann wavefront sensor, MNRAS 337(1), pp. 103-108, 2002.
4. L. Gilles and B. L. Ellerbroek, Real-time turbulence profiling with a pair of laser guide star shack-hartmann wavefront sensors for wide-field adaptive optics systems on large to extremely large telescopes, J. Opt. Soc. Am. A 27, pp. A76-A83, Nov 2010.
5. A. Guesalaga, B. Neichel, A. Cortes, C. Bechet, and D. Guzman, Using the Cn2 and wind profiler method with wide-field laser-guide-stars adaptive optics to quantify the frozen-flow decay, MNRAS 440(3), pp. 1925-1933, 2014.
6. I. Rodriguez, B. Neichel, C. Bechet, D. Guzman, and A. Guesalaga, Statistics of atmospheric turbulence at Cerro Pachon using the GeMS profiler, SPIE, 2014.
7. R. Arsenault, P.-Y. Madec, J. Paufique, et al, "ESO adaptive optics facility progress report," SPIE 8447, 2012.

Astronomical site testing and characterization of the Guadalupe Island, Mexico

D Hiriart[1],E Iñiguez-Garín[1], M Álvarez[1], R Michel[1], S Zazueta[1] and O López-Cruz[2]

[1] Instituto de Astronomía, Universidad Nacional Autónoma de México, Apdo. Postal 870, 22800 Ensenada, Baja California, México

[2] Instituto Nacional de Astrofísica,Óptica y Electrónica (INAOE), Coordinación de Astrofísica, Luis Enrique Erro No. 1 Sta. Ma. Tonantzintla, Puebla, 72840 México

E-mail: `hiriart@astrosen.unam.mx`

Abstract. We present an ongoing campaign for astronomical site testing and characterization at the Guadalupe Island at the Northern Pacific Coast of Mexico. Characterization at optical and radio wavelengths are been carried out at the site to evaluate its potential for astronomical observatories at these wavelengths.

1. Introduction

Guadalupe Island is a Mexican biosphere reserve located in the Pacific Ocean (lat: 28° 58' 24" N, long: 118° 18' 4" W) at a distance of 241 kilometers West of the Baja California peninsula between a latitude of 28°50' and 29°12' N and a longitude between 118°13' and 118°22' W. Figure 1 shows the location of the island. Its length in the North-South direction is about 35 kilometers, and its maximum wide is about 12 kilometers with a total surface of 244 km^2 [1].

The island has a volcanic origen. Two collapsed craters form the island: at the South, the highest point is Monte Picacho (975 m) and the other one at the North named Monte Augusta with 1,297 m. Guadalupe Island has two different kinds of climate: (1) between 0 and 800 meters it has a semi-warm to very arid climate with annual median temperature of 16°C and (2) over the 800 meters the wheatear is templated during the coldest month of Winter with temperatures between 3 and 18°C and hot and dry during the Summer with an average temperature of 22°C. Frequently, the marine layer rises up to a maximum high of 800 meters over the sea level.

The ecology of the island was severally damage by the introduction of non native animals brought by the first people to arrive to the island during the XVII and XVIII centuries. Cats, dogs and goats brought a lot of destruction to native plants and birds. Starting in 2002, the Grupo de Ecología y Conservación de las Islas (GECI) and other government agencies, started the first actions to preserve the ecology of the island. On April 2006, Guadalupe Island was declared a Biosphere Reserve by the Mexican Government to promote is preservation [2][5].

There has been several reports that Monte Augusta may be an excellent site for astronomical observations. During the 1960s Guadalupe Island was visited by Dr. Eugenio Mendoza from the Institute of Astronomy of National Autonomous University of Mexico (UNAM), in his search to find a new place for the Mexican Optical Astronomical Observatory at that time located in central Mexico and being very affected by light pollution and poor astronomical weather.

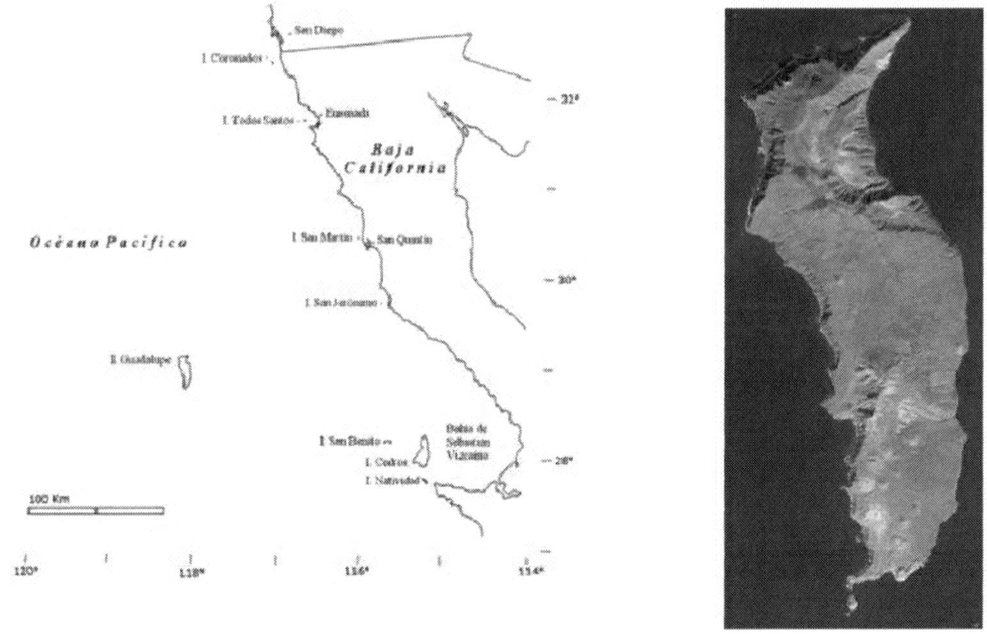

Figure 1. Macro-location of the Guadalupe Island at the West coast of the Mexican Baja California peninsula (*left*) and a satellite view of the island (*right*). In both figures North is at top and East is to the right.

This was the search that ended locating the Observatory at the San Pedro Martir sierra at the Baja California peninsula. The GECI has a biological station near Monte August since twelve years ago and its personnel have collected meteorological data and qualitative observations of astronomical quality of the site.It is claimed that the island has a very high quality, dark & transparent sky with about 240 clear nights per year (65.8 % of the nights)[3].

Since the last year a group of astronomers from Mexico and other countries started a program to quantify the astronomical quality of the site near Monte Augusta at the Guadalupe Island at optical and radio wavelengths. In this paper we present the site survey planned and carried out at the island.

2. Optical Testing

Technical staff at the Institute of Astronomy UNAM has developed and tested a robotic site testing system to assert the astronomical quality of the site near the summit Monte Augusta at the Guadalupe Island. The heart of the system is a robotic Double Image Movement Monitor (DIMM) [4]. Figure 2 shows an schematic of the system developed, it consists of:

- Robotic DIMM and protection dome
- All-sky camera
- Weather station
- web cameras.

The site survey equipment will located close to the biological station of the GECI near the summit of Monte Augusta. In that way it will be possible to the personnel from GECI to made

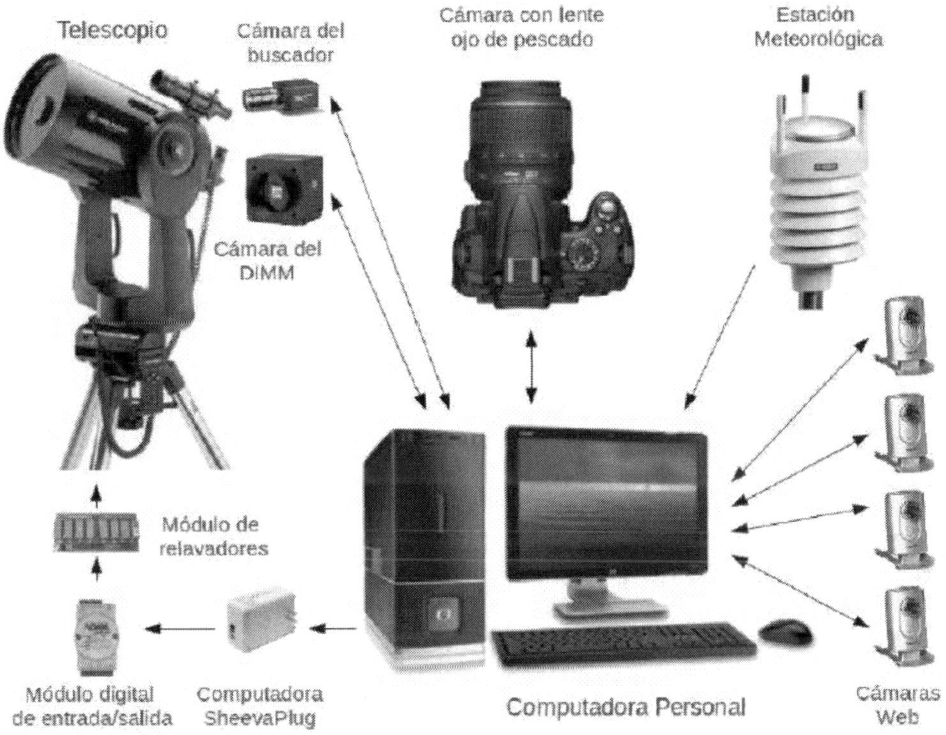

Figure 2. Equipment used to evaluate the Monte Augusta site at the Guadalupe Island.

some easy tasks. Data from all the periferical equipment will SE collected at a central computer and it will be sent thru a satellite Internet connection. Power the whole system will be provide by a set of batteries that will be charged by solar panels. The whole system was proved at the San Pedro Martir National Observatory with excellent results.

3. Radio frequency testing

As a part of the experiment "Sonda Cosmológica de las Islas para la Detección de Hidrogeno Neutro" (SCI-HI) to measure the all–sky–averaged 21 cm brightness temperature in the redshift range of $14.8<z<22.7$ the radio noise in the band from 0–250 MHz at the Guadalupe Island was measured in 2013 using a single antenna and a sampling receiver [6].

The antenna, called HIbiscus, is shown at Figure 3 deployed at the Guadalupe Island. The antenna is fixed an points to the zenith. It has a 55° beam at 70 MHz with roughly frequency independent shape on an octave of frequency [6]. Detail description will be found in J.M. Jáuregui García et al. [7].

A block diagram of the instrument is shown in Figure 4. Signal from the antenna passes through a series of amplifiers and filters to remove radio frequency interference (RFI) below 30 MHz and aliasing signals above 200 MHz. Besides the signal from the antenna, a multi-port electro–mechanical switch controlled by a computer allows collection of calibration spectra from terminators of 50Ω, 100Ω, and a short of known temperature. The signal is sampled at 500 MSamples s^{-1} with 12 bits resolution by a GE PCIe digitizer board (ICS1650) and a PC with software for power spectrum generation from 0–250 MHz. The system is placed inside a

Faraday cage about 50 m away from the antenna. The entire system is designed to run off 12 VDC batteries, allowing the system to be portable and easy to set up in remote locations [6].

The major noise contribution in the 40–130 MHz frequency band is RFI by TV and FM radio broadcasting. Even at NRAO quiet zone in Green Bank, West Virginia, the FM radio signals exceeds the sky signal by 10 dB over the entire band of 88–108 MHz. At Guadalupe Island, it is still detected some RFI from mainland although the residual FM radio signal is only about 0.1 dB above the Galactic foreground and there are no other strong RFI signals in the used band [6].

Preliminary radio frequency testing at 1.4 GHz at Guadalupe Island were completed in 2013 June. The data was cleaned to excise channels contaminated by radio frequency interference, and the system response was calibrated by comparing the measured brightness temperature to the Global Sky Model of the Galaxy and by independent measurements of Johnson noise from a calibration terminators.

Figure 3. HIbiscus antenna deployed at Guadalupe Island, Mexico [6].

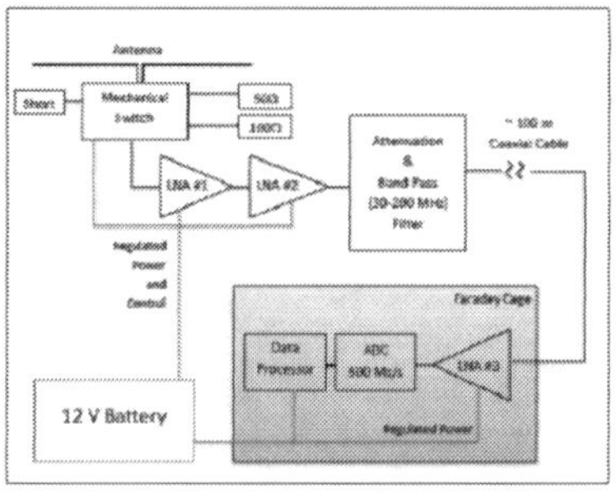

Figure 4. System block diagram [6].

4. Conclusions

An ongoing site survey campaign to assert the astronomical quality of Monte Augusta, at Guadalupe Island, Mexico, is being carried at optical and radio wavelengths. Since 1960 its as been known that this is a potential site for optical astronomy. A quantitative evaluation of the site has been asserted by the personnel of a biological station located at the site. A quantitative evaluation of the optical characteristic will be carried out by astronomers of UNAM. At radio wavelengths, Guadalupe Island has proven to be one of radio quiets sites a frequencies 40–130 MHz making it one a potential site for the radio survey of neutral hydrogen whit redshift range of 14.8<z<22.7.

Acknowledgments

D.H. and E.I.-G. acknowledge support from Consejo Nacional de Ciencia y Tecnologia (CONACyT) of México thru grant # 180817. M.Á. acknowledges support from Dirección General de Asuntos de Personal Académico (DGAPA), Universidad Nacional Autónoma de México (UNAM), thru grant PAPIIT # IT101612.

References

[1] http://www.ine.gob.mx/con-eco-ch/383-hc-isla-gpe
[2] http://islaguadalupe.conanp.gob.mx/decreto.htm

[3] GECI private communication
[4] Sarazine, M. and Rodier, F. 1990, A&A 227, 194
[5] http://www.islas.org.mx/index.php?mod=proy&op=islagua
[6] Voytek, T.C., Natarajan, A., Jáuregui García, J.M., Peterson, J.B. and López–Cruz, O. 2014, ApJL, 782, L9
[7] Jáuregui García, J.M. et al. in preparation

Measurements of the optical seeing isotropy at San Pedro Mártir Observatory

E Iñiguez-Garín[1], D Hiriart[1], M Núñez-Alfonso[1], F Lazo[1] , F Guillen[1] and T Escoboza[1]

[1] Instituto de Astronomía, Universidad Nacional Autónoma de México, Apdo. Postal 870, 22800 Ensenada, Baja California, México

E-mail: einiguez@astro.unam.mx

Abstract. We present measurements of optical seeing isotropy carried out at the San Pedro Mártir Observatory, Mexico. The orientation dependence of the seeing is analyzed based on DIMM system measurements in the East-West direction.

1. Introduction

The San Pedro Mártir National Astronomical Observatory in Mexico (OAN-SPM) has proven to be an excellent site for astronomy. Survey campaigns at the site have found a mean optical seeing of about 0.7 arc–sec [1-4].

Optical seeing depends on temperature gradient and local terrain structure, among other variables. Sierra San Pedro Mártir, in the Mexican peninsula of Baja California, runs along the NW-SE direction. The West side has a gradual elevation from the Pacific Ocean coast up to an altitude of about 3000 m over the sea level. The summit of the sierra hosts a well populated pine forest. The East side of the sierra has an abrupt decline in altitude ending at a desert at almost sea level (see Figure 1). The East rim of the sierra is the site for the 2.1-m telescope.

Because of the large variations on the surrounding terrain at the 2.1-m telescope, there was a concern on whether optical seeing was dependent on westward/eastward pointing of the telescope. Specifically, the rising thermal currents from the desert (East) were suspected to deteriorate the seeing in that direction relative to the observed seeing in the West direction. There is little variation in the terrain northward-southward of the summit, leading to the expectation that seeing should not vary too much in the N-S direction. This paper presents a study of the seeing isotropy between eastward and westward pointing on the East ridge of the San Pedro Mártir Sierra.

2. Observations

Observations were acquired using a two-aperture DIMM from LHESA Electronique described by Vernin & Muñoz-Tuñón[5]. It consists of a 20-cm Schmidt-Cassegrain telescope on a equatorial mount with automatic guiding capabilities, a diaphragm with two 60-mm apertures separated (center-to-center) by 140 mm, an optical wedge in one of the apertures, an intensified CCD camera (LHESAs LH750EIA), and a PC-compatible computer equipped with a frame grabber. An inter-calibration of the OAN-SPM DIMM to the was reported by Núñez et al [6].

Journal of Physics: Conference Series **595** (2015) 012015 doi:10.1088/1742-6596/595/1/012015

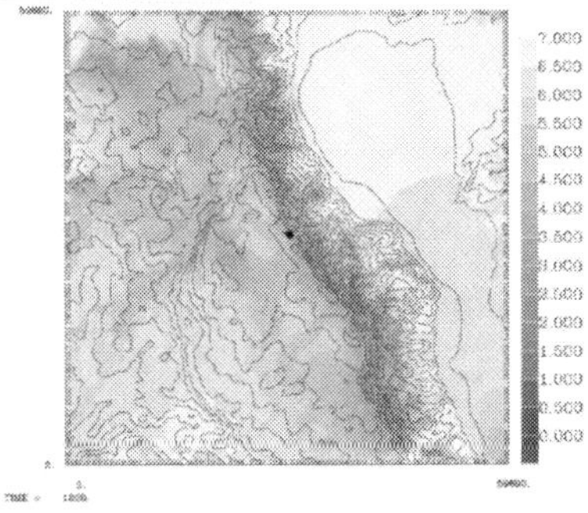

Figure 1. Topographical map of the San Pedro Mártir Sierra. The dot indicates the location of the OAN–SPM 2.1m telescope [4].

Figure 2. The OAN–SPM 2.1-m telescope at the East ridge of the San Pedro Mártir Sierra. The forest (westward) and the desert (eastward) are visible in this picture.

Data from a total of 39 observing nights were collected on 2004 September 09 through November 30 during good weather conditions. Similar air–mass stars in the East and West direction were observed at almost the same time.

3. Results

Tables 1 and 2 present the seeing measurements in the East-West direction from 2004 September 14 to September 26: Col. 1 presents the date (month,day), Col. 2 direction (West or East), Col 3 the average seeing (arc–sec), Col. 4 the rms seeing (arc–sec), Col. 5 the median seeing (arc–sec), and Col. 6 the time (minutes).

Table 3 presents the range values of all the measured seeing in the East-West direction during September and October 2004. It presents the seeing range values (column 1), the number of measurements in the range (column 2), and the percentage in that range from of the total seeing measurements (column 3). A large number of measurements (62.49%) have a seeing of less than 1.0 arc–sec.

poster2.tex

Figure 3 shows the seeing variations during the observing runs of September–October 2004. Data points representing westward or eastward telescope pointing are labeled as Forest or Desert, respectively. From Figure 3, it can be notice that seeing is not dependent on a westward versus eastward pointing direction of the telescope. Thus, seeing is isotropic for the OAN-SPM 2.1m telescope at its current location.

4. Discussion

Since it has been found the seeing is isotropic at the OAN-SPM 2.1m telescopes site, the variation in quality in seeing must be solely due to local terrain effects and thermal currents produce by differential heating. At the San Pedro Mártir Observatory the prevailing winds come from the Pacific Ocean. Simulation of wind turbulence, showed similar behavior for sites on the East and West side of the sierra [7].

The seeing isotropy between westward and eastward pointing may have the following explanation: since the Sun sets in the West, the mountains shadow the East side of the sierra starting roughly after midday. This will allow the East side to cool down and reach a thermal equilibrium with the atmosphere. (Figure 4). The West side, having a gradual elevation change,

Table 1. Measurements of the seeing: forest (West) and desert (East)

Date (mmdd)	direction	Average (")	rms	median (")	time (minutes)
0914	W	0.811	0.183	0.758	40
	E	1.124	0.164	1.115	47
	W	0.781	0.124	0.767	26
	W	0.533	0.058	0.527	16
0915	W	1.145	0.288	1.082	24
	E	0.781	0.082	0.791	10
	W	0.634	0.108	0.613	15
	W	0.477	0.044	0.480	22
	E	0.758	0.087	0.741	27
	W	0.552	0.095	0.541	25
	E	0.387	0.042	0.379	26
	W	0.409	0.042	0.411	26
	E	0.436	0.034	0.436	25
0916	W	0.741	0.206	0.696	22
	E	0.609	0.073	0.589	18
	W	0.743	0.124	0.729	24
	E	0.457	0.039	0.452	20
	W	0.605	0.086	0.597	26
	E	0.441	0.073	0.431	30
	W	0.344	0.043	0.333	27
	E	0.330	0.036	0.326	25
0917	W	0.689	0.088	0.682	26
	E	0.738	0.095	0.717	14
	W	0.750	0.149	0.723	27
	E	0.967	0.146	0.953	25
	W	0.745	0.106	0.736	25
	E	0.869	0.146	0.842	25
	W	0.639	0.119	0.635	26
0918	W	0.813	0.106	0.802	28
	E	1.126	0.130	1.124	14
	E	1.463	0.225	1.459	27
	W	0.937	0.126	0.953	28
	E	1.894	0.536	1.745	27
0921	E	1.062	0.139	1.068	23
	W	0.981	0.147	0.954	26
	E	1.010	0.354	0.909	27
	W	0.856	0.176	0.811	26
	E	1.135	0.188	1.123	25
	W	0.583	0.097	0.573	23
	E	0.742	0.073	0.745	25
	W	0.879	0.122	0.864	23

5. Conclusions

Isotropy of optical seeing in the East-West direction was confirmed at East rim of the San Pedro Mártir sierra. The expectation of having a directional dependence on seeing was motivated

Journal of Physics: Conference Series **595** (2015) 012015 doi:10.1088/1742-6596/595/1/012015

Table 2. Measurements of seeing: forest (West) and desert (East)

Date (mmdd)	direction	Average (")	rms	median (")	time (minutes)
0922	W	1.095	0.136	1.079	26
	E	1.464	0.303	1.400	22
	W	1.530	0.415	1.455	24
	E	0.885	0.266	0.799	23
	E	1.599	0.261	1.575	13
0923	W	0.540	0.110	0.507	18
	W	0.837	0.553	0.526	28
	W	0.471	0.076	0.464	23
	E	0.908	0.160	0.881	25
0924	W	1.453	0.336	1.469	17
	W	1.683	0.244	1.664	13
	E	1.205	0.163	1.173	24
	W	0.916	0.138	0.888	29
	E	0.994	0.452	0.845	31
	W	1.228	0.177	1.184	14
	W	0.898	0.127	0.869	15
	E	0.698	0.055	0.696	16
	W	0.642	0.060	0.635	17
0925	W	1.113	0.225	1.164	16
	W	0.932	0.241	0.914	17
	E	1.151	0.317	1.082	33
	W	1.142	0.233	1.092	27
	E	0.843	0.152	0.813	26
	W	0.902	0.120	0.879	23
	E	0.972	0.117	0.968	26
	W	0.850	0.145	0.829	22
	E	1.138	0.182	1.087	25
0926	W	0.592	0.069	0.584	20
	E	0.848	0.110	0.810	26
	W	0.891	0.130	0.888	25
	E	0.803	0.050	0.797	17
	W	0.921	0.124	0.903	24
	E	1.052	0.133	1.044	21
	W	0.965	0.093	0.951	24
	E	0.909	0.088	0.895	22
	W	1.134	0.118	1.119	23
	E	1.038	0.136	1.028	22

by the large variation in elevation between the eastward and westward directions at the rim site. The isotropy is likely a result of the east side of the mountain being cooling down by self–shadowing and reaching thermal equilibrium with the atmosphere during the afternoon.

Our findings are also supported by over 30 years of observations taken at the 2.1-m telescope near the same location. It has never been a report of seeing variability between eastward and westward pointing from this telescope. However, compared to the 2.1-m telescope observations,

Table 3. Measured optical seeing values.

Seeing range (arc-sec)	Measurements	Percentage (%)
Less than 1.0	13538	62.49
Between 1.0 and 5.0	5145	23.75
Larger than 5.0	2982	13.76

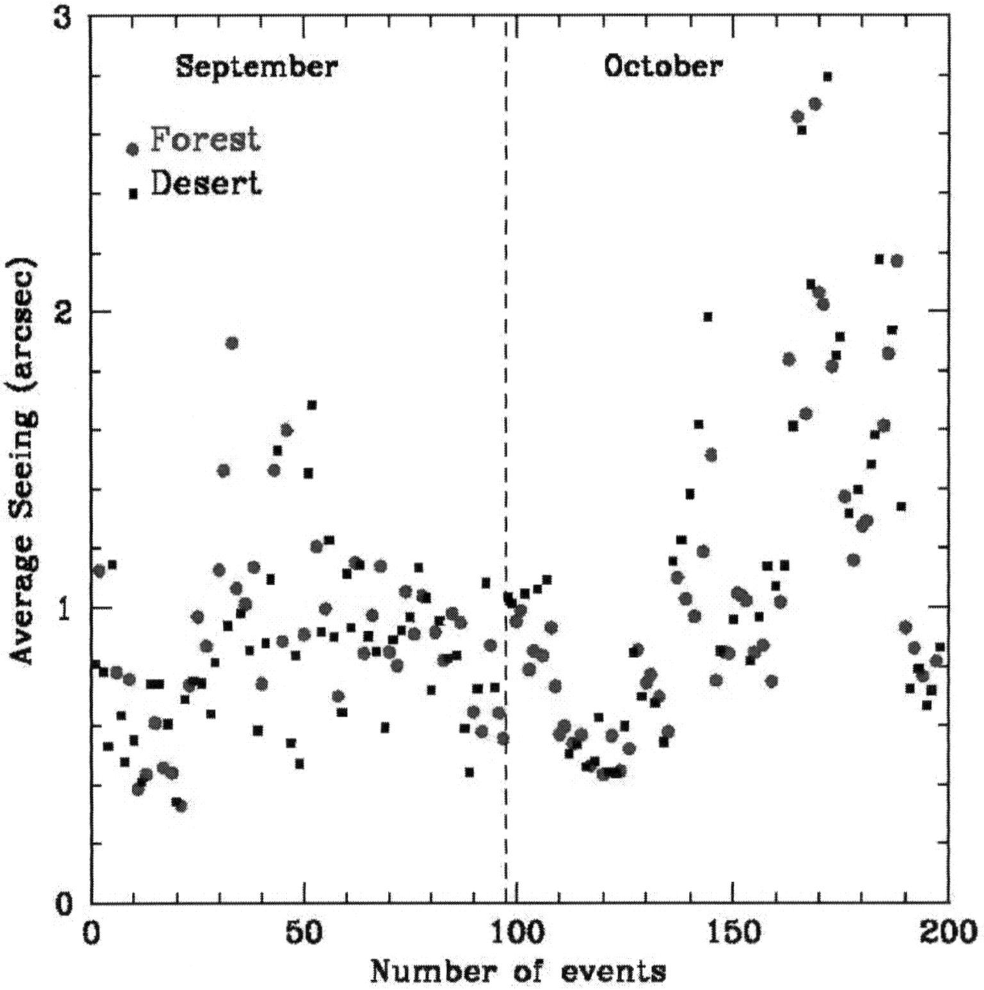

Figure 3. The variation of seeing over the September and October observing runs. Forest/Desert implies a pointing direction of West/East, respectively.

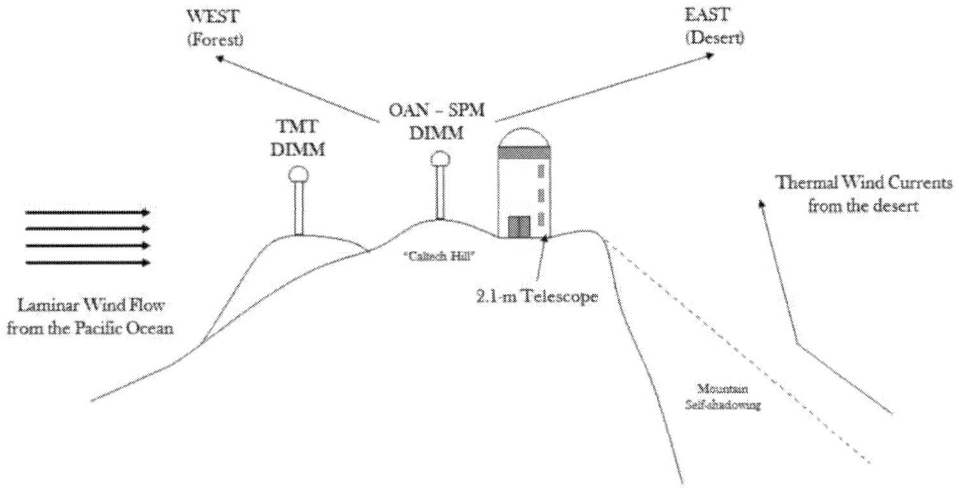

Figure 4. Diagram of the East side of San Pedro Mártir sierra ridge. It schematically shows the position of the Thirty Meter Telescope site survey DIMM (TMT-DIMM), the OAN-SPM DIMM used to made the measurements presented here, and the OAN-SPM 2.1-m telescope.

our seeing measurements were made almost simultaneous in the the E–W direction.

Acknowledgments
EI–G and DH acknowledge support from Consejo Nacional de Ciencia y Tecnologia (CONACyT) of México thru grant # 180817.

References
[1] Echeverría, J. et al., 1998, RevMexAA, 34, 47
[2] Michel, R. et al., 2003, RevMexAA, 39, 291
[3] Schoeck, M. et al. 2009, PASP, 121, 384
[4] Masciadri & Garfias
[5] Vernin, J. & Muñoz–Tuñon, C. 1995, PASP, 265
[6] Núñez, J. M. et al. 2007, RevMexAA, 43, 283
[7] Vogiatzis & Hiriart, D. RevMexAA, 34, 47

Wide-field solar adaptive optics in a layer-oriented approach

Aglaé Kellerer

University of Cambridge, Cavendish Laboratory, JJ Thomson Av., Cambridge, CB30HE, UK

E-mail: ak935@cam.ac.uk

Abstract.
We discuss a layer-oriented approach to multi-conjugate adaptive optics (MCAO) in solar imaging. The technique is a complement to the current star-oriented MCAO and appears as a necessary alternative when large field sizes are desired in solar observations. The basic procedure of the layer oriented method is indicated, and its characteristics are then illustrated in terms of numerical simulations.

1. Introduction

To obtain wide-field images of the solar surface one corrects the atmospheric turbulence through *multi-conjugate adaptive optics*. MCAO uses several deformable mirrors to correct separately the phase distortions caused by turbulence in atmospheric layers at different altitudes. To this purpose one needs to estimate the phase distortions produced at these different altitudes. This information is obtained by using not just one wavefront sensor – as in conventional AO – but by tomographic evaluation of the phase profiles measured by several wavefront sensors.

The established method in solar MCAO uses the *star-oriented* approach (SO) [1, 2] where the wavefront distortions are sensed across the telescope pupil for, say, 19 directions distributed over the field of view. The 19 phase measurements permit then a tomographic estimation of the phase fluctuations caused in the different layers conjugated to the deformable mirrors. Solar AO uses Shack-Hartmann (SH) wavefront sensors, where each SH lenslet images an element of the solar surface, rather than an unresolved star in nighttime AO. The field width must, therefore, be somewhat larger for each lenslet, which means that the effective area sensed by it increases sufficiently at high layers for the signal to be slightly washed out [3]. While the weakened signal from higher altitudes is an undesirable feature in the star-oriented method, the tomographic method does work adequately, as long as the field width remains reasonably narrow. The correlation of solar SH images requires typically 20×20 pixels of angular resolution $0.4''$. With this $8''$ viewing angle, a 0.10 m diameter sub-aperture covers then a 0.41 m diameter surface at 8 km.

Yet one may consider the opposite case where – on purpose – the viewing angle of each lenslet is made as wide as the total viewing angle (i.e. the *science field*). If conjugated to the telescope pupil, the sensor then registers predominantly the phase distortions caused near the ground, while the signal from higher altitudes is largely washed out. This suggests a fairly straight-forward alternative to the *star-oriented* approach. Using a few sensors conjugated to the most relevant altitudes one would have a method that conveniently provides the information for different layers directly. More importantly the method works best for the largest fields of view, because such fields permit the most effective zeroing-in on the specified layer. This *layer-oriented* approach (LO) has, in fact, been introduced successfully for night-time MCAO [4, 5, 6]. For solar observations it has not been utilized, and it will require different procedures there, but we propose that the technique be introduced as useful complement to the current SO-MCAO in solar imaging and, in fact, as a necessary alternative when large field sizes are desired in solar observations [7, 8]. First the basic procedure will be indicated, then the characteristics of this layer oriented method for solar observations will be illustrated in terms of a few simulations.

2. Principle of solar layer-oriented MCAO

In the star-oriented approach each sensor assesses, across the telescope pupil, the phase shift of the photons incident in the specified direction. In the layer-oriented approach each sensor assesses, across the conjugate altitude, the mean phase shift of the photons incident from the entire field of view. Since the fluctuations from other altitudes tend to average out, the sensor registers predominantly the fluctuations due to the conjugated altitude.

Night-time layer-oriented systems employ for this purpose pyramid sensors [4, 5, 6]. Each pyramid collects the light from one guide star, and the light from the different stars is then combined on a detector optically conjugated to the specified altitude. This is not a suitable solution in solar AO, where one deals with an extended source rather than multiple point sources. We have therefore suggested to implement the layer-oriented approach with SH sensors [7, 8]. The method is now briefly described.

In conventional AO the lenslet array of a SH sensor is conjugated to the telescope pupil, and the detector in the focal plane of the array registers for each lenslet an image of the observed solar area. In the layer-oriented approach, additional lenslet arrays are conjugated to dominant turbulent layers. Each lenslet images the entire field of view, but – as indicated in Fig. 1 – the image contains only those parts of the solar surface for which some of the emitted photons traverse on their way to the telescope pupil this lenslet. Fig. 1 depicts for three different altitudes the images registered behind each lenslet.

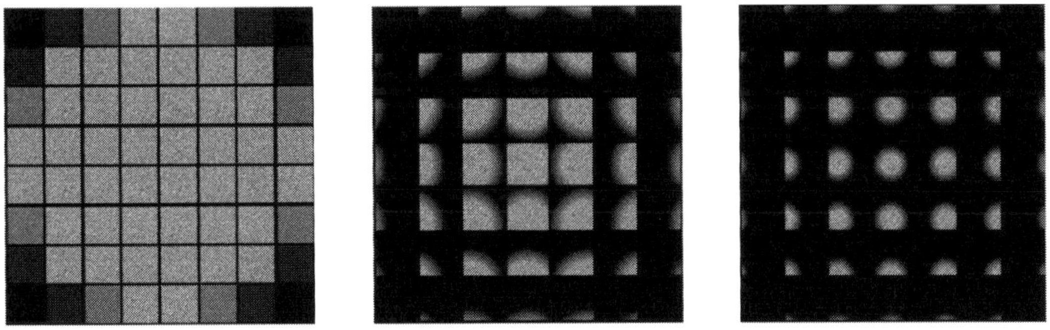

Figure 1. In a layer-oriented MCAO system the lenslet array of a SH sensor is conjugated to a turbulent layer (left, middle, right panels: 0 km, 4 km and 10 km) and the sensor registers predominantly the phase fluctuations produced at that altitude.

The vignetting of the lenslet images is no artefact of the layer oriented method, it merely reflects the fact that – at the higher altitudes – photons from only part of the observed solar area traverse the particular sub-pupil on their way to the telescope pupil; merely these photons experience the atmospheric turbulence at this location and require the corresponding phase correction.

Consider a simplified scenario where the turbulence is located at the conjugate height of a SH sensor: the image behind a lenslet is globally shifted, because the phase fluctuations are independent of the field direction. If, on the contrary, the turbulence is located far from the conjugate height, then the fluctuations vary with field direction and the SH images are distorted rather than shifted. In the layer-oriented approach, the image shift forms the sensor-signal; it is an approximation of close-by turbulence and is used to shape the deformable mirror that is conjugated to the same height as the sensor.

As an aside we note that the wavefront distortions are sensed continuously over the entire field of view: the correction loop therefore automatically converges to an optimal solution for the entire field of view. This contrasts with the star-oriented approach where the correction for the entire field is extrapolated from measurements along multiple discrete directions.

3. Reduction of the field-of-view to the relevant photon fluence

As shown on Fig. 1, SH lenslets that are conjugated above the pupil form vignetted images. Assume that the turbulence is located in two layers: at the ground and 4 km. One aims at determining the phase fluctuations introduced over a small surface S of the 4 km metapupil: see Fig. 2.

In the solar layer-oriented approach, a SH lenslet is optically conjugated to surface S. The signal measured behind this lenslet corresponds to the phase gradient introduced over S plus the gradient over a larger surface at the ground. In Fig. 2 this larger surface corresponds to the intersection between the pupil and the disc traced out by the field-of view: only the photons that end up in the intersection with the pupil (i.e. are accepted by the telescope in normal operation) traverse S. It has been suggested that the layer-oriented method is inoperable because the field-reduction restricts the effective surfaces at mis-conjugated altitudes, and thereby weakens the attenuation of mis-conjugated layers [9]. In fact field reduction affects any MCAO system, i.e. the star- and layer-oriented approaches in their night- and daytime implementation. And it affects existing MCAO approaches more strongly than the proposed solar layer-oriented method.

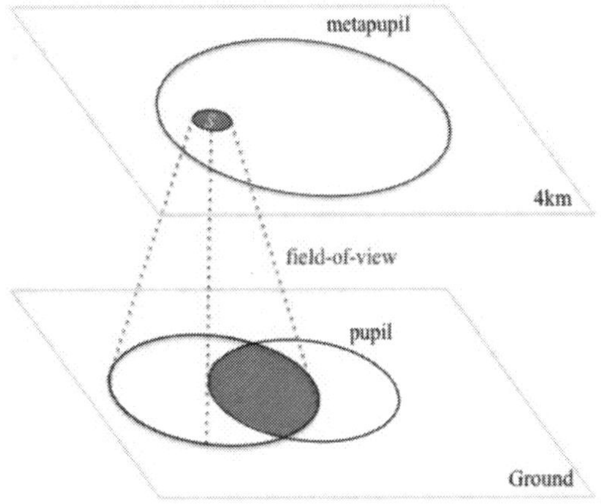

Figure 2. Solar layer-oriented MCAO: Atmospheric fluctuations at S affect only part of the field-of-view. A SH lenslet conjugated to surface S characterizes, at the ground, the phase fluctuations in the intersection between the pupil and the disc traced-out by the field of view.

In a nighttime layer-oriented system, a detector – rather than a lenslet array – is optically conjugated to a turbulent layer. The fundamental principle of both layer-oriented methods are, however, identical: signals from mis-conjugated layers are attenuated because the corresponding phase fluctuations are averaged over larger effective surfaces. In the nighttime implementation, the averaging is done over the footprints of multiple guide stars. Fig. 3 assumes three guide stars. For the detector pixels conjugated to surface S the signal from the ground layer is not attenuated because only one star lies within the unvignetted field-of-view. This example illustrates that field-reduction has more drastic consequences in nighttime than in solar layer-oriented MCAO: indeed, because of the continuous solar fields, mis-conjugated layers are always somewhat attenuated in the solar application. Since the layer-oriented approach has been successfully implemented in nighttime MCAO, field-reduction can also not be a limitation for the solar application.

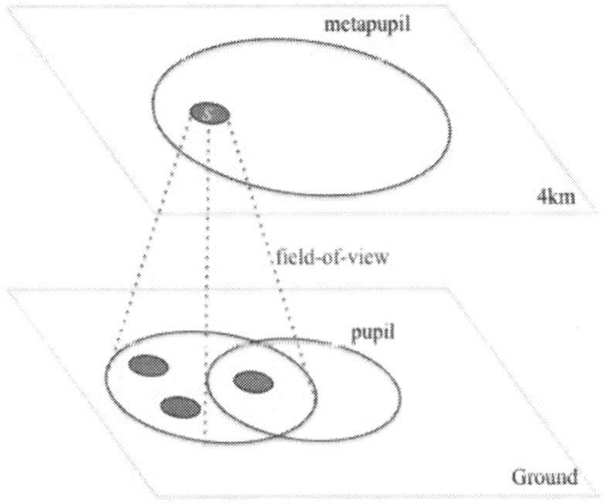

Figure 3. Nighttime layer-oriented MCAO: For the sensor elements conjugated to surface S the signal from the ground layer is not attenuated because only one star lies within the unvignetted field-of-view.

4. Simulations

Some numerical simulations in terms of *Yao*, an adaptive optical simulation code written by F. Rigaut under *Yorick* can illustrate the proposed method. We have written additional functions under *Yao* in order to simulate solar observations and wide-field Shack-Hartmann sensors conjugated above the telescope pupil.

Assume three sensor-mirror pairs conjugated to 0 km, 4 km and 10 km. The sensors (mirrors) have 8, 8 and 16 lenslets (9, 9 and 17 actuators) across the diameter. Each sensor samples a $100''$ diameter field with $0.5''$ resolution. The telescope diameter equals 1.6 m. The signal measured by a sensor is exclusively used to activate its associated mirror: in other words, once the sensor signals are determined, the tomographic reconstruction is done.

Fig. 4 shows the effect of a single layer at ground level. The Fried parameter is set to $r_0 = 0.15\,\text{m}$. As expected, the turbulence is mainly detected by the sensor conjugated to the ground, while the two other sensors measure a negligible signal: After one iteration, the mirror conjugated to the ground reproduces the shape of the turbulent screen while the two other mirrors remain nearly flat. The system has assimilated that the turbulence is introduced at the ground – the tomographic reconstruction is successful.

Fig. 5 represents the effect of a single layer at 4 km with Fried parameter $r_0 = 0.15\,\text{m}$. The system again responds correctly: the mirror conjugated to 4 km takes up the shape of the phase screen, while the two other mirrors remain approximately flat.

The effect of a single turbulent layer at 10 km with Fried parameter $r_0 = 0.50\,\text{m}$ is illustrated on Fig. 6. After one iteration, the mirror conjugated to 10 km replicates the phase screen, but the 4 km mirror likewise reproduces the lowest spatial frequencies of the screen: The sensor conjugated to 4 km has measured some signal – field-reduction has made the attenuation of distant layers less efficient. After a few iterations however, the mirror conjugated to 4 km flattens out because the sensor-mirror pair at 10 km is more efficient at correcting the phase fluctuations.

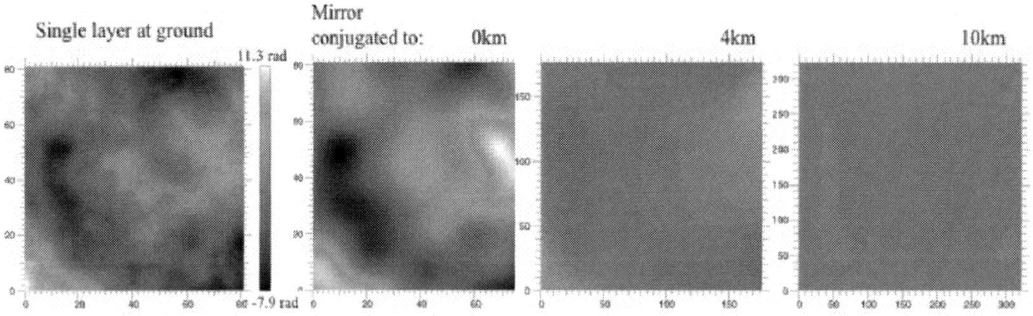

Figure 4. Effect of a single layer at ground level: After one iteration, the mirror at ground level reproduces the shape of the phase screen while the two other mirrors remain almost flat. The system has assimilated that the turbulence is located at ground level.

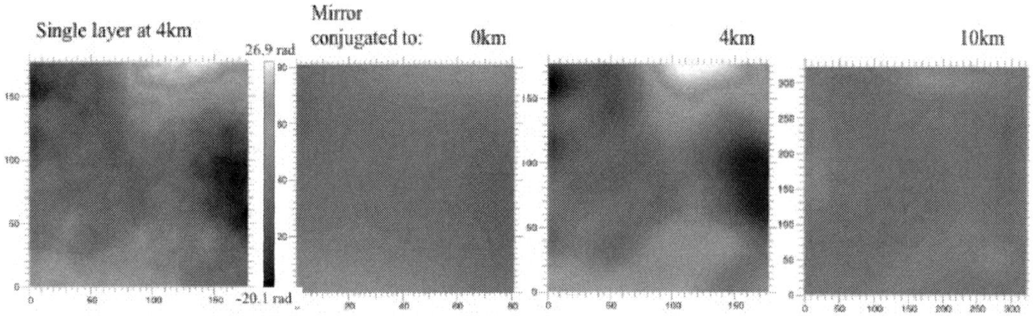

Figure 5. Effect of a single layer at 4 km: After one iteration, the mirror conjugated to 4 km replicates the shape of the phase screen while the two other mirrors remain almost flat. The tomographic reconstruction is correct.

5. Conclusion

Table 5 summarizes differences between star- and layer-oriented solar MCAO correction. A complication of the layer-oriented approach is that it requires rapid cross-correlations of large images. Further, it necessitates $\sim$ 2000 $\times$ 2000 pixel detectors that can be read-out above 1000 Hz. The availability of such detectors is currently the main challenge for the solar layer-oriented approach.

Bibliography

[1] T. Berkefeld, D. Soltau, D. Del Moro, and M. Löfdahl. Wavefront sensing and wavefront reconstruction for the 4m European Solar Telescope EST. In *Society of Photo-Optical Instrumentation Engineers (SPIE) Conference Series*, volume 7736 of *Society of Photo-Optical Instrumentation Engineers (SPIE) Conference Series*, July 2010.

[2] I. Montilla, C. Béchet, M. Le Louarn, M. Tallon, J. Sánchez-Capuchino, and M. Collados Vera. Multi-conjugate AO for the European Solar Telescope. In *Society of Photo-Optical Instrumentation Engineers (SPIE) Conference Series*, volume 8447 of *Society of Photo-Optical Instrumentation Engineers (SPIE) Conference Series*, July 2012.

[3] C. Béchet, M. Tallon, I. Montilla, and M. Langlois. Wide-field wavefront sensing in solar adaptive optics: modeling and effects on reconstruction. In S. Esposito and L. Fini, editors, *Proceedings of the Third AO4ELT Conference*, December 2013.

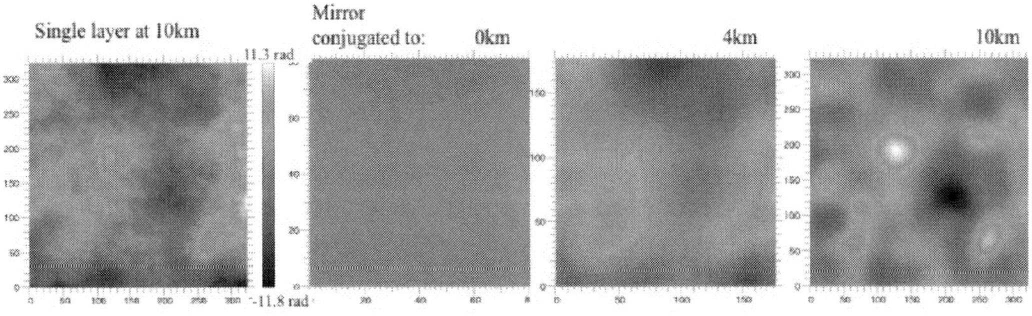

Figure 6. Effect of a single layer at 10 km. After one iteration, the 10 km mirror reproduces the shape of the phase screen, while the 4 km mirror replicates its lowest spatial frequencies: field-reduction has made the attenuation of mis-conjugated layers less efficient. This effect may be minimized by filtering out low spatial modes in the command matrix.

	Star-oriented	Layer-oriented
High-turbulence	Signal from high turbulence is attenuated	**Each sensor is most sensitive to its associated layer**
Spatial and temporal sampling	..tuned to integral turbulence	**..tuned to associated layer**
Tomographic reconstruction	..becomes more complex with increasing field size ..is done computationally	**..benefits from large field sizes** **..is done optically**
Cross-correlation	**..over ∼ 20×20 pixels**	..over ∼ 200 × 200 pixels: computationally intensive
Detector requirement	**Small detectors or large detectors with ROI read-out**	Large detectors with high read-out frequencies

Table 1. Comparison between star- and layer-oriented solar MCAO. Advantages in bold.

[4] R. Ragazzoni, J. Farinato, and E. Marchetti. Adaptive optics for 100-m-class telescopes: new challenges require new solutions. In P. L. Wizinowich, editor, *Adaptive Optical Systems Technology*, volume 4007 of *Society of Photo-Optical Instrumentation Engineers (SPIE) Conference Series*, pages 1076–1087, July 2000.

[5] E. Diolaiti, R. Ragazzoni, and M. Tordi. Closed loop performance of a layer-oriented multi-conjugate adaptive optics system. *A&A*, 372:710–718, June 2001.

[6] C. Vérinaud, C. Arcidiacono, M. Carbillet, E. Diolaiti, R. Ragazzoni, E. Vernet-Viard, and S. Esposito. Layer Oriented multi-conjugate adaptive optics systems: performance analysis by numerical simulations. In P. L. Wizinowich and D. Bonaccini, editors, *Adaptive Optical System Technologies II*, volume 4839 of *Society of Photo-Optical Instrumentation Engineers (SPIE) Conference Series*, pages 524–535, February 2003.

[7] A. Kellerer. Layer-oriented adaptive optics for solar telescopes. *Appl. Opt.*, 51(23):5743–5751, Aug 2012.

[8] A. Kellerer. Comment on: Feasibility study of a layer-oriented wavefront sensor for solar telescopes. *Appl. Opt.*, in press, 2014.

[9] J. Marino and F. Wöger. Feasibility study of a layer-oriented wavefront sensor for solar telescopes. *Appl. Opt.*, 53(4):685–693, Feb 2014.

The water vapour radiometer of Paranal: homogeneity of precipitable water vapour from two years of operations

Florian Kerber[1], Richard R Querel[2] and Bianca Neureiter[3]

[1] European Southern Observatory, Karl-Schwarzschild-Str. 2, 85748 Garching, Germany;
[2] National Institute of Water and Atmospheric Research (NIWA), Lauder, New Zealand;
[3] Ludwig-Maximilians-Universität, Geschwister-Scholl-Platz 1, 80539 München, Germany

Abstract. A Low Humidity and Temperature Profiling (LHATPRO) microwave radiometer, manufactured by Radiometer Physics GmbH (RPG), is used to monitor sky conditions over ESO's Paranal observatory in support of VLT science operations. The unit measures several channels across the strong water vapour emission line at 183 GHz, necessary for resolving the low levels of precipitable water vapour (PWV) that are prevalent on Paranal (median ~2.4 mm). The instrument consists of a humidity profiler (183-191 GHz), a temperature profiler (51-58 GHz), and an infrared camera (~10 μm) for cloud detection. We present a statistical analysis of the homogeneity of all-sky PWV using 24 months of PWV observations. The question we tried to address was whether PWV is homogeneous enough across the sky such that service mode observations with the VLT can routinely be conducted with a user-provided constraint for PWV measured at zenith. We find the PWV over Paranal to be remarkably homogeneous across the sky down to 27.5° elevation with a median variation of 0.07 mm (rms). The homogeneity is a function of the absolute PWV but the relative variation is fairly constant at 2 to 3% (rms). Such variations will not be a significant issue for analysis of astronomical data. Users at ESO can specify PWV – measured at zenith – as an ambient constraint in service mode to enable, for instance, very demanding observations in the infrared. We conclude that in general it will not be necessary to add another observing constraint for PWV homogeneity to ensure integrity of observations. For demanding observations requiring very low PWV, where the relative variation is higher, the optimum support could be provided by observing with the LHATPRO in the same line-of-sight simultaneously. Such a mode of operations has already been tested but will have to be justified in terms of scientific gain before implementation can be considered. We plan to extend our analysis of PWV variations covering a larger parameters space for temporal and spatial resolution in the future. Also for climate studies such data sets will be relevant.

[1] Email: fkerber@eso.org

1. Introduction

Water vapour is the main source of opacity in the Earth's atmosphere at infrared (IR) wavelengths. That opacity is the primary reason that ground-based astronomical observations are only made in certain wavelength regions, called windows, because of their relative lack of water absorption lines and other telluric absorption features. These windows are relatively well-defined and stable for most observations at temperate latitudes and moderate elevations above sea level. Atmospheric transmission is a function of the depth of air above the observer. This is true for almost all atmospheric constituents since gases such as O_2 are well-mixed throughout the atmosphere. In contrast, water can coexist in three phases at atmospheric temperatures, and hence, its atmospheric distribution is highly variable in altitude as well as geographical location and time. Atmospheric water vapour content can be expressed as precipitable water vapour (PWV) which is the equivalent condensed amount of water in an atmospheric column above the observer (as a depth in units of mm).

ESO's Very Large Telescope (VLT) is located on Cerro Paranal (24.6°S, 70.4°W, 2635 m.asl). Since November 2011 ESO operates a water vapour monitor in support of VLT science operations[1,2]. Paranal is a very dry site with a median PWV of ~2.4 mm offering excellent conditions for IR observations when real-time PWV information is taken into account. Before deployment of the LHATPRO monitor PWV was routinely derived from standard star observations at the 8-m unit telescopes (UTs) with the spectroscopic instruments UVES, X-Shooter, CRIRES and VISIR. The method for deriving PWV from such spectra taken at various wavelengths by fitting the observed spectrum with an atmospheric radiative transfer model is described in Querel et al[3]. The pronounced seasonal variation of PWV for sites in Northern Chile is well-documented[1,2,4]. In contrast the variation of uniformity of PWV across the sky at any given moment had not been studied; a first analysis has been presented by us[5] earlier this year and this contribution is an update of the previous work. A summary of the methods to monitor atmospheric water vapour is given in the excellent book by Kämpfer et al.[6].

2. LHATPRO

The Low Humidity And Temperature PROfiling microwave radiometer (LHATPRO), manufactured by Radiometer Physics GmbH (RPG), measures the atmosphere at two frequency ranges focusing on two prominent emission features: a strong H_2O line (183 GHz) and an O_2 band (51-58 GHz). Using 6 and 7 channels, respectively, the radiometer retrieves the profile of humidity and temperature up to an altitude of ~12 km[2,7] The spatial resolution is given by the size of the radiometer beam (1.4° FWHM). For a full calibration with absolute standards an additional external calibration target, cooled down to the boiling point of liquid nitrogen (LN2), is used. The LHATPRO offers a measurement duty cycle of >97%. Details of the radiometer are described in Rose et al[7].

The LHATPRO water vapour radiometer (WVR) was commissioned during a 2.5 week period in October and November 2011 during which the calibration was tested with respect to balloon-borne radiosondes, the established standard in atmospheric physics[2]. The LHATPRO measures PWV in the range 0-20 mm with an accuracy of better than 0.1 mm and an internal precision of about 30 μm. The 183 GHz line is intrinsically very strong and reliable readings can be obtained even in the driest of conditions encountered on Paranal. In addition, ESO's LHATPRO is equipped with an IR radiometer (10.5 μm) to measure sky brightness temperature down to -100° C. This allows for the detection of high altitude clouds, e.g. cirrus, that consist of ice crystals but contain practically no liquid water (see Kerber, Querel and Hanuschik[8]). The IR radiometer can be fully synchronized for scanning with the microwave radiometer.

The operational scheme currently set-up on Paranal was devised to meet the needs of VLT science operations and has been strictly followed for more than 2 years now:

- 2-D all-sky scan: step size of 12° azimuth, 12.6° elevation (duration 6 min, repeated every 6h; 1.7% of 24 h period)
- Cone scan at 30° elevation (Hovmöller): step size 6° (duration 2.5 min, repeated every 15 min; 16.7% of 24 h period)
- Routine observations set to zenith-staring mode (81.6 % of 24 h period)

The resulting measurements are processed by the RPG software, corrected for air mass, and so PWV values at zenith are available nearly continuously and are displayed in the Paranal VLT control room in near-real time. More details about the scanning process are given in Querel & Kerber[5].

3. Data Analysis and Results

In a previous paper we had analysed the 2D scans only[5]; now we have used data from both scanning modes. The 2D scans cover a 21 months period from June 2012 to March 2014, while the Hovmöller cone scans were acquired during a 2 year period June 2012 to June 2014. The scans contain very similar information with two important differences: the 2D scans provide better spatial coverage but worse spatial and temporal sampling, while the Hovmöller scans cover only an annulus in the sky (30° elevation) hence limited spatial coverage but with higher (factor 2) spatial and much better (factor 20) temporal sampling. We have used a total of 2588 2D scans and 67284 cone scans. The data have been treated in the same way. Only 17 scans have been excluded because of some problem with the measurement. After correcting for air mass all individual measurements in a given scan have been averaged and the standard deviation derived. The results are very similar for both kinds of scans.

Table 1. Homogeneity of PWV over Paranal. Shown in the table is the variation in PWV across the sky: Standard deviation [mm] for the 2D scans (2588 sky-scans over 21 months, representing day and night conditions); and the 30 degree cone scans (67284 scans over 24 months). The 10, 25, 50, 75, and 90 percentiles were computed. The median values are shown in boldface.

Percentiles	PWV variation SDev 2D scan [mm]	PWV Variation SDev cone scan [mm]
10	0.035	0.022
25	0.049	0.032
50	**0.066**	**0.057**
75	0.131	0.113
90	0.218	0.205

The distribution of PWV above Paranal is remarkably uniform (table 1). The variation of PWV increases as a function of the absolute PWV value (Tables 2 & 3) but is fairly constant when the fractional variation (in %) is considered. For the median variation a value of about 2-3% is found. This suggests that a PWV constraint for zenith will be sufficient to secure the correct conditions for arbitrary pointings for most science applications.

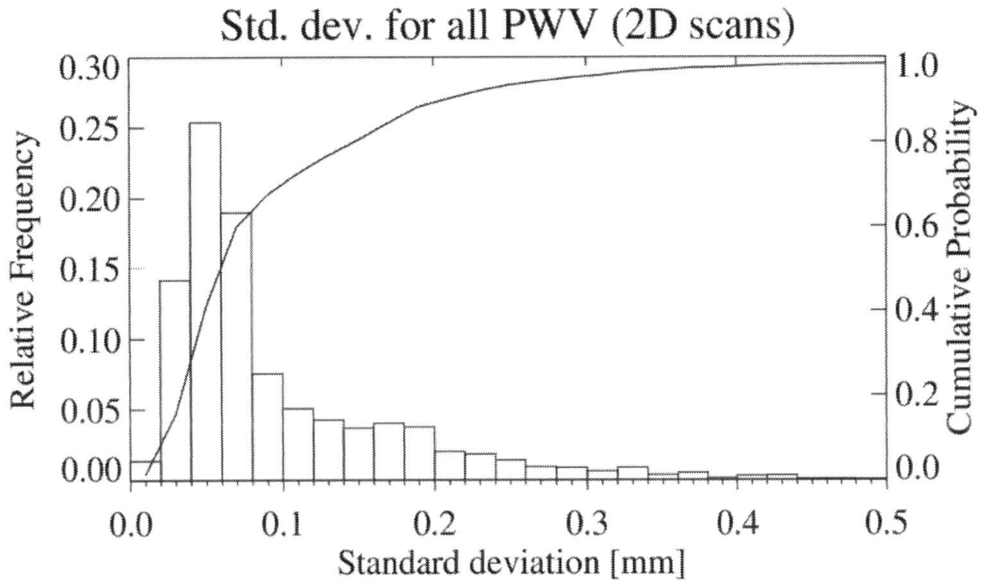

Figure 1: Histogram and cumulative probability distribution of the PWV variation across the sky down to 27.5° elevation (2D scans, 2588 samples), reproduced from Querel & Kerber[5].

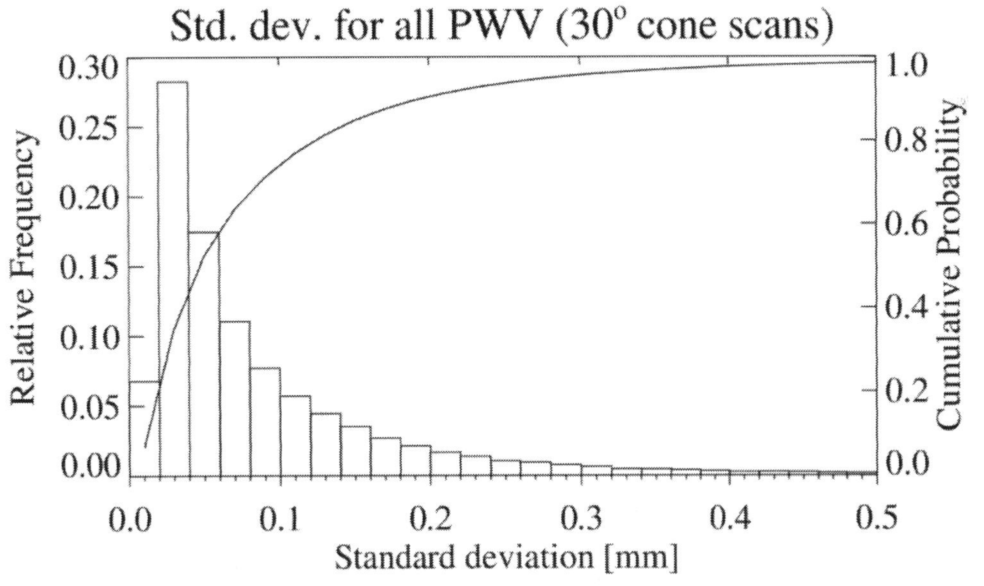

Figure 2. Histogram and cumulative probability distribution of the PWV variation of the cone scans at 30.0° elevation (Hovmöller, 67284 samples).

Variations are relatively slow in time and we have not found strong random fluctuations on the spatial and temporal scales accessible to analysis with the data available. Recently, we have obtained additional 2D data over a smaller area but at 1 degree resolution to address smaller scales. In the future we plan to use VISIR the VLT mid-IR instrument to also study PWV-induced variations of the sky

background at very high time resolution. Large variations in PWV are found to be associated with the exchange of air masses with different properties over Paranal such as large-scale frontal systems. This is consistent with the concept of a frozen flow for atmospheric processes[9, 10].
PWV can be very low on Paranal[11] enabling new science taking advantage of increased atmospheric transmission outside of established astronomical windows. This kind of science is directly dependent on the actual transmission and hence PWV along the line-of-sight towards the target. Since the variation of PWV is, relatively speaking, the highest for low PWV conditions such variations may in fact negatively impact photometric measurements. Therefore it may be prudent to consider introducing an additional constraint on the PWV representing its variation during an observation.

Table 2. The standard deviation [mm] of the PWV variations binned by zenith PWV value. Using 2588 2D sky-scans over 21 months, representing day and night conditions, the 10, 25, 50, 75, and 90 percentiles were computed. The median values are shown in boldface.

Percentiles	**0-1 mm**	**1-2 mm**	**2-3 mm**	**3-4 mm**	**4-6 mm**	**6-8 mm**	**8-10 mm**
10	0.021	0.037	0.029	0.045	0.069	0.148	0.140
25	0.031	0.048	0.037	0.056	0.091	0.173	0.163
50	**0.047**	**0.058**	**0.054**	**0.082**	**0.139**	**0.199**	**0.204**
75	0.059	0.069	0.075	0.119	0.195	0.269	0.263
90	0.068	0.087	0.114	0.184	0.261	0.345	0.325
Number	252	818	567	262	332	191	92

Table 3. The standard deviation [mm] of the PWV variations binned by zenith PWV value. Using 67284 cone scans (30 degree elevation) over 24 months, representing day and night conditions, the 10, 25, 50, 75, and 90 percentiles were computed. The median values are shown in boldface.

Percentiles	**0-1 mm**	**1-2 mm**	**2-3 mm**	**3-4 mm**	**4-6 mm**	**6-8 mm**	**8-10 mm**
10	0.015	0.019	0.024	0.041	0.058	0.080	0.078
25	0.019	0.024	0.033	0.058	0.084	0.111	0.108
50	**0.027**	**0.034**	**0.049**	**0.090**	**0.130**	**0.163**	**0.159**
75	0.041	0.052	0.079	0.149	0.204	0.250	0.250
90	0.062	0.079	0.127	0.232	0.315	0.374	0.410
Number	4569	21533	17351	6598	7513	4324	2718

Even better support to the science exploitation of such data could be given by measuring the PWV along the same air column traversed by the science light from the target (line-of-sight) simultaneously with the science observation. The LHATPRO radiometer is capable of all-sky pointing and tracking and thus line-of-sight support. The spatial resolution is limited by the size of the radiometer beam (1.4° FWHM). This mode has been technically implemented by RPG and has been tested on Paranal. Implementation of such a mode will require careful planning and would have to be justified by clear scientific gain.

The observational scheme followed over the past two years on Paranal is driven by the needs of VLT Science Operations. It is quite simple and focuses on zenith observations. The spatially resolved scans provide valuable information on the distribution of PWV acorss the sky. Since water vapour is the most important greenhouse gas and its variation in the atmopshere is highly important for climate studies we expected that we would be able to compare our findings with dedicated studies for other sites. Still when looking for similar studies of PWV homogeneity in the literature we have found that apparently only very few publications address the issue of spatial homogeneity of PWV over one specific location. In fact to the best of our knowledge our data set seems to be unique in terms of cadence and number of 2D scans over a single site obtained with a high precision microwave radiometer. Schneebeli and Mätzler[12] have obtained all-sky maps over Zimmerwald Observatory in Switzerland using the scanning all-sky radiometer ASMUWARA during the period of one week in Feb 2007 and analysed the spatio-temporal behaviour of PWV and found a high degree of spatial coherence. Morland and Mätzler[13] have studied the PWV field over Switzerland using data from a network of 31 GPS receiver stations. Specifically they have studied the passage of frontal systems and find a good correlation between the measurements at different stations again fully consistent with the frozen flow concept and our findings for Paranal.

4. Conclusion

The uniformity of PWV across the sky over Paranal is remarkably high at a level of a few percent. Variations over Paranal are consistent with the "frozen flow" concept in which most of the observed change is attributed to air masses with different properties passing over a given site. This suggests that a PWV constraint for zenith will be sufficient to secure the correct conditions for arbitrary pointings for most science applications. A more in-depth analysis of the all-sky PWV data, including diurnal and seasonal trends in the variations is in preparation. Our data set might be relevant for a better understanding of atmospheric PWV also in the context of climate change.

5. References

[1] Kerber F, et al. 2010 in *SPIE Conf. Ser.* **7733** 77331M-12M
[2] Kerber F, et al. 2012 in *SPIE Conf. Ser.* **8446** 84463N1-12
[3] Querel R R, Naylor D A and Kerber F 2011 *PASP* **123** 222-229
[4] Otarola A, Travouillon T, Schöck M et al 2010 PASP **122** 470-484
[5] Querel R R and Kerber F 2014 *SPIE Conf. Ser.* **9147** 914721-11
[6] Kämpfer N. (ed) 2013 *Monitoring Atmospheric Water Vapour, Ground-based remote sending and in-situ methods*, ISSI Report Series 10, Springer Science and Business Media, LLC
[7] Rose T, Crewell S, Löhnert U, Simmer C 2005, *Atmosph.. Res.* **75** 183-200
[8] Kerber F, Querel R R and Hanuschik R 2014 *SPIE Conf. Ser.* **9149** 91490M1-12
[9] Poyneer L, van Dam M and Veran J P 2009 *J. Opt. Soc. Am. Opt. Image. Sci. Vis.* **26** 833-846
[10] Van Baelen J, Reverdy M, Tridon F, Labbouz L, Dick G, Bender M and Hagen M 2011 *Q. J. R. Meteorol. Soc.* **137** 204-223
[11] Kerber F, Querel R R, Rondanelli R et al. 2014 *MNRAS* **439** 247-255
[12] Schneebeli M and Mätzler C 2011 *IEEE Geoscience and Remote Sensing Letters* **8** 948-952
[13] Morland J and Mätzler C 2007 *Meteorol. Appl.* **14** 15-26

LOTTTTUCE: Layer Oriented Tip-Tilt Turbulence Tomography using Covariance and Elevation

Olivier Lai[1,2], Mark Chun[3], Jessica R. Lu[3], Yutaka Hayano[1], Shin Oya[1], Douglas Toomey[4]

[1]Subaru Telescope, 650 North A'ohoku Place, Hilo HI 96720, USA
[2]Gemini Telescope, 670 North A'ohoku Place, Hilo HI 96720, USA
[3]Institute for Astronomy, University of Hawaii, 640 North A'ohoku Place, Hilo HI 96720, USA
[4]Mauna Kea Infrared, 21 Pookela Street, Hilo HI 96720, USA

lai@naoj.org, olai@gemini.edu

Abstract. LOTTTTUCE is based upon the fact that turbulence at the pupil produces correlated tip-tilt motion over the entire field (averaging the tip-tilt across the widest field possible gives the strength of the turbulence at the telescope), while the on-axis (any axis) image motion measures the integrated tip-tilt over the line of sight (single stars provide the variance of the tip-tilt, which allows to infer the integrated seeing). Between these two extremes, the amount of correlation across a given field size is the integral of the turbulence from the ground to the altitude where the tip-tilt decorrelates over the meta-pupil. Differentiating the altitude-integrated tip-tilt with respect to altitude generates an estimate of tip-tilt (hence turbulence, assuming Kolmogorov properties) at each altitude. Alternately, the 3D Fourier transform of a data cube containing the time evolution of the tip (or tilt) across the field allows to determine the amount of energy for "field" frequencies (in other words, the integrated seeing across each same size patches) and the temporal spectrum of each of these features. Differentiating the spectrum with respect to spatial frequency would provide the amount of energy, as well as speed and direction, of each layer. The LOTTTTUCE method is a novel method of measuring the vertical turbulence profile that uses wide field tip-tilt information such as that provided by Pan-STARRS. However, the method also has limitations due to tip-tilt decorrelation as a function of meta-pupil overlap, finite outer scale, and non-Kolmogorov power spectrum.

1. By way of an introduction

Although this method *should* work (see below), simulations have not yet yielded any satisfactory results. Nonetheless, these simulations, as well as handling new forms of data (tip-tilt data cubes across a wide field are not commonly available and manipulating such data reveals correlations and relationships which are not immediately apparent) have revealed much more information than the originally simple algorithm described in the abstract. It would have been easy to withdraw this paper from publication, but null results are an important aspect of the scientific process.

The term "null result" comes from the latin *nullus resultarum* which does not imply no result, but rather that the results do not support the underlying hypotheses, that the experimental outcome does not show an otherwise expected effect.

Figure 1: LOTTTTUCE - Layer oriented Tip-Tilt Turbulence Tomography using Covariance&Elevation. Demonstrating that a good acronym and a logo are no guarantee of success.

The reason this method should work in principle is that it is an extension of a demonstrated method originally developed by Flicker&Rigaut [1]: They used the adaptive optics corrected isoplanatic angle to determine the altitude integrated r_0, as well as its first moment, i.e. its mean height. The LOTTTTUCE method proposes to extend higher orders of isoplanatism (though using tip-tilt, i.e. isokinetism only), to obtain the higher moments, and hence a vertical profile, of the turbulence.

The general context in which this method is relevant is the importance of turbulence profiling. Any sensor measuring tip-tilt across wide fields (such as GLAO or MOAO but may be even more appropriately, Pan-Starrs [2]) will generate data that contains information about the turbulence profile. For instance in GLAO, the ground layer is given by the commands to the deformable mirror, while the residual isokinetism from the wavefront sensor measurements could be used to determine the turbulence. This will be useful for PSF reconstruction. Being able to separate turbulent layers and speeds would also have applications in predictive algorithms for MOAO.

2. Simulations

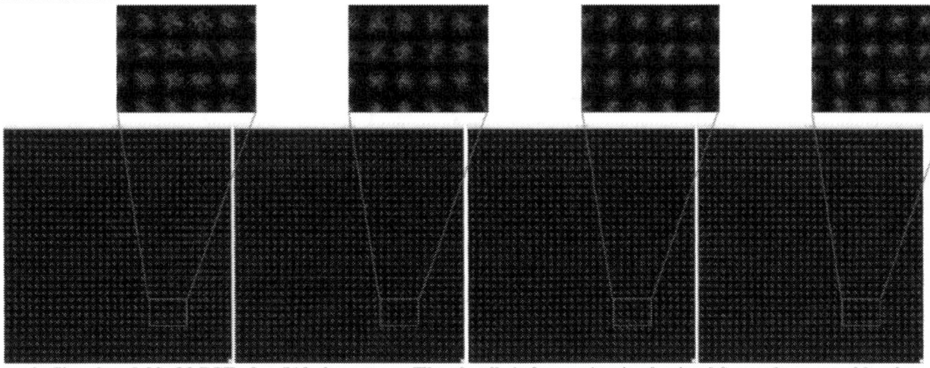

Figure 2: Simulated 32x32 PSFs for 512 time steps. The tip-tilt information is obtained from the centroids of each of these PSFs. 2 layers are present, implying that there should be two different correlations sizes for the tip-tilt.

The ideal data to test out the LOTTTTUCE method would be a pair of data cubes containing `[tip(x,y,t)]` and `[tilt(x,y,t)]`. A GLAO code developed to simulate very wide field AO systems (e.g. Imaka [3]) named `instant_GLAO` was modified to produce a grid of instantaneous PSF stars seen through a layered atmosphere. Originally, `instant_GLAO` was developed to simulate complex effects and implementation issues of very wide field AO, such as a tiled or misconjugated DM, field aberrations seen by different wavefront sensors, etc. All these features were turned off, as well as the adaptive optics correction (i.e. a non AO, sky only simulation). Instead, each instantaneous PSF was recorded and the tip and tilt were determined by measuring the shift in centroid.

Using a 1.8m telescope with a field of view of 1°, a grid of 32x32 PSF were generated for 1024 time steps at 200Hz. In the simplest case, the model atmosphere typically consisted of two layers, one at a few hundreds of meters altitude and the other a few kilometers high moving perpendicular to each other at a few ms^{-1}. These PSFs are shown in Figure 2. Because only two layers are present, two different scales of correlation should be present in the centroid data. Centroid cubes are shown on Figure 3. Based on such data, different methods were developed to try to retrieve the vertical turbulence profile, which are described below (section 3.

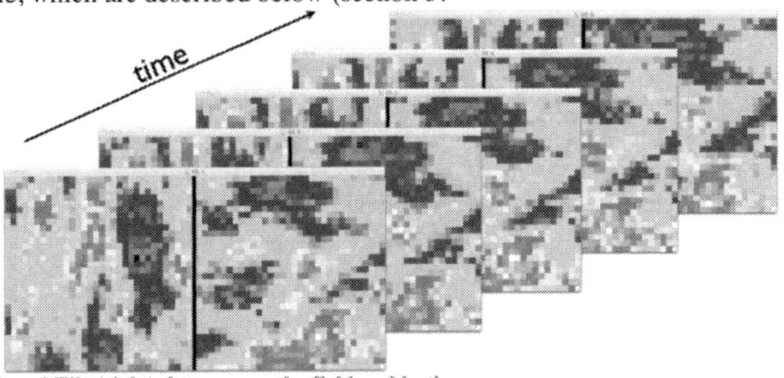

Figure 3: Tip (left) and Tilt (right) data across the field and in time.

3. Methods to retrieve the turbulence profile

3.1. Multi-scale tip-tilt averaging

The simplest approach to retrieve the turbulence profile consists in averaging the instantaneous tip-tilt into different size boxes (if N is the size of a box, there will be N^2 boxes of 32/Nx32/N pixels each) and measuring the time variance of the signals. The largest box (the variance of the average) is a measure of the variance of the tip-tilt at the ground and the average of the $(32)^2$ smallest boxes of single star/pixel measurements (the average of the variances) provides the altitude integrated seeing. We can write an equation of the form:

$$\sigma_{tip}^2(N) \propto 1.03\int_0^h \left(\frac{D}{r_0(h)}\right)^{\frac{5}{3}} \cdot dh$$

with the size of the box-averaged field N proportional to the altitude h of the layer. Plots of the field-size dependent isokinetism (Figure 4) indeed show a monotonic increase of the variance with N, but it is surprisingly smooth and differentiating with respect to N does not retrieve the layers introduced in the simulations.

There seems to be a faint trend in the results, apparent when changing the number of layers or their height but this is not sufficient to reliably retrieve the input turbulence height or strength. This may be due to the fact that the tip-tilt angular correlations are in fact much larger than the meta-pupil at most altitudes. Also, the tip-tilt angular correlation function is most likely very smooth, translating into a very broad altitude response. The tip-tilt angular correlations are also very sensitive to the outer scale of the turbulence, L_0. It is expected that the outer scale close to the ground varies with altitude and there is no reason to assume that the free atmosphere outer scale is correlated to the ground. For instance, we can assume that the dome seeing outer scale has to be smaller than the diameter of the dome itself. All of this could readily be included in the simulations, but the disappointing results shown in Figure 4 are unfortunately just for the simplest of cases...

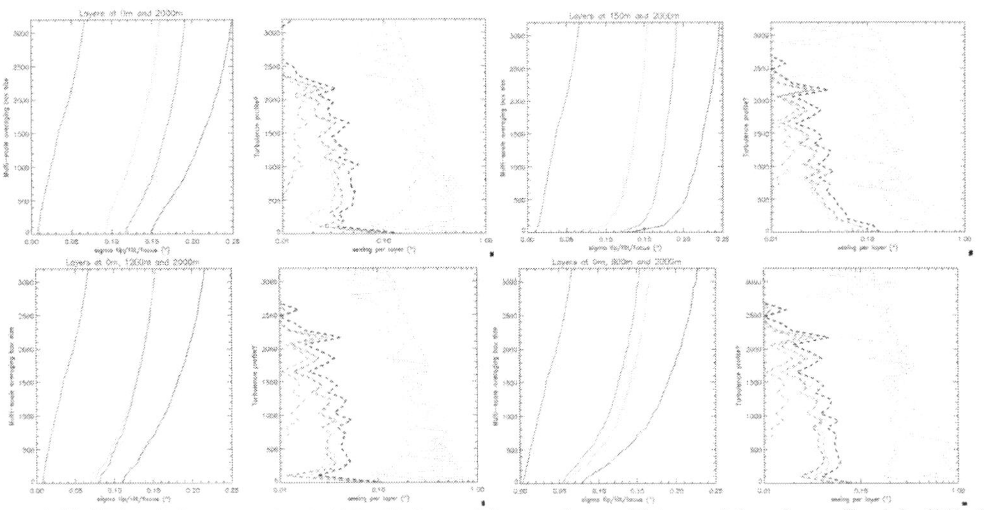

Figure 4: Multipl-scale box-averaging field tip-tilt for model atmospheres with two and three layers. Top left: 40% of turbulence at 0m and 60% at 2000m; top right, 70% at 150m and 30% at 2000m. Bottom left: 40% at 0m, 30% at 1200m and 30% at 2000m; bottom right, 20% at 0m, 40% at 800m and 40% at 2000m. In each case, the cumulative seeing obtained from $\sigma_{tiptilt}(N)$ is shown (left, tip=green, tilt=red, focus=blue, r.m.s. tip-tilt=black) as well as its derivative with respect to altitude, which should be the vertical turbulence profile.

3.2. Covariance method

The multi-scale averaging method only uses the temporal statistical information, but we have do access to the full time series of tip tilt measurements, so it may be possible to extract more detailed information by studying the tip-tilt spatio-temporal covariances. Because these are field covariances (and not pupil covariances as one is more accustomed to in SLODAR work), it is worth pointing out that the effect of altitude on layers is to broaden the covariance peak for low-altitude turbulence and to narrow the covariance peak as the altitude increases (turbulence at the ground produces a constant plane). Furthermore, at t=0, all the covariance peaks associated with each layer will be superposed; it is only by looking at temporal covariance that the wind speeds and directions can be extracted, although this is non-trivial as it ideally requires fitting multiple Gaussian component. Because turbulence at the pupil introduces a constant plane, it may not be possible to extract the direction of the ground layer, though its speed can possibly be recovered from the decay time of the constant.

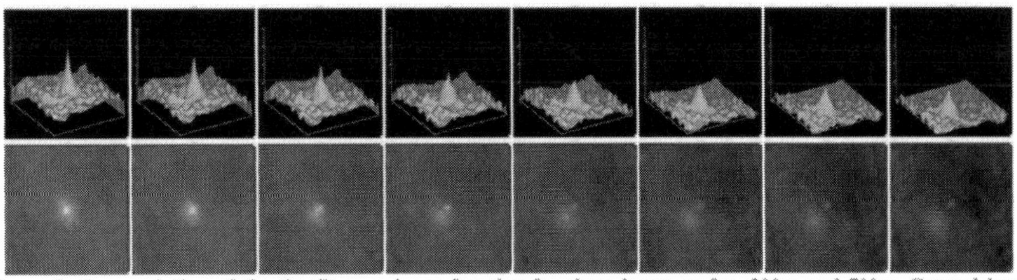

Figure 5: time evolution of the tip-tilt covariance function for three layers at 0m, 200m and 700m. Ground layer turbulence produces a constant plane, which decreases in time. Higher layers produce sharper peaks, which move across the covariance as a function of their speed, height and direction.

Although this method appears to contain more information, it is not trivial to retrieve the turbulence profile, as quantification would require fitting Gaussian components to the data. More information also seems to be available for example from one time step to the next (e.g. being able to track a correlation peak through the time dimension) as shown in the 3D projection of the temporal data

(Figure 6). This is known as the autocovariance and is work in progress, as at least conceptually, it may open the path for a truly layer-oriented version of the turbulence profiling (see Section 4.

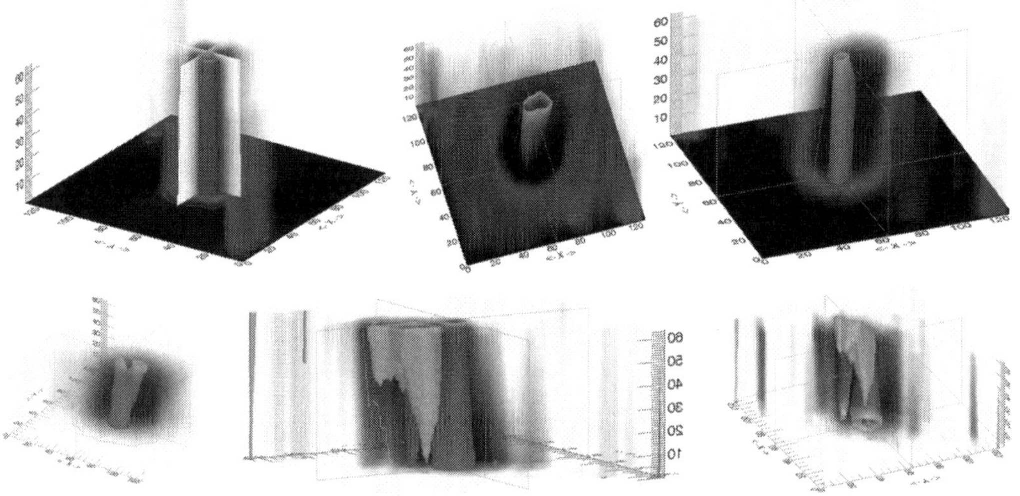

Figure 6: 3D plots of autocovariance maps. The top row shows three different views/cuts of the simple case of 60% at 250m and 40% at 2000m moving perpendicular to each other. The bottom row (left) shows a clear example of two separated layers (70% at 150m, 30% at 2000m), but the autocovariance plots are not always so easy to decipher (middle and right: 40% at 150m, 60% at 2000m).

3.3. 3D FTT

The last method that was developed used a 3-dimensional Fourier transform of the tip and tilt cubes. The result is the temporal and spatial power spectrum of the tip and tilt. Each spatial frequency corresponds to a given spatial scale of field variations (for example the average tip-tilt is the frequency zero, corresponding to the DC bias; the highest spatial frequency corresponds to very small scale variations of the signal, in other words, small isokinetism introduced by high altitude layers). The temporal dimension of the spectrum should then provide information related to the speed of the layers/spatial frequencies.

One difficulty of this method is that the phase screens don't evolve, but simply translate. A temporally translating phase screen will produce a plane of constant phase in the plane orthogonal to the 3D direction of motion. Examples of cuts through the amplitude and phase of the 3D FFT of the tip and tilt cubes are shown in Figure 7.

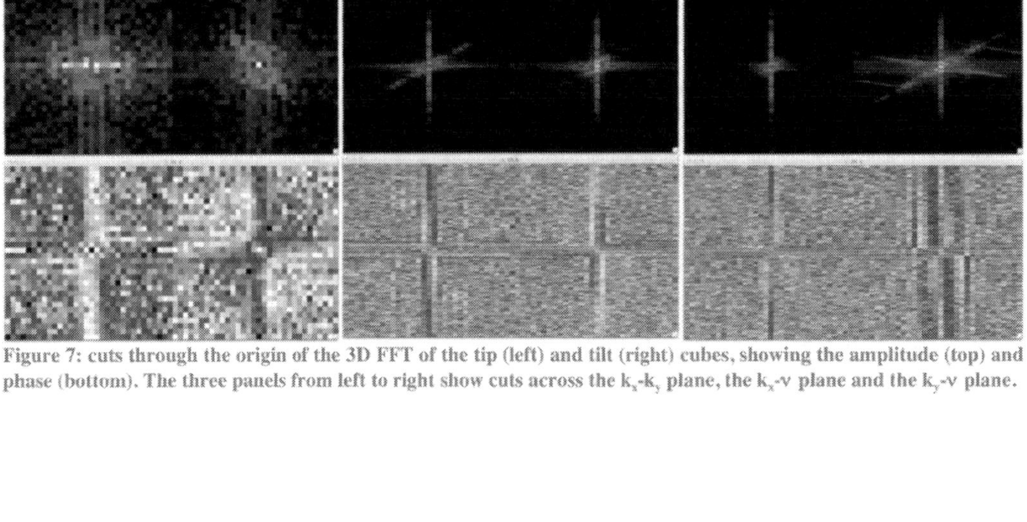

Figure 7: cuts through the origin of the 3D FFT of the tip (left) and tilt (right) cubes, showing the amplitude (top) and phase (bottom). The three panels from left to right show cuts across the k_x-k_y plane, the k_x-v plane and the k_y-v plane.

Journal of Physics: Conference Series **595** (2015) 012018 doi:10.1088/1742-6596/595/1/012018

Another apparently trivial difficulty is being able to visualize the phase in 3 dimensions. Iso-phase surfaces might be useful, but because the phase wraps modulo 2π at high frequency, it hides the coherent phase information at lower frequencies. It can again be seen that the data is very rich, but in this case it is the difficulty in extracting the relevant information (profile, wind speed), which makes this method impractical.

Taking the inverse Fourier transform of the modulus squared of the previous step produces the spatiotemporal autocorrelation of the tip and tilts across the field. In theory, individual layers at different heights and speeds produce autocorrelation peaks that narrow with layer altitude (similar to the covariance). Different attempts at visualizing dual and triple layers are shown in Figure 8.

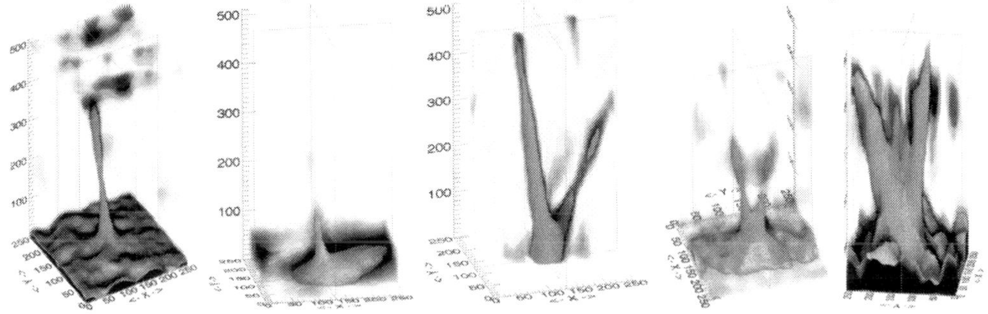

Figure 8: FFT spatiotemporal autocorrelation of two and three layer atmosphere. Left to right: 60% at 0m and 40% at 2000m; 60% at 1000m and 40% at 2000m; 40% at 15000m and 60% at 2000m; 40% at 0m, 30% at 1200m and 30% at 2000m; 20% at 0m, 40% at 800m and 40% at 2000m. As the altitude of the layer increases, the correlation peaks narrows (it is a constant plane for ground turbulence). The direction and angle of the correlation peaks give the direction and angular velocity components respectively of the velocity vector of each layer.

4. Why it doesn't work and future work

Although the tip-tilt correlations across the field do contain information regarding the vertical turbulence profile, the response function is most likely too smooth to be able to reconstruct useful profiles. The information is encoded in the data (and correlations can be seen "by eye"), but a different method is needed for this concept to work quantitatively on realistic turbulence profiles. For example, instead of tip and tilt, it might be possible to use higher orders of aberrations (because they smaller correlation lengths). This requires looking at a different observable than the centroid (for example, focus is encoded in the second moment of the PSF, shown in the dashed blue line of Figure 4); or by using low order wavefront sensors on field stars - but this is no different than SLODAR!

The conference was very stimulating at generating new ideas and discussions: a suggestion which has not yet been tried out was made by Miska LeLouarn to build the equivalent of an interaction matrix between field covariances and single layer altitudes and then invert it, to retrieve the altitudes from the measured field covariances. Another very interesting discussion with Aglae Kellerer indicated that the layer oriented MACO approach to solar observations[4] might be applicable if the notion of extended object is applied to star fields. How to translate measurements on partially illuminated high-altitude meta-pupils into useful tip-tilt correlations remains to be determined and will be the focus of this ongoing work.

References

[1] R.Flicker&F.Rigaut. "Hokupa`a Anisoplanatism and Mauna Kea Turbulence Characterization." 2002, *Publ. Astron. Soc. Pac.* **114**, 1006-1015.

[2] K.Hodapp, N. Kaiser, H. Aussel etal. "Design of the Pan-STARRS telescopes." 2004, Astronomische Nachrichten, **325**, 636-642.

[3] O. Lai, M. Chun, J-C. Cuillandre, "IMAKA: Imaging from Mauna KeA with an atmosphere corrected 1 square degree optical imager." 2008, SPIE **7015**, Adaptive optics systems.

[4] A. Kellerer, "Layer-oriented multi-conjugate AO for solar observations", *these proceedings*.

Multi-instrument characterization of optical turbulence at the Ali observatory

L.-Y. Liu[1], Y.-Q. Yao[1], J. Vernin[2], H.-S. Wang[1], J. Yin[1], X. Qian[1]

[1]National Astronomical Observatories, Chinese Academy of Sciences, Beijing 100012, China
[2]Université de Nice-Sophia Antipolis, Observatoire de la Côte d'Azur, CNRS-UMR7293, Laboratoire Lagrange, 06108 Nice Cedex 2, France

E-mail: `liuly@nao.cas.cn`

Abstract. In order to characterize the atmospheric optical turbulence at Ali observatory, we have deployed multi-instruments, which are able to continuously monitor the optical turbulence for site evaluation. These instruments include the DIMM, MASS, Single Star SCIDAR and Polaris seeing monitor, and we also plan to install SNODAR and Micro-thermal sensors for the turbulence on surface layer by the end of this year. This configuration allows us to collect a substantial database and make cross-comparison of the results. We have successfully obtained the profiles of optical turbulence and wind speed with Single Star SCIDAR, as well as the key parameters for adaptive optics application, such as seeing, coherence time, and isoplanatic angle. The DIMM seeing measurements are also carried out simultaneously. The median seeing measured by the DIMM and SSS in 2013 is 0.69 and 0.79 arcsec, respectively.

1. Introduction

The Ali observatory is located at 32°19'(N), 80°01'(E), with an altitude of 5100m, near the central town of Ali area, Tibet (Fig.1). The remote studies and local site testing have been performed, the first results show that Ali observatory is promising for very large optical-infrared telescopes [1]. The Ali observatory has begun construction since 2010, and up to date, there are nine domes already be installed for site testing. The site testing instruments including automatic weather station (AWS), cloud monitor, water vapor monitor, and multi-instruments for atmospheric optical turbulence characterization. Due to optical turbulence seriously limits the performance of high angular resolution instruments, the characteristics of optical turbulence is important to evaluate astronomical observatory. In this paper we introduce the instruments for turbulence measurement at Ali observatory, and present preliminary results of the site testing campaign in 2013.

2. Instruments

In this section, a brief description is given to the principles and observations of multi-instruments in Ali observatory. DIMM is a portable instrument for detection of optical turbulence, but it only senses the integrate atmosphere. The Single Star SCIDAR (SSS) is a portable version of the SCIDAR, which provides the profile of optical turbulence and velocity of the whole atmosphere from ground to 25 km. Microthermal sensors, equipped on a 40m mast, can detect the surface layer turbulence. Although Polaris seeing monitor is fully automated, the measurements easily

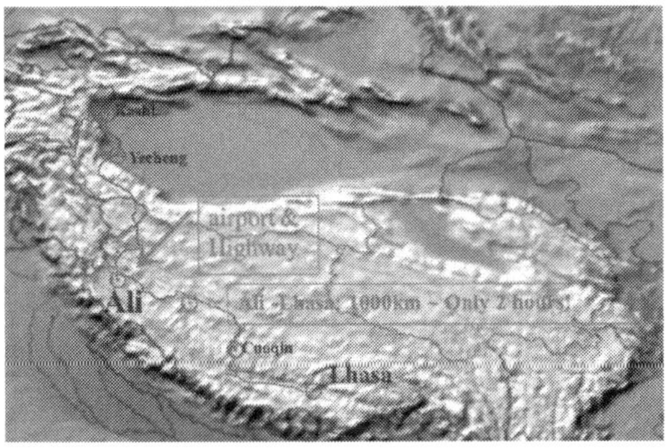

Figure 1. The location of Ali observatory.

Figure 2. The layout of multi-instruments.

be affected by gusty winds. MASS can subdivide the whole atmosphere into 6 thick layers, yet it tends to detect the turbulence in the free atmosphere. SNODAR is a profiler for the atmospheric boundary layer with a vertical resolution of 1 m. Fig.2 shows the layout of multi-instruments for turbulence measurements at the Ali observatory.

2.1. DIMM

The DIMM is a small transportable instrument [2], with the differential technique to accurately measure the seeing conditions. The seeing is measured via the differential positions of two sub-aperture images of a star, and therefore, DIMM seeing monitor is almost insensitive to tracking errors. This instrument yields the integrated seeing from telescope pupil to the top of the atmosphere.

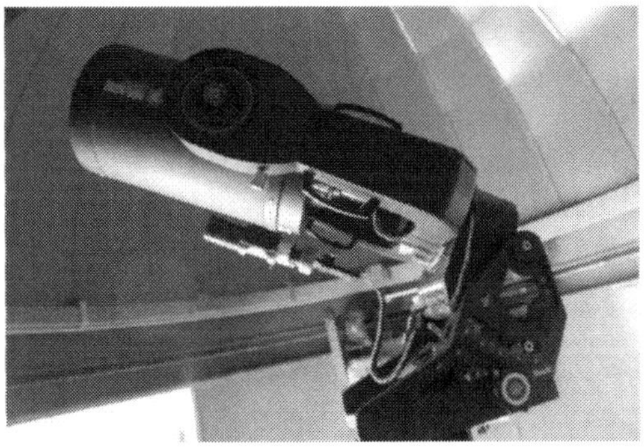

Figure 3. The DIMM instrument at Ali observatory (The position C in Fig.2).

Figure 4. The simultaneous measurements with 2 sets DIMM for calibration.

The DIMM technique is widely employed in site testing. We have developed a DIMM for site testing [3] and installed it at the position C on Fig.2. The DIMM instrument is shown in Fig.3. The Fig.4 is the simultaneous measurements with 2 sets of DIMM for calibration. According to the principle mentioned above, our DIMM instrument has the following configurations:

(i) a Meade LX200GPS telescope, with the entrance pupil of 20 cm, and the focal length of 200 cm;

Journal of Physics: Conference Series **595** (2015) 012019 doi:10.1088/1742-6596/595/1/012019

(ii) a mask with two sub-aperture, with diameter of 5 cm, and a separation of 15 cm;
(iii) an optical wedge installed on one sub-aperture, to adjust an appropriate separation between the two star images at focus;
(iv) a Lumenera SKYnyx 2.0M CCD camera, attached to the telescope focal plane for fast sampling star images ($\Delta t = 0.5ms$),with a 640 × 480 format and the pixel size of 7.4 μm;
(v) two computers, one calculating seeing values and closed-loop guiding, and the other remotely monitoring the status via internet.

2.2. Single Star SCIDAR
SSS is a remote sensing technique to measure the vertical structure of the atmospheric turbulence by analyzing the scintillation patterns of single star. SCIDAR technique has been first proposed by Vernin in 1973 [4], which analyzing the double star scintillation in order to separate the effect of each optical turbulent layer. The SSS is a new member in SCIDAR family, and this instrument can deliver the distribution of $C_n^2(h)$ with height, by using a small telescope.

Figure 5. The SSS instrument at Ali observatory (The position A in Fig.2).

Figure 6. The MASS instrument at Ali observatory (The position D in Fig.2).

We have developed a SSS for remotely sensing atmospheric turbulence profiles [5]. Fig.5 shows the SSS location at the Ali observatory (The position A in Fig.2). The SSS consists of a 40 *cm* telescope and a CCD camera for fast sampling of stellar scintillation pattern. The single star images are analyzed, providing vertical profiles of optical turbulence intensity $C_n^2(h)$ and wind speed $V(h)$, and allowing the determination of seeing, isoplanatic angle and coherence time. Although the SSS is a low resolution optical turbulence profiler, it needs only 40 cm aperture telescope, showing a practical advantage of being transportable for field site testing. The main specifications of our SSS are the following.

(i) a 40 cm Meade M16 tube, on an Astro-Physics 1200 equatorial mount, with the focal length of 400 cm;
(ii) a collimating lens after the focus of the telescope, with 10 mm focal length, to make the beam parallel;
(iii) a CCD camera, after the collimating lens for fast sampling the star scintillation pattern. The fast readout Pixelfly CCD-200XS, 640×480 format with pixel size of 9.9×9.9 μm^2, allows continuous acquisition with a high frequency rate without loss of any image. The exposure time is usually taken as 1 ms, every 5.6 ms, the pixel scale is 0.51 arcsec/pixel;
(iv) an auto-guiding system, made with a cube splitter;

(v) two control computers, one capturing scintillation images from the CCD camera and computing the spatial auto-correlation and cross-correlation images, the other serving as real time tracking through a CCD guider.

2.3. *Micro-thermal sensor*

The refractive index variations arise from density variations in the air, so the refractive index fluctuation is related to its thermal fluctuations. The first micro-thermal measurements are made with resistance thermometers by Lynds [6]. The temperature structure coefficient C_T^2 is defined as: C_T^2=<|T(x)-T(x+r)|^2> $\times r^{(} - 2/3)$, where r is a separation of two temperature sensors in meter. Assumed that fluctuations are passive under the constant pressure, the refractive index variations arise directly from temperature fluctuations and can be measured by using the temperature fluctuations as following:

$$C_N^2(h) = \left(7.9 \times 10^{-5} \times P \div T^2\right)^2 \times C_T^2(h) \quad (1)$$

where C_N^2 is the refractive index structure coefficient, P is pressure in hPa, T is temperature in K degree, and h is height above ground in meter.

We have built a 40m tower for the micro-thermal measurements (The position B in Fig.2), and we plan to measure optical turbulence of the surface layer by cooperating with Japanese team at Ali observatory [7]. A total of 10 C_T^2 sensors on 5 levels of support stays will be installed on the 40m tower at the height of 36m, 19m, 10m, 6m and 4m above ground. A separation of a pair of C_T^2 sensors is set up to be 1 meter, and a temperature sensor is attached near the center of each stay.

2.4. *Polaris Seeing Monitor*

We have employed a Polaris seeing monitor (Made by SBIG) to determine the quality of seeing in the early site testing, because it is fully automated, and virtually no maintenance. The Polaris seeing monitor is fixed on the position F in Fig.2.

The seeing values can be calculated automatically according to the Polaris trail [8]. The monitor is developed using ST-402ME CCD with TDI readout mode, and the exposure time is set up to 5ms. The optical configuration is a set of F/5.3 lenses, providing a field of view of 2.7° × 1.6°. Due to the atmospheric turbulence, the line below Polaris image is deviated left and right as Polaris's position is perturbed by seeing, and the SBIG software measures this perturbation and automatically calculates FWHM at the zenith. The comparison with SBIG seeing monitor and DIMM shows that SBIG seeing monitor is easily affected by gusty wind, resulting in larger seeing values [3]. Therefore, the SBIG seeing monitor just provide a trend of seeing variation.

2.5. *MASS*

MASS is an instrument to measure the vertical distribution of turbulence in terrestrial atmosphere by analyzing the scintillation of bright stars [9]. MASS can subdivide the whole atmosphere into 6 thick layers, e.g. 0.5, 1, 2, 4, 8, 16 km, and the turbulence intensity of each layer is measured. However, the MASS is not sensitive to the turbulence of the ground layer.

The turbulence of the ground layer in the first 0.5 km can be measured by combining MASS and DIMM. We have obtained a set DIMM-MASS from CTIO in 2009, and we installed it at Ali observatory in 2013 afer some experiments. Fig.6 shows the SSS location at the Ali observatory (Position D in Fig.2). The DIMM-MASS will run the turbulence measurements next year.

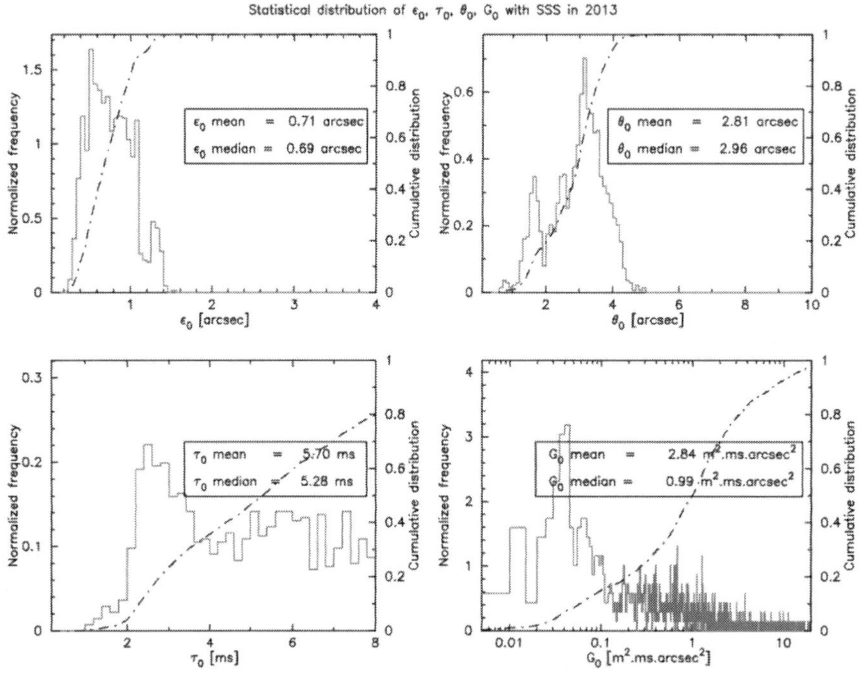

Figure 7. Histogram (solid line) and cumulative distribution (dashed line) of atmospheric parameters computed with the SSS measurements at Ali observatory in 2013. The parameter distributions are seeing ε_0 (top left), the isoplanatic angle θ_0 (top right), the coherence time τ_0 (bottom left), and coherence étendue G_0 (bottom right).

2.6. SNODAR

We have ordered 2 sets of SNODAR from UNSW. SNODAR is a high resolution (5 kHz) acoustic radar designed specifically for profiling the atmospheric boundary layer, which can measure the atmospheric turbulence within the first 200m atmosphere with a vertical resolution of 1 m [10]. We plan to install the SNODAR in the position E in Fig.2, which is a platform without surrounding shelter.

3. Preliminary Results

We have performed site testing campaigns at Ali observatory to characterize the optical turbulence with the instruments of DIMM and SSS. One of the campaign are carried out the turbulence measurements from June 4 to 6 in 2013. One DIMM seeing value is calculated using a set of 100 images, and the final seeing corrected to the zenith angle. The SSS spatiotemporal cross-correlation functions are obtained every 11.2 *sec* corresponding to a set of 2000 images. Due to measurement errors caused by continuously changing sky background, especially during moon nights, the sky background is checked every hour during the SSS measurements.

Fig.7 presents the statistical distributions of the high angular resolution parameters with SSS. The seeing values (top left) span over the range [0.3 and 1.5 arcsec], with a half of the seeing better than 0.69 arcsec, and 75% of the seeing better than 0.9 arcsec. The isoplanatic angle (top right) follows a log-normal distribution, with an average of 2.81 arcsec, and a median of 2.96 arcsec, respectively. The coherence time (bottom left) also follows a log-normal distribution, and the average and median values are 5.70 and 5.28 ms, respectively. For the coherence étendue

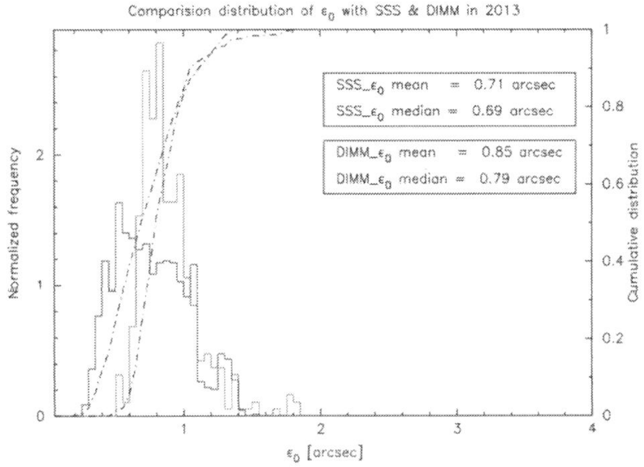

Figure 8. The comparison distribution of seeing ε_0 with SSS (in blue) and DIMM (in red) at Ali observatory in 2013.

G_0 (bottom right) [11], the average and median values are 2.84, and 0.99 m^2.ms.arcsec^2.

Fig.8 shows the statistic comparison of the histogram and the cumulative distribution of the entire DIMM (in red) and SSS (in blue) data set. The DIMM mean seeing value is 0.85 arcsec, and the DIMM median value is 0.79 arcsec. There are 25% of the DIMM seeing values better than 0.65 arcsec, and 75% better than 0.95 arcsec.

4. Conclusions

Multi-instruments have been installed or planned at Ali observatory to investigate atmospheric optical turbulence properties. The set of instruments can obtain the turbulence profiles of surface layer, boundary layer, free atmosphere and the whole atmosphere. These instruments can provide optical parameters involved in high angular resolution astronomy, including seeing, isoplanatic angle, coherence time, coherence étendue, etc. The preliminary results in the site testing campaign in 2013 shows the median seeing with DIMM and SSS is 0.69 and 0.79 arcsec, indicating that Ali observatory is promising for large optical-infrared telescopes.

Acknowledgements

This work is supported by the National Natural Science Foundation of China (NSFC Grant Nos.10903014, 11373043, 11103042, 11203044, 11303055 and 11473041), and the Young Researcher Grant of National Astronomical Observatories, CAS.

References

[1] Liyong Liu, Yongqiang Yao, et al. 2012, Proc.of SPIE, Volume 8444, pp. 844464-844464-7
[2] Sarazin, M., & Roddier, F. 1990, A&A, 227, 294
[3] Li-Yong Liu, Yong-Qiang Yao, et al. 2010, Research in Astronomy and Astrophysics, 10, 1061
[4] Vernin J, Roddier F. 1973, J. Opt.Soc.Am, 63, 270
[5] Liu LiYong, Yao YongQiang, Jean Vernin, et al. 2013, Chinese Science Bulletin, 58, 4566
[6] Lynds, C. R. 1964, IAU Symposium 19 (ed. Jean Rosch, Gauthier-Villars, Paris), 126-131
[7] Sasaki, T., Yao, q., et al. 2011, The proceeding of Asian-Pacific Regional IAU Meeting 2011, 428-429
[8] Harlan, E.A., & Walker, M.F. 1965, PASP, 77, 246
[9] Tokovinin, A. & Kornilov, V. 2007, MNRAS, 381, 1179
[10] Bonner, TC.S., Ashley, M.C.B., et al. 2009, Acoustics Australia, 37, 47
[11] Lloyd J P. 2004, Proc. of SPIE, 5491, 190L

Dealing with the forecast of the optical turbulence as a tool to support astronomy assisted by AO facilities

Elena Masciadri, Franck Lascaux, Luca Fini

E-mail: masciadri@arcetri.astro.it

Abstract. In the context of the research activities related to the forecast of the optical turbulence and the atmospherical parameters being relevant for ground-based astronomy we focus here our attention on two specific topics: **1.** pros and cons of different solutions to supply wind speed and direction stratification on the whole atmosphere all along a night to support AO facilities; **2.** the necessity of instrumentation for optical turbulence monitoring (vertical profiles on the whole atmosphere) to be used operationally. In the last two decades the development and the use of different vertical profilers covering the whole atmosphere or part of it in application to the astronomy took place. Several instruments based on different principles (with associated pros and cons) have been applied in different contexts in astronomy with a main use in the site characterization and site selection. Time changed and the necessity of the astronomy supported by AO facilities is much more demanding with a diversification of applications. Recently, motivated by a precise necessity related to the identification of an absolute reference to carry out studies on optical turbulence forecasts (MOSE project), we carried out a verification of the reliability of a few instruments that lead us to put in evidence some limitations for a few of them. At the same time such a detailed analysis permitted us to clarify the nature of some astroclimatic parameters. The main conclusion at which we arrived is two-fold. From one side we could trace a list of warnings related to different uses of such instruments. On the other side we could identify open problems that indicate that there is still space for research in the field of turbulence monitoring in application to the astronomy. Some suggestions are proposed.

1. Introduction

In this contribution we concentrate our attention on two specific topics both related to the forecast of the atmospheric parameters and optical turbulence with atmospherical models. From one side we present a method for the estimate of the wind speed stratification on the whole atmosphere and its performances; the advantage of this method with respect to other techniques; the usefulness in using such a method to validate wind speed measurements estimate obtained with the MCAO system GeMS [1] and in general with whatever other wide field AO system; the perspectives for an optimized method of wind speed stratification estimation based on the simultaneous use of the atmospherical model Meso-Nh and the MCAO system GeMS. Besides we present the state of art of optical turbulence vertical profilers to be used operationally to support AO systems and useful to be able to monitor the performances of atmospherical models for the turbulence forecasts. This study permitted us to achieve a better understanding of some atmospherical parameters such as the wavefront coherence time.

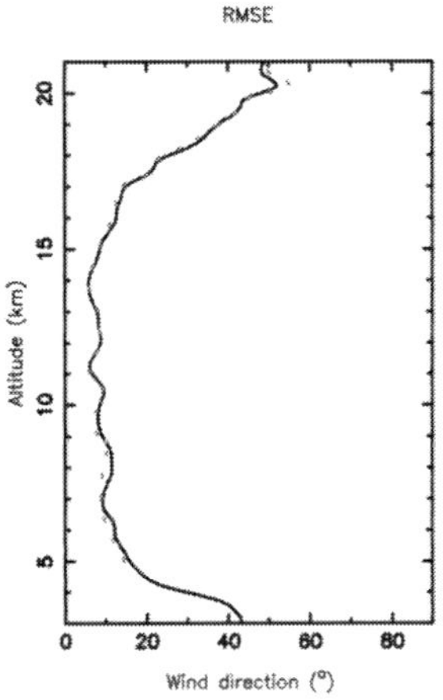

Figure 1. RMSE of the wind direction between radio-soundings and Meso-Nh (full black line) and radio-soundings and analyses from the ECMWF (red dots).

2. Wind speed vertical stratification on 20 km

The method we propose consists of a non-hydrostatic atmospherical model at mesoscale called Meso-Nh [2] that, used with the Astro-Meso-Nh packages [3], permits to calculate all the atmospheric parameters relevant for the ground-based astronomy and the optical turbulence (OT). The most solid validation of this model in terms of wind speed stratification [4]-Fig.2 is supported by the fact that we proved that the model is very well correlated to measurements related to 50 radiosoundings launched on 23 different nights in summer and winter periods. Results indicate that the correlation is good in statistical terms as well as on individual launches. In statistical terms simulations and measurements of wind speed (WS) and direction (WD) have been analyzed giving a bias for WS typically of the order of 1 ms^{-1} and not larger than 2 ms^{-1} along the whole 20 km; the RMSE of WS is typically of the order of 3 ms^{-1}. For the wind direction in the central part of the atmosphere the bias is just a few degrees. In the low part of the atmosphere and the top the bias is not larger than -/+ 10 degrees. The RMSE of WD in the central part of the atmosphere is as small as 10 degrees. Its value increases in the range 3-5 km above sea level and for h $>$ 16km achieving a maximum value of 40 degrees. In this new calculation with respect to [4] we filtered out all data associated to a wind speed $\leq$ 3 ms^{-1} (Fig.1).

The residual 40 degrees is due to the fact that when the wind speed is weak the direction is much more variable. In the same paper [4] we proved that the residual biases are due to the initialization data and not the mesoscale model. If we look at the model performances in terms of the correlation between measurements and calculation obtained on the individual flights results are even more interesting. In many nights we had radiosoundings launches at different time and this permitted us to investigate the ability of the model in reconstructing the real-time modification of the wind stratification during the same night. We could proved that the model adapts perfectly to the temporal evolution of the wind speed stratification on 20 km ([4]-Fig.3), changing the wind speed values as they are observed at all the heights above the ground. In the same paper (Fig.B1) the model ability in matching the measured wind speed stratification on all the individual 50 radiosoundings is shown indicating excellent model performances. A maximum relative error of just 23% has been observed in the whole sample. This tells us that we can reasonably use the wind speed predicted by the model for the calculation of the τ_0 (with C_N^2 measured or simulated) with the temporal sampling that we wish. We typically use a sampling of 2 minutes but we could go down to a few seconds. Meso-Nh solves therefore important criticisms in the field of the wind speed characterization. The calculation of τ_0 using wind speed obtained by General Circulation Models (GCMs) can induce to important errors because these models are not reliable in the low part of the atmosphere and, moreover, they provide informations only at synoptic hours that means each six hours (00, 06, 12 and 18) UT.

The wind speed stratification can be very useful in application to the ground-based astronomy. Here are a few of these applications:

1 to manage observational scheduling of scientific program (leading application);
2 to support the use of the AO systems in real time (see for example an operational application to GeMS [1]);
3 to support studies of simulations of AO performances and new AO techniques validation (this is a work to be done offline and not in real time);

How can we use the wind estimates ?

1 the simplest output is the forecast of the temporal evolution of the wind speed at different heights (wind speed stratification).
2 if the wind speed from the model is joint to the C_N^2 predicted by the same model we can calculate the forecasted τ_0.
3 if the wind speed from the model is joint to the C_N^2 measured by an instrument we can have a real time measurement of τ_0. For such a high level of model reliability for the wind speed stratification it is indeed useless to use an instrument even more because, in many cases, uncertainty and limitations for instruments are more severe.

The applications of points 1, 2 and 3 can be obtained with very a high temporal sampling.

2.1. Advantage of the Meso-Nh model with respect to other techniques
The numerical technique, and more precisely the use of the Meso-Nh model presents some advantages with respect to other techniques that are available at present:

1: the atmospheric model is by far the cheapest and fastest technique;
2: there are no logistical constraints;
3: instruments such as the Generalized SCIDAR (GS) [13] and the Stereo SCIDAR [14] need a telescope with a pupil size D of at least of 1 meter. This implies that they can hardly be used for operational applications;
4: Stereo SCIDAR has an automatic procedure for the detection of the wind speed but to cover all the wind spectrum it is necessary a pupil size of the telescope of at least 2 m otherwise the maximum measured wind speed is of the order of 36 ms^{-1}. Wind speed can reach values well above this threshold in the free atmosphere;
5: the model has a better spatial and temporal coverage than instruments based on remote sensing principle. It is enough to compare [4]-Fig.4,top with [5]-Fig.2;
6: the model has basically no constraints on the temporal sampling.

2.2. Meso-Nh and GeMS wind speed estimates

In a recent preliminary study [1], it has been possible to investigate and validate the ability of the MCAO system GeMS [6, 7] in reconstructing reliable wind speed stratification. Vertical wind speed stratification estimates obtained with the Meso-Nh model on 11 nights above Cerro Pachon have been compared to measurements obtained with GeMS during those nights ([1]-Fig.3 and Fig.4). Results of this study indicate that GeMS is able to reconstruct in most cases a good wind speed vertical stratification. Besides that, it has been possible to observe that, in some cases, it might produce not negligible errors. However, in those cases of failure, it has been possible to observe that the Meso-Nh estimates are determinant to correct the GeMS estimates. Results indicate therefore that a simultaneous (and preferably automatic) estimate of the wind speed done simultaneously by Meso-Nh and GeMS might give an exhaustive answer to the necessities of this MCAO system. Of course such a solution might be used for other WFAO systems such as AOF [8] of ESO.

3. Optical turbulence vertical profilers for operational applications

A critical element for the studies on the OT forecast done with atmospherical models is the verification (and quantification) of a good absolute estimate of the OT vertical stratification provided by the vertical profilers (instruments) that have to be used as a reference. In [9]-Fig.3 are shown the scattering plots of the seeing as measured by the Generalized Scidar (GS) and the MASS in all the six layers of the MASS. The C_N^2 profiles of the GS have been therefore projected on the six triangle shape weighting functions of the MASS to perform the comparison. The sample analyzed is part of the data related to the site test campaign of PAR2007 [10, 11] at Cerro Paranal. In [9]-Fig.4 is shown the integral of the seeing in all the 6 layers as seen by the MASS and the GS. The comparison puts in evidence that, in many layers (with exception of the layer 3 and 6), there is an evident discrepancy between the two instruments that is translated in a underestimation of the total seeing in the free atmosphere of the MASS. If we look at the relative errors in each layer ([9]-Table 4) we realize that these discrepancies are not negligible, particularly for the layers 1, 2 and 5 and for the total seeing (order of 50-60%). This in spite of the fact that the difference of the seeing in the free atmosphere is of roughly 0.2 arcsec. In other words, we should not be induced in error by the weak absolute differences. The determinant thing is the relative error. In the same paper we proved that the problem is in the MASS and not in the GS because the correlation between GS and DIMM is consistently good.

3.1. Seeing and J

If we look at the temporal evolution of J (related to the seeing by [9]-Eq.8) on all the 12 nights in which simultaneous measurements from the two instruments are available ([9]-Fig.A1), we observe that the trend of the seeing measured by the two instruments is substantially good but the underestimation by the MASS is systematic. If we look at the temporal evolution of J in the first layer the underestimation is quite important ([9]-Fig.A2). If we look at the temporal evolution of J of the highest layer we observe that the correlation is mostly very good ([9]-Fig.A7). The trend remains in general satisfactory.

3.2. Isoplanatic angle

In [9]-Fig.6 is shown the scattering plot of θ_0 as measured by MASS and GS. It is possible to observe a good correlation with a small bias (~ 0.07 arcsec). Looking at the temporal evolution for the different nights ([9]-Fig.A8) also the trend and the correlation seem good. This is not in contradiction with results obtained for the seeing. We know, indeed, that θ_0 is mainly driven by the turbulence in the high part of the atmosphere and the correlation of seeing between the two instruments in the highest layer was satisfactory (see [9]-Fig.A7). In other words, even if there are discrepancies in the spatial distribution of the OT in the middle of the atmosphere these have negligible impact on θ_0. It is a pity that the statistics in this case is a bit limited. For θ_0 we have only 12 nights with simultaneous measurements and 4 nights with very few values. However it seems that θ_0 estimates from MASS are mainly reliable.

3.3. Wavefront coherence time

In the scattering plot of [9]-Fig.7 it is possible to observe a good correlation between τ_0 estimates obtained from the two instruments with limited BIAS and RMSE. The correlation is better for small values of τ_0, for large values the MASS shows a tendency in providing larger values with respect to the GS. The same can be said for the temporal evolution of τ_0 ([9]-Fig.A9). In the first part of the campaign the correlation is somehow better but it is in general good. It is interesting to investigate the reasons for such a general good behavior. We should expect an overestimation of the τ_0 from the MASS because of the underestimation of the seeing in the free atmosphere. In the same paper we formulated a thesis to explain how it is possible that the τ_0 from MASS is satisfactory also if the seeing is not.

A τ_0 MASS measurement in reality implies the use of three different instruments: a MASS, a DIMM and an automated weather station (AWS) because τ_0 is given by the addition of two contributions. One coming from the MASS (associated to the free atmosphere) and the other coming mainly from the DIMM (associated to the ground layer). The hypothesis is that, if τ_0 is substantially well correlated with the GS that means that the ground layer contribution dominates. We proved this thesis in [9]-Fig.A10 where the three contributions of τ_0 provided by the ground, the free atmosphere and the whole atmosphere are displayed. Details of the demonstration can be found in [9]. We conclude that, differently from what is frequently assumed, the τ_0 is mainly driven by the ground contribution and not by the free atmosphere.

4. Conclusions

The main conclusions of this contribution can be summarized as:

1: Our recent studies [9] permitted us to establish a list of warnings with respect to the use of MASS (the unique monitor used operationally at present): (a) θ_0, τ_0 appear reliable, the seeing in the free atmosphere is underestimated by the MASS. (b) all major conclusions on site selection for TMT and E-ELT are not affected by our results because the analysis in those context is done in relative terms. (c) we proved for MASS the existence of important discrepancies on individual layers with respect to GS estimates. At the same time the GS revealed to be in good agreement with DIMM and the assumption of an overestimation of the GS (in alternative to an underestimation of the MASS) would lead to the absurd result that GS would not be well correlated to DIMM any more. (d) as a consequence it is not suggested to use MASS in applications to the AO and in general in all applications in which is critical the knowledge of the turbulence stratification in individual layers.

2: We put in evidence that the discrepancies on OT absolute estimate obtained with different instruments under conditions of fair comparison can be important and we identified specific limitations in some cases. Also we could identify contexts in which the level of accuracy is acceptable. The problem of the OT absolute estimate is the king topic in the turbulence field in application to the astronomy.

3: We promoted an action (multi-instrument campaign) that goes in the direction to solve this problem [12]. The goal of the site-testing campaign is to quantify the absolute estimate of the OT vertical distribution obtained by independent different instruments.

4: The OT absolute estimate is fundamental to quantify the accuracy. We proved [9] that, when the BIAS in seeing estimates in the free atmosphere between MASS and GS is equal to 0, the bias corrected RMSE σ_{FA} = 0.13". This tells us that the accuracy in seeing estimates in the free atmosphere is at present of +/- 0.13".

Journal of Physics: Conference Series **595** (2015) 012020 doi:10.1088/1742-6596/595/1/012020

References

[1] Neichel, B., Masciadri, E., Guesalaga, A., Lascaux, F., Bechet, C., 2014, *SPIE 2014*, 9148, id. 914863
[2] Lafore, J.-P. et al., 1998, *Annales Geophysicae*, **16**, 90
[3] Masciadri, E., Vernin, J., Bougeault, P., 1999a, *A&ASS*, **137**, 185
[4] Masciadri, E., Lascaux, F., Fini, L., 2013, *MNRAS*, **436**, 1968
[5] Avila, R., Carrasco, E., Ibanez, F., et al., 2006, *PASP*, **118**, 503
[6] Rigaut, F., Neichel, B., Boccas, M., et al., 2014, *MNRAS*, **437**, 2361
[7] Neichel, B., Rigaut, F., Vidal, F., et al., 2014, *MNRAS*, **440**, 1002
[8] Arsenault, R., Madec, P.-Y., Paufique, J. et al., 2014, *SPIE*, **9148**, id. 914802
[9] Masciadri., E., Lombardi, G., Lascaux, F., 2014, *MNRAS*, **438**, 983
[10] Dali Ali, W., et al., 2010, *A&A*, **524**, A73
[11] Masciadri, E., Lascaux, F., Fuensalida, J.J., Lombradi, G., Vazquez-Ramió, H., 2012, *MNRAS*, **420**, 2399
[12] Masciadri, E., Lascaux, F., Fini, L., AO4ELT 3rd Edit., 2013, DOI: 10.12839/AO4ELT3.13219
[13] Avila, R., Vernin, J., Masciadri, E., 1997, *Applied Optics*, **36**, 7898
[14] Sheperd, H.W., Osborn, J., Wilson, R. W., Butterley, T., Avila, R., Dhillon, V.S., Morris, T. J., 2013, *MNRAS*, **437**, 3568

Adapting to the Atmosphere Conference 2014 IOP Publishing
Journal of Physics: Conference Series **595** (2015) 012021 doi:10.1088/1742-6596/595/1/012021

Tomographic Adaptive Optics and Turbulence Profiling

Tim Morris

Centre for Advanced Instrumentation, Department of Physics, Durham University, South Road, Durham, DH1 3LE. UK.

E-mail: t.j.morris@durham.ac.uk

Abstract. The use of tomographic adaptive optics is fundamental to fulfilling scientific goals for many proposed instruments at major observatories. Tomographic AO uses knowledge of the atmospheric C_n^2 profile and to date, the majority of the profiles used to design and simulate these systems have come from external turbulence profilers. The C_n^2 profile resolution required for accurate predictions of ELT instrumentation exceeds that of existing instrumentation and here we define the requirements on these profilers for ELT support. However, tomographic AO systems can also measure C_n^2 profiles and we highlight several cases where external profilers can provide critical functionality to support on-sky operations.

1. Introduction

The next generation of facility-class Adaptive Optics (AO) systems will make use of tomograhpic wavefront sensing techniques to provide correction over wider fields of view or improve the performance of laser guide star AO systems [1]. Through the use of multiple Wavefront Sensors (WFSs) these tomographic systems are capable of mapping the full volume of turbulence above the telescope and thus determining a full C_n^2 profile (see e.g. Vidal 2010 [2] and references therein). Profiles defined using WFSs that are part of the AO system are generally called *internal* profiles as opposed to an *external* profile derived from a profiling instrument mounted on an auxiliary telescope.

Common with many types of existing turbulence profilers the vertical resolution and maximum altitude of the C_n^2 profile is dependent on telescope diameter and the angular separation of sources being observed. Tomographic AO system performance is C_n^2 profile dependent and the sensitivity of the AO system to changes in the profile is also defined by the WFS configuration. The tomographic configuration of next generation of wide-field AO systems on Extremely Large Telescopes (ELTs) defines a profile sensitvity that exceeds the capabilities of existing external profilers. However, the internal profile measured using the AO system WFSs will be sensitive enough to determine a C_n^2 profile at sufficient resolution to for example, analyse AO correction [3] or use as an input to PSF reconstruction algorithms [4]. At first sight this makes the role of the external profiler redundant once a tomographic AO system is operational.

Recent on-sky tests of tomographic AO systems [5] have provided insight into how external profilers could be used in the era of facility-class tomographic systems. In this paper I aim to identify both requirements and potential roles of external profiling systems as facility support instruments. I also argue that external profiling, whilst not essential for AO operation, may provide significant operational advantages and improve observing efficiency. First I introduce the various types of tomographic AO system that will be in use in the future, then describe their sensitivities and how profiling could be used with these systems. Finally I present several uses for external profilers that would support tomographic AO system operations.

2. Tomographic AO

There are several 'flavours' of tomographic adpative optics system which cover a range of performance levels, sky coverage and corrected fields of view. In general there is a trade-off between the fidelity of correction and the corrected field of view, but all tomographic AO systems rely on the use of multiple wavefront sensors. These wavefront sensors have to date been typically of Shack-Hartmann (SH) type, therefore most tomographic AO systems therefeore resemble multi-baseline SLODAR [6] systems.

The sensitivity of tomographic AO performance to changes in C_n^2 is defined primarily by the sepration of the guide stars (natural or laser) in use, and the diameter of the WFS sensing elements (subapertures in the case of a SH-WFS) although it is complicated by the number of independent baselines between guide stars within the asterism. The impact of a given layer changing altitude on AO performance is then also dependent upon the strength of that layer. Simulations of E-ELT AO systems have shown that for the most demanding cases, the C_n^2 profile vertical resolution must approach 100-200 m [7]. At this resolution, the AO system may start to become sensitive to changes in the thickness of the turbulent layer also, particularly for Ground Layer AO (GLAO) [8].

Tomographic AO systems can be separated into those that provide 'fixed' and 'floating' tomographic correction. A 'floating' tomographic system in general uses a single Deformable Mirror (DM) that is optically conjugated to the telescope pupil. The shape the DM should assume is calculated by first defining a number of virtual layers within the atmosphere at which the incoming wavefront will be reconstructed. The DM is then mathematically projected through these virtual layers to find an optimal DM shape that will provide the correction along the desired line of sight to the target object [2]. This is one reconstruction approach that can be adopted for both Laser-Tomographic and Multi-Object AO, although several others exist. As the C_n^2 profile changes the altitude of the virtual layers can be adjusted within the reconstructor and AO performance remains optimised for the observed turbulence profile.

For 'fixed' tomographic systems (such as Multi-Conjugate AO or GLAO), the DM(s) are optically conjugated to one or several fixed altitudes within the atmosphere. This configuration means that if the strongest turbulent layers in the atmosphere do not match the conjugated altitude of the DM(s), AO system performance can be degraded [9]. It should be noted however that both GLAO and MCAO systems can also make some use of floating tomographic control techniques to optimise performance (see e.g. [10] and references therein).

The number of reconstructed layers in the atmosphere is dependent on the number of subapertures within the WFSs. For a low-order system such as CANARY on a 4.2m telescope, 3-4 reconstructed layers are required [3]. Simulations of ELT-scale AO systems have shown that AO performance is even more sensitive to the number of layers within the atmosphere [11]. To optimise performance, these N layers should be continually adjusted within the reconstructor to track the strongest N layers in the atmosphere. However, weaker layers within the atmosphere can also have a large impact on performance, particularly for wide-field observations. Simulations of E-ELT AO have shown that the altitude of the strongest 40 layers should be known up to altitudes of 20-30 km.

3. Profiling requirements

Irrespective of the system architecture, it is clear that knowledge of the profile is critical to operate a tomographic AO system on-sky. As stated in Section 1 however, the internal WFSs can achieve sufficent resolution to satisfy these requirements. Initially where knowledge is required before the internal WFSs will be available is during the instrument design phase. This is particularly true for MCAO where the conjugate altitude of the DMs must be fixed. To allow estimates of the number of nights an instrument could be used or to optimise the system design for robustness, requires not only a median profile but example profiles over an extended time period. Representative profiles of night-to-night profile variation would be useful over the course of at least a year to fully describe the expected AO system performance.

Measurements of the variability of the layer altitude at shorter timescales is also required. Adjustment of the reconstructed layers and subsequent recalculation of the tomographic reconstructor can carry a significant computational overhead. However, all that is required is the altitude of the layers and not an absolute measure of their strength.

It should be noted that telescope observations have shown that there is always a significant ground/boundary layer component and therefore a reconstruction layer is always conjugated to the telescope pupil. Although the ground layer is often the dominant layer within the atmosphere, it does not affect the performance of the tomography because it is common to all WFS signals. Whilst a measure of atmospheric strength in terms of seeing or r_0 is required to estimate overall system performance, individual high layer strengths need not be calibrated absolutely.

In addition to C_n^2 profiles, wind velocity profiles can also be used with some reconstructors that can make use of the temporal statistics of the atmosphere. Whilst these techniques have not been implemented to date within any facility class AO systems, initial on-sky tests have shown they provide a significant improvement in performance particularly in the presence of high wind speeds or telescope vibrations (e.g. [12], [13]). Limits on the required accuracy of the wind velocity profile have not been fully calculated at this stage.

4. External profilers for operational support

There are many situations where an external profiler supporting tomographic AO systems could very rapidly return any initial investment required for its installation. For major facilities such as the E-ELT this statement holds true even if only a few minutes were saved per observation/target due to the high cost of observing time. Here I present several possible ways in which an profiler could assist with tomographic AO operations:

- **Queue scheduling** Tomographic AO systems with multiple WFSs and DMs are complex to setup for observation. Whilst existing tomographic AO systems have not in general been developed to facility-class level, it could take 2-3 hours to prepare an instrument for observations. An early indication of the high-altitude profile (remembering that the ground layer is not required), either through profile forecasting or direct measurement near sunset would help to ensure that the optimal tomographic instrument is used for the predicted tomographic conditions.
- **Reconstructor pre-computation** If an internally measured profile is used to build the reconstructor, then the target must be acquired and an on-sky dataset recorded before the compuation of the reconstructor can begin. As already stated in section 3, there can be a significant overhead in the calculation of the tomographic reconstructor. Building the reconstructor from externally measured profiles whilst (e.g. the telescope is slewing) could save significant amounts of time.
- **AO Performance optimisation** The variability of the atmospheric turbulence profile is such that it is highly unlikely that the system will have been exposed to a wide range of

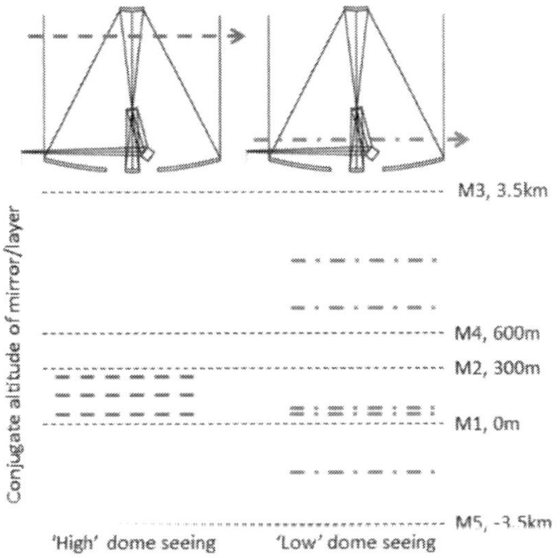

Figure 1. Sketch of E-ELT (not to scale) optical layout showing conjugate altitudes of each mirror in main optical train and obseved conjugate layers caused by passage of two idealised layers passing through the telescope structure at different heights.

tomographic configurations during the instrument commissioning phase. Due to this, much of the initial performance optimisation tasks will have to be performed off-line using recorded WFS measurements taken during science observations. An unbiased measurement of profile at high vertical resolution will be required as inputs for developing both simulations and PSF reconstruction algorithms.

- **Instrument comparisons** Each AO instrument will use its own WFSs, correction elements and real-time control optimisations. The only unbiased atmospheric profile reference between instruments would therefore be an external profiler. Not only will this be crucial during commissioning, but also to allow performance optimisations developed for one AO instrument to be applied to another.
- **Automation** For existing tomographic AO systems, the reconstructor is currently crafted by hand through fitting of the number of layers, layer altitudes and wide range of other parameters (such as misalignments due to flexure etc.) that can affect AO performance. The number of degrees of freedom required can result in a reconstructor fit which can contain errors that are not always obvious even to the trained observer, let alone an automated system. The use of an external profiler as an automated check of the internally measured profile could provide a robust check and form a critical part of the on-sky operating procedure.
- **Targeted line of sight profiling** An external profiler need not point along the line of sight of main science target after initial acquistion. Profiling along alternate lines of sight, such as that of the next science target, or 'upwind' from the current target would once again provide an estimate of how the profile could change in the coming minutes. This could potentially provide early-warning of upcoming profile variations to allow reconstructor optimisation before AO performance becomes degraded.

- **Dome-seeing and structure-induced turbulence** One aspect of ELT design that could be of particular importance is the impact of dome-seeing, or structure-induced turbulence [14]. For example, due to the optical design of the E-ELT, a single turbulent layer flowing through the telescope could be observed at several conjugate altitudes up to ±3.5 km from the telescope pupil (see Fig. 1) Turbulence caused by wind-flow through the telescope structure and localised heat sources could also contribute to the observed profile. In this manner, the ground layer could appear to the internal profile as conjugated layers. An external profiler would not be observe these layers, providing a potentially important diagnostic tool to improve telescope performance.

5. Conclusions

In this paper I have addressed the requirements that future tomographic AO systems will place on external profilers, based primarily on on-sky experience gained using CANARY and the external turbulence profilers supporting the CANARY project. I have summarised requirements on the required vertical resolution of turbulence profiles required for instrument design based on simulations of several E-ELT AO systems where a 100-200 m vertical resolution profile up to 20 km in altitude is required for the most demanding wide-field AO instruments for the E-ELT. The absolute strength of the individual layers is not required, but at least the strongest 11 layers must be identified. More layers are required in the measured profile to fully characterise AO performance. The strength of the ground layer is not required although its thickness can be used to estimate/characterise GLAO performance.

Whilst noting that tomographic AO systems are in general self-calibrating with respect to the turbulence profile, I have identified several cases in which external profilers could be used to support tomographic AO observations that hold the potential to improve observational efficiency and AO performance. These applications of external profilers in this manner will become more critical for E-ELT operations, where observing time is at a premium.

References

[1] Davies R, and Kaspar M, 2012, *Ann. Rev. Astron. Astrophys.* **50** 305
[2] Vidal F, Gendron E and Rousset G, 2010 *J. Opt. Soc. Am. A* **27** A253
[3] Vidal F *et al.*, 2014 *A&A* **569** A16
[4] Villecroze R, Fusco T, Bacon R, Madec PY, 2012. *Proc. SPIE* **8447** 84475W
[5] Myers R *et al.*, 2008 *Proc. SPIE* **7015** 7015E
[6] Wilson R, 2002. *MNRAS* **337** 103
[7] Basden A, Myers R and Butterley T, 2010. *Appl. Opt.***49** G1
[8] Tokovinin A, 2004. *PASP* **116** 941
[9] Ragazzoni R, Diolaiti E, Farinato J, Fedrigo E, Marchetti E, Tordi M and Kirkman D, 2002. *A&A* **396** 731
[10] Tokovinin A, Le Louarn M, Viard E, Hubin N and Conan R, 2001 *A&A* **378** 710
[11] Le Louarn M, 2012 *Proc. SPIE* **7736** 773618
[12] Sivo G *et al.*, 2014 *Opt. Express* **22** 23565
[13] Poyneer L, Macintosh B, Veran JP, 2007 *J. Opt. Soc. Am. A* **24** 2645
[14] Farahani A, Kolesnikov A, Cochran L, Hull C and Johns A, 2012. *Proc. SPIE* **8444** 84440U

Characterising atmospheric optical turbulence using stereo-SCIDAR

James Osborn, Tim Butterley, Dora Föhring and Richard Wilson

Department of Physics, Centre for Advanced Instrumentation, University of Durham, South Road, Durham DH1 3LE, UK

E-mail: `james.osborn@durham.ac.uk`

Abstract. Stereo-SCIDAR (SCIntillation Detection and Ranging) is a development to the well known SCIDAR method for characterisation of the Earth's atmospheric optical turbulence. Here we present some interesting capabilities, comparisons and results from a recent campaign on the 2.5 m Isaac Newton Telescope on La Palma.

1. Introduction

The Earth's turbulent atmosphere has a detrimental effect on astronomical observations. For imaging random phase aberrations blur the image reducing the attainable spatial resolution. For photometric observations, scintillation induced by the turbulence can be a dominant noise source [1].

Several techniques have been implemented to estimate the atmospheric turbulence profile, measured as the refractive index structure constant, $C_n^2(h)$, and wind velocity, both as a function of altitude. The most widely exploited are MASS (Multi Aperture Scintillation System, [2]), SCIDAR (SCIntillation Detection And Ranging, [3]) and SLODAR (SLOpe Detection And Ranging, [4]). MASS is not intended as a high vertical-resolution technique. It has a limited logarithmic vertical resolution and the high altitude response is very broad [2]. Here, we only address high altitude-resolution techniques and therefore we will only discuss SLODAR and SCIDAR. Both SLODAR and SCIDAR are triangulation techniques in which the atmospheric turbulence profile is recovered from either the correlation of wavefront slopes in the case of SLODAR, or scintillation intensity patterns in the case of SCIDAR, for two target stars with a known angular separation

Section 2 briefly describes stereo-SCIDAR instrument and how it works, a full description can be found in [5, 6]. Section 3 states the parameters which can be derived and gives short examples of how they can be applied, and section 4 shows that turbulence profiles are important photometric measurements as well as imaging.

2. Stereo-SCIDAR

Stereo-SCIDAR is a stereoscopic system in which the distance to a turbulent layer is estimated by measuring the offset of a covariance peak in the cross-correlation function of the scintillation pattern from two stars separated by angle θ. The strength of the turbulence is proportional to the magnitude of the covariance (figure 1). The velocity of the turbulent layers can also be

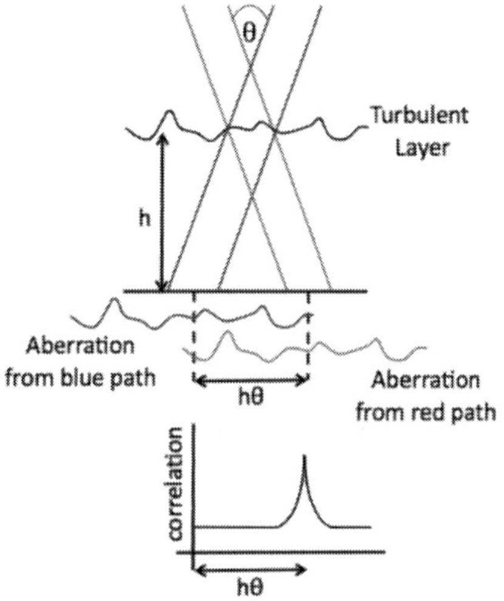

Figure 1. If a turbulent layer at height, h, is illuminated by two stars of angular separation, θ, then two copies of the aberration will be made on the ground separated by a distance $h\theta$. By cross correlating either the centroid positions from a Shack–Hartmann wavefront sensor (SLODAR) or the intensity patterns (SCIDAR) we can triangulate the height of the turbulent layer and the amplitude of the correlation peak corresponds to the strength of the layer.

estimated optically by tracking the covariance peaks through the spatio-temporal covariance function. The reader is referred to Shepherd *et al.* [6] for a full description stereo-SCIDAR including how it works and the opto-mechanical design.

By separating the scintillation patterns onto separate detectors instead of overlapping them on a single camera (as with traditional SCIDAR instruments) we reduce the noise in the profile estimation. This is because, in conventional SCIDAR instruments, the intensity speckles lose contrast in the overlapping patterns, reducing the visibility of the covariance peaks (figure 2).

3. Atmospheric Parameters at La Palma

Stereo-SCIDAR provides measurements of the refractive index structure parameter, $C_n^2(h)$, as a function of altitude, h, (figure 3) and the turbulence speed, $|V(h)|$, and direction $V_\theta(h)$ (figure 4). Using these parameters it is possible to derive several other optical parameters:

$$r_0 = \left(0.423\left(\frac{2\pi}{\lambda}\right)^2 \cos(\gamma)^{-1} \int C_n^2(h)\mathrm{d}h\right)^{-3/5}, \quad (1)$$

$$\theta_{\mathrm{FWHM}} = \frac{0.98\lambda}{r_0}, \quad (2)$$

$$\theta_0 = \left(2.914\left(\frac{2\pi}{\lambda}\right)^2 \cos(\gamma)^{-8/3} \int C_n^2(h)h^{5/3}\mathrm{d}h\right)^{-3/5}, \quad (3)$$

$$\tau_0 = \left(2.914\left(\frac{2\pi}{\lambda}\right)^2 \cos(\gamma)^{-8/3} \int \frac{C_n^2(h)}{V(h)}\mathrm{d}h\right)^{-3/5}, \quad (4)$$

$$\sigma_I^2 = 10.7D^{-4/3}t^{-1}\cos(\gamma)^{\alpha(V_\theta(h))} \int \frac{C_n^2(h)h^2}{V(h)}\mathrm{d}h, \quad (5)$$

where r_0 is the Fried parameter, θ_{FWHM} is the Full Width at Half Maximum (FWHM) of the point spread function (PSF) or seeing, θ_0 is the isoplanatic angle, τ_0 is the coherence time and σ_I^2 is the scintillation variance. Other required parameters for the above calculations are the zenith angle, γ, the wavelength of the observation, λ, the telescope diameter, D and the air

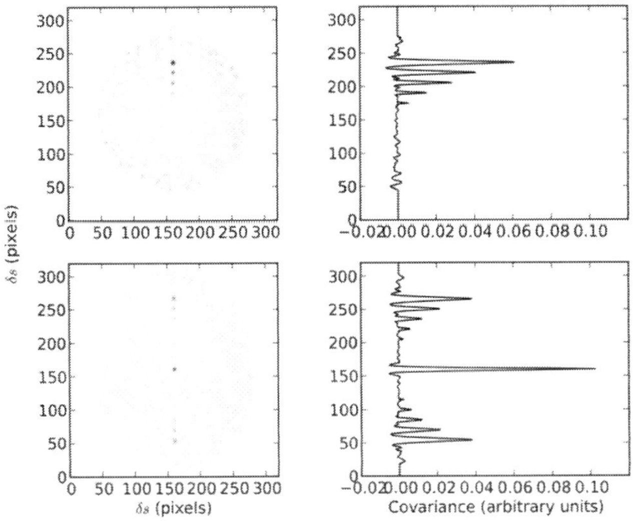

Figure 2. 2D covariance plots for Stereo-SCIDAR (upper) and single camera Generalised-SCIDAR (lower) for an atmospheric simulation containing six equal strength turbulent layers at 2 km spacing between 0 and 10 km, inclusive. A vertical cut through each covariance function is shown on the right. δs is the position in the covariance function. We see that for single camera SCIDAR we have two sets of spatially separated peaks and one set of overlapping peaks at the centre. For Stereo-SCIDAR we only have one set. Both plots have the same contrast scale, the correlation peaks for Stereo-SCIDAR are larger in magnitude than that of single camera Generalised-SCIDAR.

mass exponent, α. Note, that the value of the airmass exponent, α, will depend on the wind direction and vary between -3 for the case when the wind is transverse to the azimuthal angle of the star, up to -4 in the case of a longitudinal wind direction. This is a geometric correction. In the case where the wind direction is parallel to the azimuthal angle of the star, the projected pupil onto a horizontal layer is stretched by a factor of $1/\cos(\gamma)$, which changes the projected wind speed.

Each of these parameters has its own influence for particular applications. r_0 and θ_{FWHM} are usually used interchangeably and are a measure of the effect on an image caused by a wavefront which has propagated through the complete atmosphere. The isoplanatic angle defines the angular extent over which the atmospheric effects are correlated. It is this parameter that defines the angular size of an Adaptive Optics (AO) corrected field. Multi-Conjugate Adaptive Optics (MCAO) systems can be used to increase the isoplanatic angle and hence increase the corrected field of view. The coherence time defines the update rate that an AO system must function at in order to minimise residual wavefront errors due to the temporal lag between the wavefront measurements and correction on the deformable mirror (DM). All of this information can be used in real time for AO support, PSF reconstruction, observatory and observation management.

In addition to the above, statistical data on the typical profiles and variability of each of the profiles can be used for instrument development and performance analysis [7].

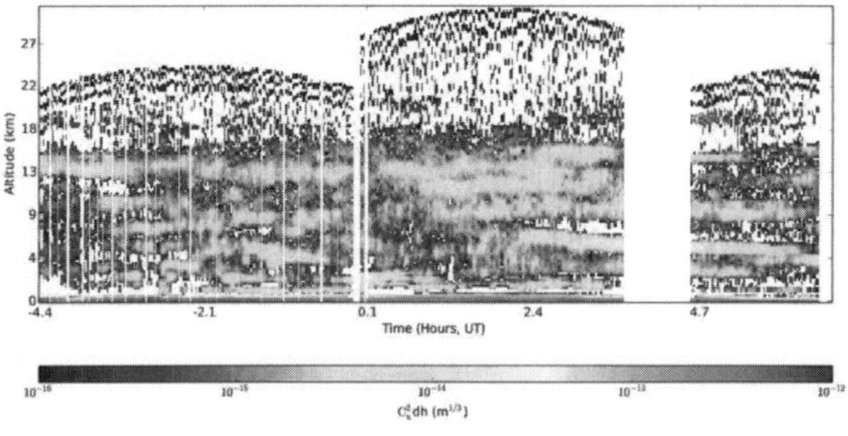

Figure 3. Example C_n^2 profile from the 2.5 m Isaac Newton Telescope, La Palma, on the 9th October, 2014.

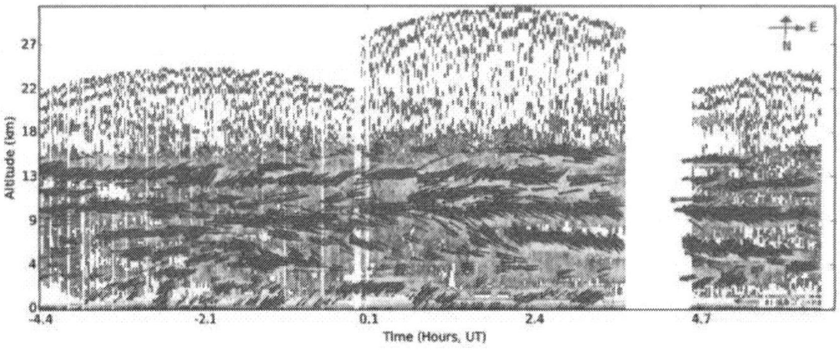

Figure 4. Example wind velocity profile from the 2.5 m Isaac Newton Telescope, La Palma, on the 9th October, 2014. The length and direction of the arrow indicating the speed and direction of the optical turbulence.

4. Scintillation noise

The scintillation index (equation 5) can be used to estimate the scintillation noise for photometric observations. As scintillation is a second order propagation effect depending on the local curvature of the wavefront, high altitude optical turbulence is dominant in causing scintillation at the ground. Therefore, scintillation noise is dominated by high-altitude turbulence [8] and it can be assumed that the scintillation noise is isotropic. The data from a turbulence monitor can then be used to estimate the scintillation noise applicable to photometric measurements made at nearby telescopes [9]. The fractional scintillation noise is defined as,

$$\sigma^2 = \frac{\left\langle (I - \langle I \rangle)^2 \right\rangle}{\langle I \rangle^2}, \tag{6}$$

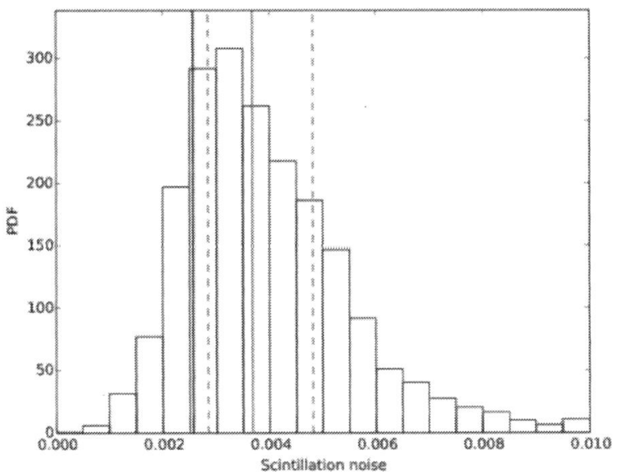

Figure 5. Scintillation noise distribution for a 1 m telescope and 1 second exposures, calculated from recent Stereo-SCIDAR results on the INT, La Palma. The red solid and dashed lines denote the distribution median, and first and third quartiles respectively. The black solid line indicates the value from Young's approximation.

where I is the measured intensity and $\langle\rangle$ indicates the expect value.

A standard equation for calculating the expected fractional scintillation variance at any particular site in the long exposure regime was first suggested by [10] and is given by,

$$\sigma_Y^2 = 9 \times 10^{-6} D^{-4/3} t^{-1} (\cos\gamma)^{\alpha} \exp\left(-2h_{\mathrm{obs}}/H\right) \tag{7}$$

where h_{obs} is the altitude of the observatory and H the scale height of the atmospheric turbulence structure, which is generally accepted to be approximately 8000 m. Equating Young's equation with equation 5 we see that Young is effectively assuming a turbulent layer of strength $\sim 80 \times 10^{-15}$ at 8 km. All previous work on the detection limits of fast photometry have used this equation to estimate the scintillation noise in their measurements.

Recent results from stereo-SCIDAR on the 2.5 m INT, La Palma, show that the conventional estimation for scintillation noise (Young's equation) [10] underestimates the median fractional scintillation noise by a factor of approximately 1.5 (figure 5).

Real profiles from Stereo–SCIDAR shows that the optical turbulence profile (both turbulence strength and velocity) can evolve rapidly. We have seen periods of strong high layers which survive for the duration of several hours and we have seen small bursts of activity lasting only a few minutes [6]. The altitude and magnitude of these layers varies with time, having significant impact on observations. Therefore, no single number metric can be used to estimate the scintillation noise at any particular time.

Knowing the contemporaneous scintillation noise will enable performance assessment, calibration and optimisation of photometric instrumentation. It will also help to explain and constrain model fits to photometric data (for example, extra-solar planet transit light curves) [11], and to help develop scintillation correction concepts [1, 8, 12].

5. Conclusions

Stereo-SCIDAR provides sensitive measurements of the optical atmospheric turbulence variables such as the refractive index structure function and wind velocity profile. From these variables several other useful parameters can be estimated, for example, seeing, isoplanatic angle, coherence time and scintillation index. All of this information can be used in real time for AO support, PSF reconstruction, observatory and observation management and statistical data on the typical profiles and variability of each of the profiles can be used for instrument development and performance analysis. We have shown from recent stereo-SCIDAR data on La Palma that standard scintillation noise estimates, as used by astronomers, tends to underestimate the actual scintillation noise. This has ramifications on understanding the model fits to photometric data.

Acknowledgments
We are grateful to the Science and Technology Facilities Committee (STFC) for financial support (grant reference ST/J001236/1). FP7/2013-2016: The research leading to these results has received funding from the European Community's Seventh Framework Programme (FP7/2013-2016) under grant agreement number 312430 (OPTICON). The Isaac Newton Telescope is operated on the island of La Palma by the Isaac Newton Group in the Spanish Observatorio del Roque de los Muchachos of the Instituto de Astrofsica de Canarias.

References

[1] Osborn J 2015 *MNRAS* **446** 1305 – 1311
[2] Tokovinin A and Kornilov V 2007 *MNRAS* **381** 1179–1189
[3] Vernin J and Roddier F 1973 *J. Opt. Soc. Am.* **63** 270–273
[4] Wilson R W 2002 *MNRAS* **337** 103 – 108
[5] Osborn J, Wilson R W, Shepherd H, Butterley T, Dhillon V S and Avila R 2013 Stereo scidar: Profiling atmospheric optical turbulence with improved altitude resolution *Adaptive Optics for Extremely Large Telescopes 3*
[6] Shepherd H, Osborn J, Wilson R W, Butterley T, Avila R, Dhillon V and Morris T J 2013 *MNRAS* **437** 3568–3577
[7] Morris T J 2014 Tomographic adaptive optics and turbulence profiling *Proc. of IOP: conference series: Adapting to the Atmosphere*
[8] Osborn J, Wilson R W, Dhillon V, Avila R and Love G D 2011 *MNRAS* **411** 1223–1230
[9] Föhring D, Wilson R W, Osborn J and Dhillon V 2014 Scintillation noise in exoplanet transit photometry *Proc. of IOP: conference series: Adapting to the Atmosphere*
[10] Young A T 1967 *AJ* **72** 747–753
[11] Föhring D 2014 *The Effect of Scintillation on Ground-Based Exoplanet Transit Photometry* Ph.D. thesis Durham University
[12] Viotto V, Ragazzoni R, Bergomi M, Brunelli A, Dima M, Farinato J, Magrin D, Marafatto L, Nascimbeni V and Piotto G 2012 Mcao: Wavefront sensing only as a tool for high precision photometry? *Proc. SPIE*

Atmospheric transmission and thermal background emission in the mid-infrared at Mauna Kea

A Otárola[1], M Richter[2], C Packham[3] and M Chun[4]
1111 South Arroyo Parkway, Pasadena, CA 91105, USA
1 Shields Ave, Davis, CA 95616, USA
1, UTSA Circle, Dept. of Physics & Astronomy, UTSA, San Antonio, TX 78249, USA
640 N. A'ohoku Place, Hilo, HI 96720, USA

E-mail: aotarola@tmt.org

Abstract.
We present results of a preliminary study intended to quantitatively estimate the atmospheric transmission and thermal background emission in the mid-infrared (MIR), 7 μm - 26 μm, at the 13N TMT site in Mauna Kea. This is in the interest of supporting the planning of MIR instrumentation for the posible second-generation of astronomical instruments for the Thirty Meter Telescope (TMT) project. Mauna Kea, located at high altitude (4,050 m above sea level), enjoys natural conditions that make it an outstanding location for astronomical observations in the mid-infrared. The goal of this work is to produce a dataset and model that shows the atmospheric transmission and thermal emission for two cases of precipitable water vapor (PWV), a low value of 0.3 mm, and at 1.5 mm which represent near median conditions at the site. Besides, and driven by the interest of the MIR community to exploit the daily twilight times, we look at the specific atmospheric conditions around twilight as a function of season. The best conditions are found for cold and dry winter days, and in particular the morning twilight offers the best conditions. The analysis of PWV data, shows the median value for the site (all year conditions between 6:00 PM and 7:30AM) is 1.8 mm and that periods of water vapor lower than 1.0 mm are common, these supports the opportunity and discovery potential of the TMT project in the mid-infrared bands.

1. Introduction

The Thirty Meter Telescope (TMT) Project contemplates the development of MIR instrumentation (imaging and spectroscopy) for its second generation instruments. In general, the MIR allows to probe the dynamics and chemistry of regions highly obscured by surrounding dust, such as is found in the central region of galaxies and star forming regions. Observations in the MIR enable interesting science, such as: evolution of gas and chemistry in protoplanetary discs, the study of star and planet evolution (kinematics and physical & chemical evolution), study of accretion and outflows around protostars, characterization of exoplanet atmospheres as well as potential for the detection of biomarkers from transmission spectral studies in exoplanet atmospheres, evolutionary connections of black holes & galaxies. Also, the MIR enables the study of moderate redshifted objects, the [SIV] line (10.5 μm) is an excellent substitute with superior ground-based sensitivity instead of the [OIV] line. In our solar system, near- and mid-infrared astronomy allows the study of the thermal structure, chemistry, photo-chemical products, and dynamics in the thick planetary atmospheres of the giant gaseous plane [1].

One important limiting factor in the observation at MIR wavelengths (7 μm to 26 μm) from ground-based observatories is the water vapor in our atmosphere. This arises from the water vapor absorption, due to the complex ro-vibrational spectrum of water vapor molecules, and added noise in the detectors from atmospheric thermal emission [2]. Consequently, the biggest difference between the MIR and visible spectral regions, from the point of view of instrument design, is the large amount of thermal background radiation in the mid-infrared [3].

The overall effect of such limiting factors vary strongly from site to site.

Here we model the atmospheric transmission and thermal background emission in the MIR at the 13N site; the place chosen for the construction and operation of the Thirty Meter Telescope [4]. Ultimately, once including information on the thermal emission of the telescope optics (3 mirrors), that of the MIR adaptive optics system as well as the thermal emission in the optics of a potential MIR instrument, a more general model will be obtained to estimate the signal-to-noise ratio of astronomical observations in this spectral band and estimate the performance of the MIR camera and spectrograph when operated as part of the TMT suite of instruments.

Other goals of this study are to summarize the cumulative functions and statistics of precipitable water vapor at the 13N site, as well as provide information on the median conditions of surface temperature and ground-layer atmospheric turbulence at the site. All these parameters are relevant for the modeling of atmospheric transmission in the mid- and far-infrared, especially in the Q (17 μm to 26 μm) and Z (28 μm to 40 μm) spectral bands. The N mid-infrared band 7.0 μm to 14.5 μm isn't that sensitive to the water vapor levels in the atmospheric column, but the wings of this spectral band are limited by atmospheric O_3 in the middle of the band while CH_4 is the major feature at the short end, and CO_2.

Section 2 presents the datasets and methodology used in this study, section 3 shows the main results for the site, in terms of main atmospheric characterization and the atmospheric transmission and thermal background emission in the (7 μm to 26 μm) spectral region, and finally Section 4 includes the main conclusions of this work.

2. Datasets & Methodology

2.1. Datasets

The most relevant variables on which information is needed to model the atmospheric transmission and thermal background emission in the MIR includes the atmospheric temperature and absolute humidity. The best conditions for astronomical observations in the mid- far-infrared, specially for the Q infrared band are obtained, from very high elevation sites which are cold and dry [5]. The 13N site in Mauna Kea, with its altitude of 4,050 m above sea level, is one of the driest places [6] and is already hosting MIR instruments such as TEXES [3] used in the Gemini North and the IRTF telescopes and COMICS [7] at the Subaru telescope.

For characterization of the PWV of the 13N site, we used PWV data derived from the monitoring of optical depths at 225 GHz (τ_{225}) using a tipping radiometer[1] performed at the Caltech Submillimeter Observatory (CSO) in Mauna Kea and nearby the 13N TMT site[2]. The relation used in this study for deriving PWV from τ_{225} is $PWV(mm) = 21.422 \times \tau_{225} - 0.2950$[3].

Also important is the vertical distribution of absolute humidity in the atmosphere, since low-level water vapor is responsible for larger collisional broadening of the atmospheric absorption lines, and so is a source of higher thermal emission. For the vertical profile of absolute humidity we learned the median profile, for winter conditions at Mauna Kea, from radiosonde soundings launched from the PHTO 91285 (HILO) National Weather Service station[4].

[1] ALMA Memo Series #49, http://science.nrao.edu/facilities/alma/aboutALMA/Technology/ALMA_Memo_Series
[2] 225 GHz optical depth database at Caltech Submillimeter Observatory, http://cso.caltech.edu/tau/
[3] S Radford, Private Communication
[4] Radiosonde soundings database, Hilo station ID is PHTO: http://weather.uwyo.edu/upperair/sounding.html

Journal of Physics: Conference Series **595** (2015) 012023 doi:10.1088/1742-6596/595/1/012023

The surface temperature data come from the site study[5] conducted by the TMT site testing team in the period 2004-2008 [4].

2.2. Methodology

PWV at the exact location of the 13N site ($\phi = 19.8339^o$ N; $\lambda = 155.4810^o$ W), was derived from 2-minute averages of surface measurements of air temperature (T) and relative humidity (RH) obtained during the TMT site testing campaign in the period 2005/06/29 - 2007/12/31. For the exact procedure for estimating PWV from surface weather data see [6].

As for the determination of the atmospheric radiance for the MIR, the following steps were performed:

The radiosonde soundings, obtained from the HILO National Weather Station, were used to obtain a typical profile of atmospheric temperature and absolute humidity for the Mauna Kea site. Best conditions of observations in the MIR are achieved in winter time, at low PWV and cold atmospheric temperatures. The profiles, in the upper troposphere and up to 90km above sea level, were complemented with data from the US standard atmosphere for winter conditions [8]. The absolute humidity profile was normalized, and subsequently adjusted for any desired level of integrated precipitable water vapor.

A line-by-line, layer-by-layer radiative transfer model, implemented in Matlab, making use of HITRAN [9] spectral line information, was used to derive the layer and total atmospheric transmission spectra. The radiative transfer model includes the atmospheric composition as shown in Table 1 of [5], and was run for two cases of PWV, a very dry case (with PWV=0.3 mm) and a near median condition (PWV=1.5 mm).

The physics formulation of the radiative transfer model is out of the scope of this conference paper. However, suffice is to say that the model output produces the absorption spectra ($\alpha_{mol_j,layer_i}$), for the following molecular species mol_j =N_2, O_2, CO_2, CH_4, CO, N_2O and H_2O, and for each and all of the atmospheric layers ($layer_i$) defined by the temperature and water vapor profiles.

The transmission $Tr_{layer_i}(\lambda)$, of each of the atmospheric layers is obtained from the first expression in the set of equations shown below. The layer emissivity $\epsilon_{layer_i}(\lambda)$, for each corresponding layer, is obtained from the complement to unity as follows: $\epsilon_{layer_i}(\lambda) = 1 - Tr_{layer_i}(\lambda)$. The radiance of each layer, as a function of wavelength, is obtained from multiplying the layer's emissivity and the Planck energy density function evaluated at the physical temperature of the layer. Finally, the total atmospheric radiance is the sum of all the atmospheric layers. This form of computing the overall atmospheric radiance is more accurate than simply multiplying the emissivity of the lowest most atmospheric layer by the corresponding Planck density function.

$$\mathrm{Tr}_{layer_i}(\lambda) = \exp(\sum_{mol_j} \alpha_{mol_j,layer_i} \cdot dH_{layer_i})$$

$$Rsky_{layer_i}(\lambda) = \epsilon_{layer_i}(\lambda) \cdot \{\frac{2\pi hc^2}{\lambda^5}\frac{1}{exp(\frac{hc}{\lambda kT_i})-1}\}$$

$$Rsky_{total}(\lambda) = \sum_{i=1}^{N} R_{layer_i}(\lambda) \quad (1)$$

Multi-annual time series of precipitable water vapor data, surface temperature and surface turbulence data was analyzed to obtain an statistical characterization of those parameters, in terms of hourly medians, for the course of the day. This allow us to obtain a perspective on

[5] TMT Site testing database: http://sitedata.tmt.org

how those parameters change through the day and learn their behavior in the period from the afternoon twilight, through the night and into the morning twilight. The information presented this way will bring light to the proposal of the MIR community about using one or two hours of observing time at around morning twilight for additional observing time.

3. Results & Analysis

3.1. Atmospheric transmission and atmospheric radiance in the mid-infrared

Table 1 (insert to Fig. 1) shows the statistical percentiles of PWV out of 352534 values of PWV (only data in the period 6:00 PM to 7:30 AM local time are used for the statistics) and Fig. 1 shows the cumulative function of PWV for every month in 2012. The cumulative functions in Fig. 1 illustrate that, when looking at the driest conditions, the driest period is that of January through April, as shown by the first quartile with a PWV $\leq$ 0.75 mm. These are also cold months, as shown in Fig. 4, and together low precipitable water vapor in the atmospheric column and low low-level temperatures, are the best combination for astronomical observations in the MIR bands.

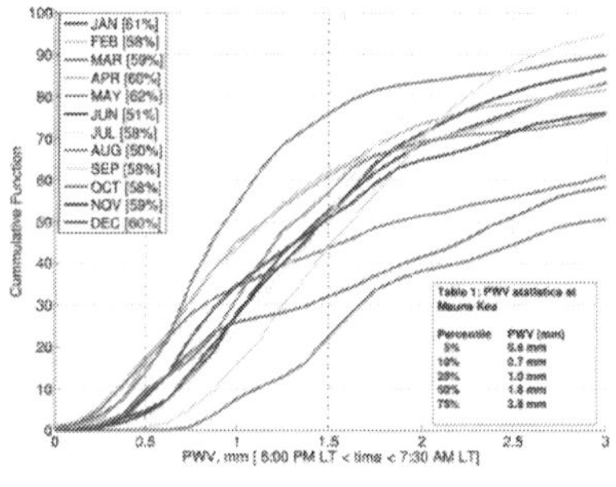

Figure 1. PWV monthly cumulative functions (2012).

In particular, we computed the atmospheric irradiance, following the procedure outlined in the previous section, for two conditions of PWV: a very dry atmosphere, with PWV=0.3mm, and a near median condition with PWV=1.5 mm. Figure 2(a) shows the atmospheric transmission, while Fig. 2(b) shows the atmospheric irradiance. The atmospheric molecular species responsible for the main absorption bands are CH_4 below 8 μm, O_3 in the vicinity of 9 μm, CO_2 with a strong absorption band in the range 13 μm - 17 μm, and various absorption lines resulting from the vibrational spectrum of the H_2O molecule in the range from 18 μm to 30 μm. The N band shows little change with water vapor, but the transmission of the various spectral windows beyond 18 μm show an increasingly dependance on the water vapor content.

The information shown in Fig. 2(b) can be used to compute the noise, at a particular spectral window in the detector of a mid-infrared instrument, that originates in the thermal emission of the atmosphere. For instance, the lowest radiance levels in the region around 11.5 μm wavelength is of about 2×10^{-7} W sr^{-1} cm^2 μm^{-1} (or 1.16×10^{-16} W arcsec^{-2} cm^2 μm^{-1}).

3.2. A proposal for using the twilight time for observations in the mid-infrared bands

The MIR community has proposed that one way of maximizing the scientific output of the telescope could be done by using about 1-2 hours of observing time available in the evening and/or morning twilights. One example of such approach is the use of TEXES instrument at the IRTF in Mauna Kea. This has motivated the community to look at how the PWV, air temperature and turbulence behaves as a function of time through the day at sites of interest.

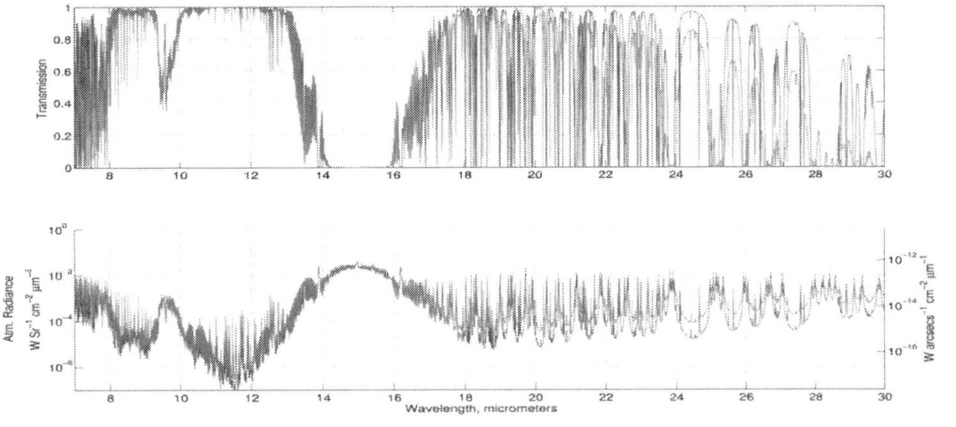

Figure 2. a) Atmospheric Transmission Spectrum for two cases of PWV (0.3 mm in blue, and 1.5 mm in read) and for a winter temperature profile condition; b) Atmospheric Radiances for the same cases of PWV.

In particular, for the 13N site, the PWV data measured at the CSO site in the period of four years (204-2007), and inferred from observations of optical depth at 225 GHz, was analyzed to compute the percentage of the time when the PWV was at a given PWV level as a function of local time. The calculations were done computing the fraction of the number of cases when the PWV falls in a bin of 0.125 mm size (as a function of absolute PWV level) for a time interval of 15 minutes (through the day) with respect to the total number of observations performed within that time window in the 4 years of data. Such map is therefore a robust way to show what is the statistical behavior of PWV as a function of time of day. The results are shown in Fig. 3 together with the 25%, 50% and 75% quartiles. This clearly illustrate that the maximum of PWV levels is about 0 hours UT (2:00 PM LT). The trend shows that PWV levels decreases through the afternoon reaching the lowest levels at about 5 UT (7:00 PM LT) and it remains at low levels until about 19 UT (9:00 AM LT). This is an interesting result. First, PWV in the atmospheric column is dominated by advection of air masses of different PWV content, but at low levels there is also a contribution from evaporation and/or sublimation depending on season and soil moisture condition. This in itself might lead to some variability as a function of time of day. However, this result is likely also the effect of a near surface temperature inversion that changes in altitude during daytime, when subsiding then it traps the water vapor below and at some point the temperature inversion goes below the altitude of the CSO site, allowing a dryer atmospheric column.

Another important parameter is near surface atmospheric temperature. For this purpose, the monthly median, as a function of time of day, was computed from data obtained at the 13N site during the TMT site testing campaign in 2006. The results are shown in Fig. 4. The daily and monthly temperature cycles are evident, with the coldest months being January, February and March. Regarding the daily cycle, the near surface atmosphere cools down by radiative processes starting with sunset and reaching the lowest temperature (near equilibrium) by 15:00-16:00 UT (5:00 - 6:00 LT). The temperature starts warming up with the sunrise. Stable temperature conditions are found during the morning twilight.

In this analysis we have also looked at the behavior of near surface atmospheric turbulence. The turbulence strength (C_n^2), was derived from fast temperature fluctuations observed with a Sonic anemometer operating at 30 Hz. The monthly median of C_n^2 as a function of time of

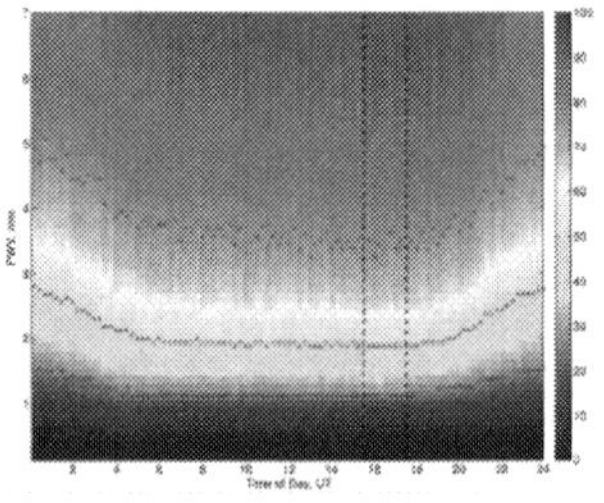

Figure 3. PWV probability of occurrence matrix as a function of time (UT) at Mauna Kea (CSO site).

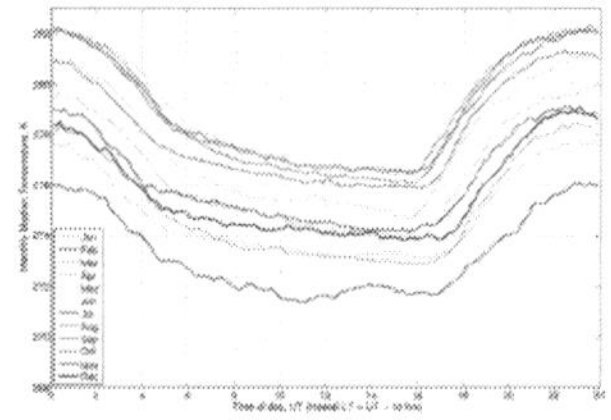

Figure 4. Monthly medians of near surface air temperature at the 13N site in Mauna Kea.

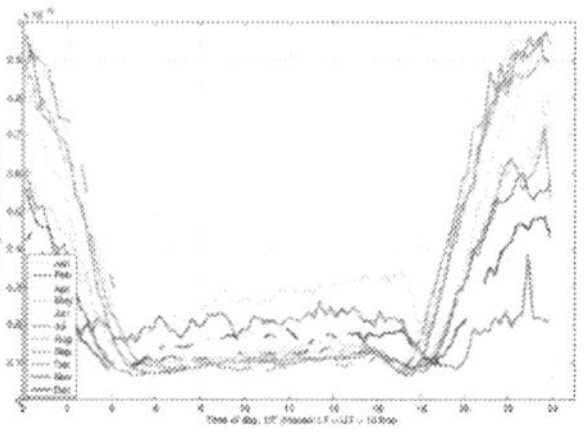

Figure 5. 2006 Monthly medians of surface turbulence strength (C_n^2)) at Mauna Kea.

day (with a time resolution of 15 min) are shown in Fig. 5. Again, the lowest conditions of near surface turbulence are observed towards the end of the night into the morning twilight, this as a result of lower variability in the air index of refraction due to the fact the atmospheric temperature by this time of day has reached near equilibrium. However, in terms of absolute magnitude of the turbulence Fig. 5 also shows that the months of lowest absolute turbulence -at night and towards the early morning- are the spring and summer months. The reason for this is that the optical turbulence is proportional to the index of refraction fluctuations, which in turn have origin in fractional changes of air temperature. Therefore, in the summer months with higher atmospheric temperatures, the fractional fluctuations are lower and this translates into lowest C_n^2.

So what is best, the afternoon or the morning twilight for MIR astronomy? Based on the findings shown in here, is no doubt that the morning twilight, with a drier atmospheric column (due to the local temperature inversion layer located at altitudes lower than the observatory sites), and colder near surface atmospheric conditions (due to the radiative cooling of the site) which have also allowed the telescope structures to reach their lowest temperature, together with lower near surface turbulence (because of atmospheric temperatures show lower fractional variability), make the morning twilight a great temporal window for astronomical observations in the mid-infrared.

3.3. Would it be possible to predict periods of low PWV?

The atmosphere is highly unsaturated with respect to water vapor, which is limited by the air temperature through energy considerations. Local sources of water vapor, originating from sublimation of snow/ice, and evaporation of surface waters as well as diffusion of water vapor from sub-surface, are to be considered as important sources of humidity at very dry sites. However, the water vapor field is highly dominated by synoptic and mesoscale dynamics, where the location of high and low pressure centers with respect to a site location become very relevant in the advection of wet and or cold air. Therefore, PWV can be forecasted by means of a mesoscale model [10, 11]. PWV in the Hawaiian Islands is strongly influenced by the location of the Hawaiian High pressure center, which could either bring cold and dry air from the northern latitudes, or warm and wet from the southern latitudes, respectively.

How often and for how long do the dry atmospheric conditions remain? This is an important question. Without generalizing, but as a way to show an example of PWV conditions at Mauna Kea we looked at the PWV gathered at the CSO site (with 10 min time resolution) in 2012. With this the nightly median value of PWV was calculated for each day in 2012, results are shown in Fig. 6. This illustrate that dry conditions, with PWV lower than 1 mm, are usual at

Mauna Kea with advection of very dry air (PWV lower than 0.5 mm) that extend for various days. See for instance, the dry conditions at around day 35 (early February), day 80 (second half in March), day 120 (beginning of May) and day 180 (end of June/beginning of July). This helps to illustrate that when dry conditions occur they tend to stay for several days in agreement with conditions controlled by synoptic scale atmospheric conditions. By the same argument the upper plot in Fig. 6 shows various occasions of advection of wet mass air.

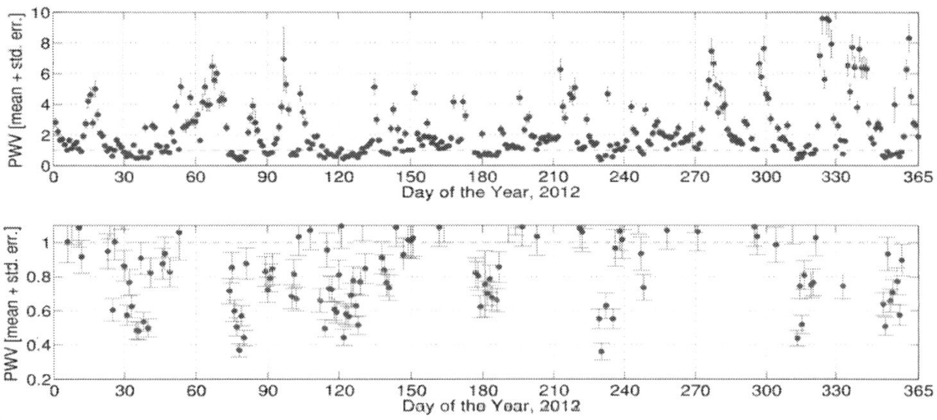

Figure 6. 2012 Nightly mean and its uncertainty of PWV (in mm) at Mauna Kea. The bottom figure limits the vertical scale to 1.1 mm of PWV. The red line shows the 1 mm PWV level and helps as a visual reference to identify all the cases when PWV was lower than 1 mm at the CSO site.

4. Conclusions & Future Work

Astronomical observations in the MIR bands, specially in the Q band, require a dry and cold site to minimize the noise arising from thermal emission of the atmosphere as well as minimize the atmospheric absorption of the astronomical signals. The 13N site, location for the Thirty Meter Telescope, was studied through a site testing campaign effort between 2004 and into 2008. Continuous measurements of PWV have been conducted at the location of the Caltech Submillimeter Observatory for many years. The atmospheric radiance under median conditions of the site in Q band is around 2×10^{-15} W arcsec^{-1} cm^{-2} μm^{-1} and down to less than 10^{-15} W arcsec^{-1} cm^{-2} μm^{-1} in very dry conditions (0.3 mm of PWV). In the N band, the atmospheric radiance get as low as 10^{-17} W arcsec^{-1} cm^{-2} μm^{-1}.

The analysis of site testing data, including temperature and near surface turbulence strength, indicate that optimum conditions are found around morning twilight (between 5:30 AM until 7:30 AM Local Time) when the atmospheric column is usually the driest, colder and more stable. By morning twilight, it is expected that the telescope structures have reached equilibrium with its environment minimizing noise arising from thermal emission scattered from the telescope structures into the mid-infrared detectors.

The atmospheric radiance information, as calculated in this study and shown in Fig. 2 can be used as input for the determination of the thermal emission arriving at the detector of a MIR instrument, a source of noise in the detector. The overall thermal radiance into the detector is a function of, the number of mirrors in the telescope, the optical throughput of the the telescope optics, the thermal emission of the Adaptive Optics instrument, and the quality of the mid-infrared instrument optics. Equation 2, in a generic way, includes the terms relevant for the determination of the thermal background radiation arriving at the mid-infrared detector), the results in this work can be used to determine that quantity provided all relevant parameters are know.

$$R_{total}(\lambda) = t_{tel}(\lambda) \cdot t_{AO}(\lambda) \cdot t_{midIR}(\lambda) \times Rsky_{total}(\lambda)$$
$$+ t_{AO}(\lambda) \cdot t_{midIR}(\lambda) \times R_{telescope}(\lambda)$$
$$+ t_{midIR}(\lambda) \times R_{AO}(\lambda)$$
$$+ R_{midIRt}(\lambda) \ (2)$$

Where;

$R_{total}(\lambda)$ is the total thermal background radiation reaching the mid-infrared detector. $t_{tel}(\lambda)$, $t_{AO}(\lambda)$, and $t_{midIR}(\lambda)$ are the optics throughput of the telescope, AO instrument and of the mid-infrared instrument, respectively. $R_{telescope}(\lambda)$ is the thermal emission of the telescope's mirrors, and this can be taken as the summation of each of the mirrors' emissivity multiply by the Planck function at the physical temperature of the mirrors (considered the ambient temperature), i.e. $\epsilon(\lambda) \cdot Planck(\lambda, T)$. Similarly, $R_{AO}(\lambda)$ is the thermal emission from the AO system that goes into the mid-infrared instrument. Finally, $R_{midIR}(\lambda)$ is the internal thermal emission of the mid-infrared instrument itself.

4.1. Acknowledgments
We thank Dr. Simon Radford (Caltech Submillimeter Telescope) for sharing of the 225 GHz tipping radiometer data, and Dr. Hanae Inami (National Optical Astronomy Observatory) for valuables comments regarding science topics of interest for mid-infrared astronomy.

5. References
[1] de Bergh C, Bézard 2003 *Probing Thick Planetary Atmospheres with High Resolution Infrared Spectroscopy* (Proceedings of ESO High Resolution Infrared Spectroscopy in Astronomy Workshop, Springer Science & Business Media) eds H U Käufl, R Siebenmorgen, A F M Morrwood, pp. 514-526.
[2] Kendrew S, Jolissaint L, Brandl B, Lenzen R, Pantin E, Glasse A, Blommaert J, Venema L, Siebenmorgen R, Molster F 2010 *SPIE* **vol. 7735-201**
[3] Lacy J H, Richter M J, Greathouse T K, Jaffe D T & Zhu Q 2002 *Pub. of the Astr. Soc. of the Pac.* **114**, pp.153-168.
[4] Schöck M *et al.* 2009 *Pub. of the Astr. Soc. of the Pac.* **121**, pp.384-395.
[5] Bustos R, Rubio M, Otárola A & Nagar N 2014 *Pub. of the Astr. Soc. of the Pac.* **126**, No 45.
[6] Otárola A *et al.* 2010*Pub. of the Astr. Soc. of the Pac.* **122**, pp.470.
[7] Kataza H, Okamoto Y, Takubo S, Onaka T, Sako S, Nakamura K, Miyata T & Yamashita T 2010 *Proc. SPIE 4008* (Optica and IR Telescope Instrumentation and Detectors) eds M Iye & F M Moorwood.
[8] Seinfield J H & Pandis S N 1998 *Atmospheric Chemistry and Physics* (John Wiley & Sons, New York), pp.1326
[9] Rothman L *et al.* 2008 *J. Quant. Spec. Radiat. Transf.* **110**, pp.533.
[10] Chacón A *et al.* 2011 *Rev. Mex. of A. & A., Serie de Conferencias* **41**, pp.20.
[11] Pozo D *et al.* 2011 *Rev. Mex. of A. & A., Serie de Conferencias* **41**, pp.55.

Ground-layer turbulence evaluation project at Subaru Telescope

Shin Oya
Subaru Telescope, NAOJ,650 North A'ohoku Place, Hilo, Hawaii 96720, USA
E-mail: oya@naoj.org

Abstract. A candidate of the next-generation adaptive optics (AO) system at Subaru Telescope is a ground-layer AO (GLAO) using an adaptive secondary mirror. The performance of GLAO depends on the turbulence profile because only ground-layer turbulence is corrected. At Mouna Kea, the profile data have been obtained at the summit ridge site and TMT site. However, the height difference of the Subaru site from these sites is about 100 m, close to the scale height of the ground-layer turbulence. There is a possiblity that the topographical difference affects the ground-layer turbulence property, and then the performance of GLAO. In this paper, the activity to evaluate the ground-layer turbulence at the Subaru site is introduced.

1. Introduction

A plan for the next infrared instrument with improved spatial-resolution by use of adaptive optics (AO), is on discussion at Subaru Telescope [1]. The instrument will be installed at Cassegrain focus, because the current infrared instrument with AO (LGSAO188) has been completed commissioning and in open-use operation at Nasmyth focus. A candidate is a wide-field (10' or more) camera-and-spectrograph with a ground-layer AO (GLAO) using an adaptive secondary mirror, called ULTIMATE-SUBARU as the project name (Hayano et al. 2014 [1]). The wide-field infrared instrument will be a natural upgrade of current seeing-limited instrument, Multi-Object Infrared Camera and Spectrograph (MOIRCS), and also has a good synergy with seeing-limited optical wide-filed instruments of Subaru, Hyper Suprime Cam (HSC; commissioned) or Prime Focus Spectrograph (PFS; on-going project) used in dark nights. Also the capability of the infrared wide-field instrument will be complement of the Thirty-Meter-Telescope (TMT) at Mauna Kea. Studies have been done for the ULTIMATE-SUBARU project on technical feasibility, performance estimation by simulation, science cases and the instrument, the budget plan and time-line, and so on. According to the preliminary simulation (Oya et al.2012 [2], 2014 [3]), the expected performance of full-width at the half maximum (FWHM) is 0.2 arcsec in the K-band over 10 arcminute of the field of view. under moderate seeing condition. However, the result of performance simulation depends much on the seeing condition at the telescope. This is why the seeing evaluation at the Subaru Telescope is needed.

[1] http://www.naoj.org/

Journal of Physics: Conference Series **595** (2015) 012024 doi:10.1088/1742-6596/595/1/012024

2. Impact of ground-layer turbulence on GLAO performance

GLAO realizes the widest working field-of-view (FoV) by correcting only the ground-layer turbulence, which is common for objects over wide FoV (Fig.1). The aimed performance is improvement of the seeing rather than the diffraction-limit, because the turbulence in the upper free-atmosphere is not corrected. The performance is determined by the uncorrected turbulence in the upper free atmosphere and affected by the turbulence profile. Interestingly, GLAO performance changes by the height distribution of turbulence even with the same integrated strength (r_0) of the turbulence (Gemini Ground Layer Adaptive Optics Feasibility Study Report, Appendix F1; Oya et al. 2012 [2]). Thus, it is quite important to know how much of the turbulence is attributed to the ground-layer.

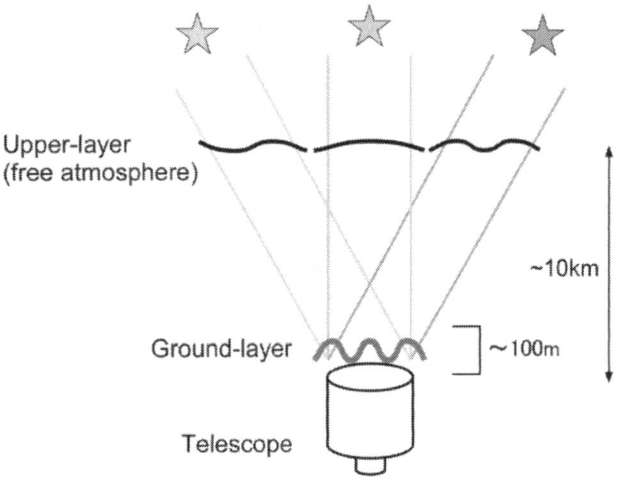

Figure 1. Schematic diagram explaining the principle of GLAO

3. Preceding campaigns at Mauna Kea

3.1. Mauna Kea Ridge

The campaign was done by Chun et al. 2009 [4] over 18 month from October, 2006 to July 2008. Two optical turbulence profilers using double star were used for the measurement. SLODAR (SLOpe-Detection And Ranging) detects the the turbulence height by spatial cross-correlation between the centroids measured by Shack-Hartmann WFS subapertures and obtains data up to 640 m with 80 m bins. LOLAS (LOw-Layer Scintillation Detection and Ranging) determines the turbulence height by calculating the spatial auto-correlation between scintillation pattern generated by double star and obtains data up to 60 m by 15 m bins. In the SLODAR results, the most of the turbulence is concentrated in the first 80 m bin and the LOLAS reveals that all of the turbulence is below 60 m and most of time below 30m.

3.2. TMT site, 13N

The campaign was done by Els et al. 2009 [5] from Jun, 2005 to February, 2008 as a part of TMT site testing for five candidate sites. One optical profiler and one sonic profiler have been used for the measurement. MASS-DIMM (Multi-Aperture Scintillation Sensor - DIfferential Motion Monitor) consists of two optical detectors. DIMM measures the integrated total strength of the turbulence (r_0). MASS determines the height profile at 0.5, 1, 2, 4, 8, 16 km, by spatial auto-correlation of scintillation at different separation. The ground-layer (0 m) strength is subtracting the sum of the upper layer strength measured by MASS from the total strength by DIMM. SODAR (SOund Detection And Ranging) is a sonic profiler which measures the

Journal of Physics: Conference Series **595** (2015) 012024 doi:10.1088/1742-6596/595/1/012024

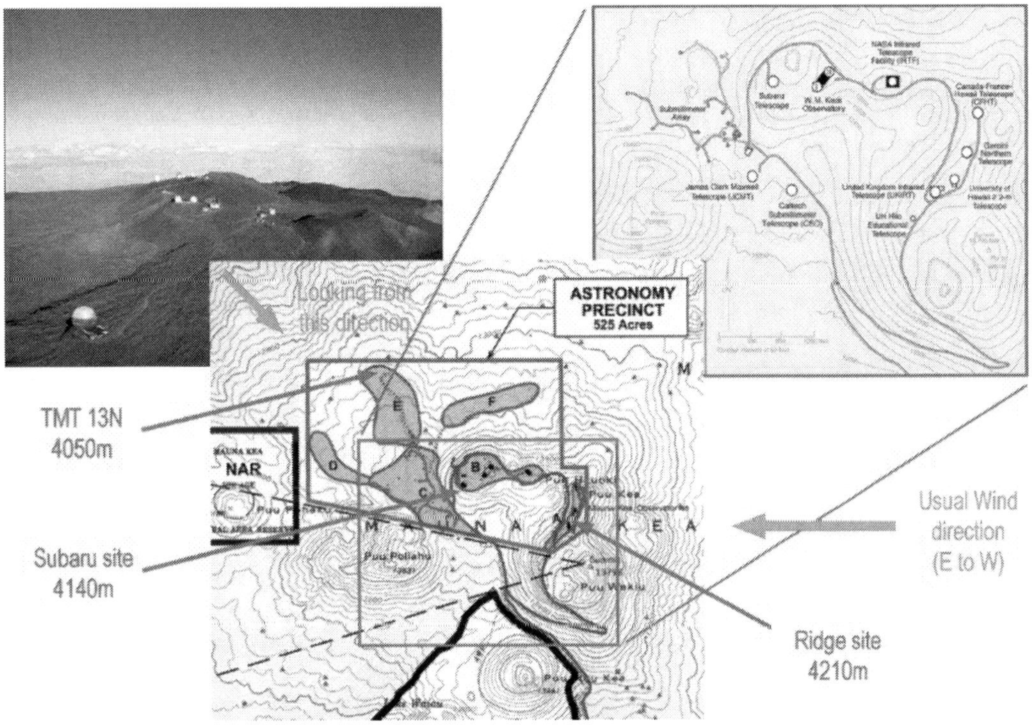

Figure 2. The topography of the Mauna Kea summit. The upper-left panel is an aerial photograph looking the summit from the north-west. The computer graphics of TMT is synthesized at 13N location (lower-left corner in the photograph). The location of the ridge site and 13N relative to the Subaru site is displayed in the lower-middle panle. The upper-right panel is the close-up of the area for major observatories, including Subaru site and the ridge site. The original source of the upper-left panel is IfA web site (http://www.ifa.hawaii.edu) and the others are from TMT web site (http://www.tmt.org).

turbulence from 10 m to 200 m with 5 m resolution and is used between October, 2005 to December, 2007 The MASS-DIMM data show that the turbulence is well concentrated at the ground and SODAR data indicate about half of the seeing is dominated by the first 100m.

3.3. Topographical difference of Subaru site

The data obtained by previous works indicate the ground-layer is concentrated below 100 m. However, the elevation of these site-testing locations is different from Subaru about 100 m. Moreover the local terrain is not smooth due to cinder-cone structure (Fig.2). The wind predominantly blows from the east. The Subaru site is on the leeward side of the ridge site and at the lower elevation. The behavior of the ground-layer turbulence after blowing over the ridge is not clear, if it blows down along the surface and keeps laminar flow or blows off at the ridge and goes rather horizontally. The ground-layer turbulence height at the Subaru site is also expected to be below 100m in the former case, while the height will be around 200m due to the elevation difference in the latter case. Therefore, the local measurement of the ground-layer turbulence at Subaru site is necessary, in addition to the preceding campaigns at other sites of the Mauna Kea.

4. Activity at Subaru Telescope

To evaluate the local ground-layer turbulence at the Subaru site, two profilers (one optical and one sonic) have been introduced. Additionally, the actual correction by GLAO mode through the telescope has been tested, which indicates the performance of GLAO at the Subaru telescope is promising as expected by the simulations.

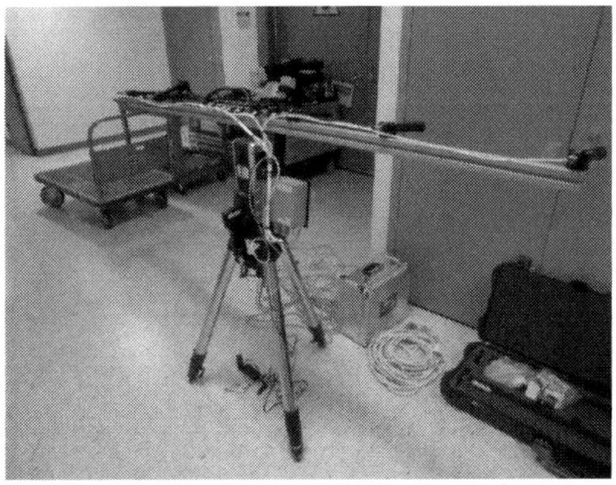

Figure 3. Portable turbulence profiler 2 (PTP2)

Figure 4. Surface layer NOn-Doppler Acoustic Radar (SNODAR)

4.1. PTP2

PTP2 (Portable Turbulence Profiler 2) is a mobile type of lunar scintillometers developed at University of British-Columbia (e.g., Hickson et al. 2013 [6]). PTP2 has twelve photodiodes (PDs) placed at various separation on a horizontal bar (Fig.3). A PD detects the scintillation of the entire moon. The height of the turbulence is determined by correlation calculation between PDs with different baselines. The height resolution is high at ground-layer and detectable height ranges up to several hundreds; hence, suitable for ground-layer measurement. The initial test has been finished and detection of lunar scintillation has been confirmed. After improving the tracking of the moon and reducing 60 Hz noise on PDs, data accumulation will be started on regular basis.

4.2. SNODAR

SNODAR (Surface layer NOn-Doppler Acoustic Radar) is developed at University of New South-Wale (e.g., Bonner et al. 2010 [7]) and used for boundary layer evaluation in Antarctica. SNODAR has quite high resolution (1 m at best) up to 100 m. An advantage of sonic profiler is that measurement is possible even in day time or cloudy night. A standard type sonic profiler, SODAR, had been introduced once at the Subaru site, but did not work properly due to the acoustic noise from the ventilator fan of the Keck Telescope. In addition to the use of single frequency, SNODAR has a large sound cone (Fig.4) which is effective to successfully reduce the influence of the acoustic noise. Figure 5 displays a quick result for a night obtained during the initial test in July, 2014. The result indicates that the thickness of the boundary layer is less than 50m also at the Subaru Telescope. Although this is only a result for a night, it is quite encouraging for the GLAO activity at the Subaru Telescope. Further long-term monitoring under various weather condition is desired and will be started after installation of SNODAR on the roof of the control building.

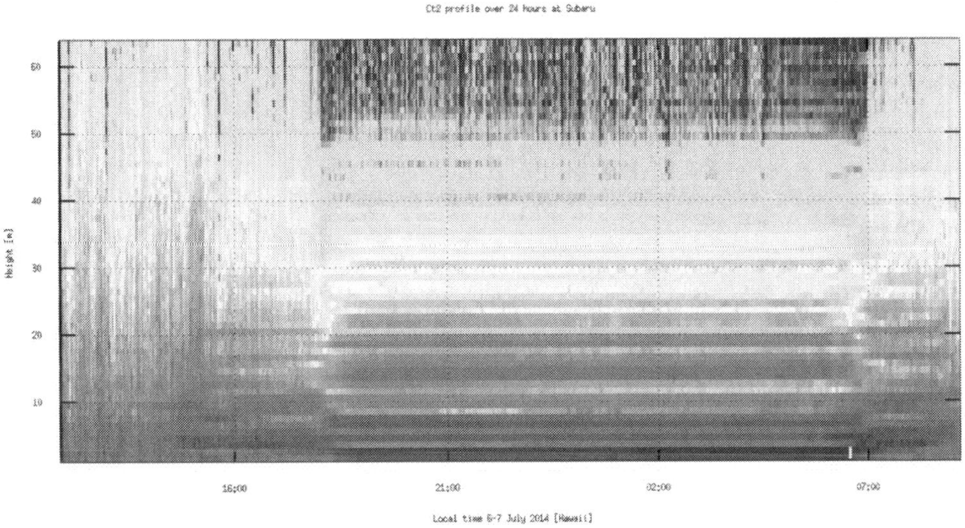

Figure 5. An example of SNODAR result obtained during initial test (by courtesy of Colin Bonner).

4.3. GLAO correction by RAVEN

RAVEN is an MOAO (Multi-Object Adaptive Optics) demonstrator developed at University of Victoria and carried-in to the Subaru Telescope (Lardière et al. 2014 [8]). RAVEN also works in an open-loop GLAO mode. During the first engineering run in 2014, better correction than 0.13" has been achieved in the H-band under good seeing condition (0.32"). In Fig.6, the quick

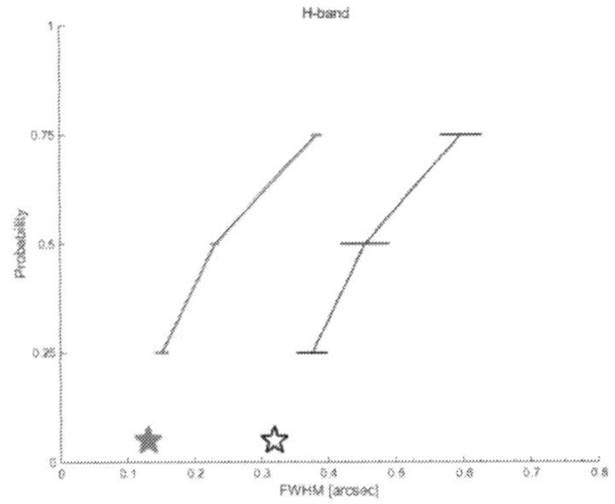

Figure 6. A quick result of GLAO mode obtained during the first engineering run by RAVENN, is overplotted on GLAO simulation for ULTIMATE-SUBARU.

result obtained during the first engineering run is overplotted on the expected performance of Subaru GLAO system (ULTIMATE-Subaru; Oya et al. 2012 [2], 2014 [3]) in the H-band. The observed value by RAVEN is in fairly good agreement with the expected performance, even with the difference of detailed system configuration from ULTIMATE-SUBARU (e.g., open-loop operation and narrower GS configuration in the RAVEN case). This is probably because, once the ground layer is corrected enough, the image size of GLAO is determined by the rest of uncorrected turbulence in the upper free-atmosphere regardless of the details of the system.

4.4. Summary
A ground-layer turbulence evaluation project is on-going at the Subaru Telescope. An optical profiler, PTP2, and a sonic profiler, SNODAR, have been introduced. Initial tests for both of the profilers have been finished. Especially, the quick result by SNODAR indicates the ground layer is below 50m, which is encouraging for GLAO plan at the Subaru Telescope. The actual GLAO corrected image through the telescope obtained by the open-loop GLAO mode of RAVEN, shows better FWHM size than 0.2" as expected by GLAO simulations.

Acknowledgments
I would like to thank all staff of Subaru Telescope, National Astronomical Observatory, Japan, who kindly support this research. Especially, Naruhisa Takato, an expert of site-testing, always aid my activity and Hirofumi Okita, who has experience of seeing evaluation at Antarctica, helped the initial test of SNODAR. I am grateful to Paul Hickson for development of PTP2 and to Colin bonner and Michael Ashley for development of SNODAR. Thanks are also to all members of RAVEN team from the University of Victoria, NRC Herzberg-Astronomy, Tohoku University and other institutes. This work is supported by JSPS KAKENHI 24654050.

References
[1] Hayano Y, Akiyama M, Hattori T, Iwata I, Kodama T, Lai O, Minowa Y, Ono Y, Oya S, Takiura K, Tanaka I, Tanaka Y and Arimoto N 2014 ULTIMATE-SUBARU: project status *Adaptive Optics Systems IV* (*Proc. SPIE* vol 9148) ed Marchetti E, Close L M and Véran J P p 2S:8pp

[2] Oya S, Akiyama M, Hayano Y, Minowa Y, Iwata I, Terada H, Tomonori U, Takami H, Nishimura T, Kodama T, Takato N, Tomono D and Ono Y 2012 A preliminary simulation results of the next-generation wide-field AO at Subaru Telescope *Adaptive Optics Systems III* (*Proc. SPIE* vol 8447) ed Ellerbroek B L, Hart M, Hubin N and Wizinowich P L p 3V:11pp

[3] Oya S, Hayano Y, Lai O, Iwata I, Kodama T, Arimoto N, Minowa Y, Akiyama M, Ono Y H, Terada H, Tomonori U, Takami H, Nishimura T, Takato N and Tomono D 2014 ULTIMATE-SUBARU: simulation update *Adaptive Optics Systems IV* (*Proc. SPIE* vol 9148) ed Marchetti E, Close L M and Véran J P p 6G:8pp

[4] Chun M, Wilson R, Avila R, Butterley T, Aviles J L, Wier D and Benigni S 2009 *Mon. Not. R. Astron. Soc.* **394** 1121

[5] Els S G, Travouillon T, Schöck M, Riddle R, Skidmore W, Segufl J, Bustos E and Walker D 2009 *Publ. Astron. Soc. Pacific* **121** 527

[6] Hickson P, Gagné R, Pfrommer T and Steinbring E 2013 *Mon. Not. R. Astron. Soc.* **433** 307

[7] Bonner C S, Ashley M C, Bradley S G, Cui X, Freng L, Gong X, Lawrence J S, Luong-van D M, Shang Z, Storey J W V, Wang L, Yang H, Ji, Yang, Zhou X and Zhu Z 2010 Snodar: 2009 performance at Dome A, Antarctica *Ground-based and Airborne Telescopes III* (*Proc. SPIE* vol 7733) ed Stepp L M, Gilmozzi R and Hall H J p 4A:6pp

[8] Lardière O, Andersen D, Blain C, Bradley C, Gamroth D, Jackson K, Lach P, Nash R, Venn K, Véran J P, Correia C, Oya S, Hayano Y, Terada H, Ono Y and Akiyama M 2014 Multi-object adaptive optics on-sky results with Raven *Adaptive Optics Systems IV* (*Proc. SPIE* vol 9148) ed Marchetti E, Close L M and Véran J P p 1G:14pp

Adapting to the Atmosphere Conference 2014 IOP Publishing
Journal of Physics: Conference Series **595** (2015) 012025 doi:10.1088/1742-6596/595/1/012025

Characterization of sites for Indian Large Optical Telescope Project

Padmakar Parihar, T.K. Sharma, A. Surendran, P.M.M. Kemkar, Dadul, Urgian Stanzin and G.C. Anupama

Indian Institute of Astrophysics, 2nd Block, Koramangala, Bangalore, Karnataka, India

E-mail: psp@iiap.res.in

Abstract.
Growing Indian astronomical community require access to large observing facility preferably located within the country. Site surveys carried out in the post indicates that the best sites for optical-NIR astronomy can better be found in Indian part of trans-Himalayan region. This very high-altitude, extremely dry and cold region is far from any artificial light pollution having low atmospheric aerosols, not much affected by monsoon and has got clear sky all through year. In the year 2000, as a pathfinder to large telescope, one 2m size optical telescope was commissioned at Indian Astronomical Observatory(IAO) Hanle Ladakh. However, before initiating a project to install a large 8-10m size Optical-NIR telescope it is essential to thorough characterize few candidate sites in and around IAO. Keeping this in consideration, once again we have started a detailed site survey program. In this paper, we will briefly present our effort to indigenously develop site survey instruments as well as results obtained by using them.

1. Introduction

While great strides have been made in observational astronomy all over the world, the Indian astronomers have to solely rely on 2 m class telescopes in all these years. To keep up with the emerging trends and explore the new horizons in astronomy, there is proposal by Indian astronomical community to install a modern, state-of-the-art 8-10m class telescope some where in India. In eighties, its was found by Indian astronomers that the best place for optical NIR observation can only be found in Himalayan region. A national project headed by A Bhatanagar was initiated to search for sites in in Ladakh region. The survey carried out during 1984-1989 near Nimmu (4100 m above msl) on Leh-Kargil road northwest of Leh found this place very suitable for astronomy, except relatively less number of clear nights. Few year later, under separate initiative to identify suitable sites for large Optical-NIR observatory, Indian Institute of Astrophysics (IIA) started surveying Himalayan and trans-Himalayan regions. By making use of available topographical maps, weather data and satellite imagery, six potential sites all more than 4000m above from mean sea level were identified. Based preliminary survey carried out at sixes places, Digpa-ratsa Ri, Hanle (mount Saraswati), was chosen for further detailed studies. Hanle is a high-altitude, extremely dry and cold place. This place is far from any artificial light pollution having low atmospheric aerosols, not much affected by monsoon and has got clear sky all through year. This place provides an excellent opportunity for developing astronomical facilities at a variety of frequency and an initiative has already been taken to establish astronomical facility over there. As a pathfinder to large telescope, one moderate

Table 1. Three candidate sites for the NLOT

Place	Altitude(M) (M)	Longitude (HH:MM:SS)	Latitude (DD:MM:SS)	Wind speed(night) (m/s)	RH %	Extreme Temp (C)
IAO Hanle	4500	78:58:29	32:46:48	4.0	33	-25.0/26.0
Kalak-Tartal	5486	79:00:44	32:38:23	6.2	45	-23.8/18.7
Randog	5055	78:55:20	32:50:46	7.2	50	-27.7/17.9

size Optical telescope was installed during 2000 at one of the peak called mount Sarswati in Hanle. Since then, slowly all required infra-structure is being created. More than a decade long observing experience with 2m size telescope, it has been found that Hanle region thus has the required characteristics of a good astronomical site and could be a natural candidate site for any future large aperture Indian optical-infrared telescope (Cowsic et. al. 2002). However, setting up 8-10 meter size telescope requires huge amount of investment and the manpower, therefore, a thorough understanding and characterization of the intended sites are necessary. In the year 2007-08, a detailed site survey program was initiated to characterize the site where current 2m telescope is installed as well few more potential sites within few km from it. In this paper we present our effort being carried out in this regards.

2. Meteorological Data

In order to record meteorological parameters, several weather stations have been installed at different candidate sites. Weather installed at IAO is continuously used to collect data over more than 18 years. Whereas, weather stations in other two peaks are operating since 2007. These weather station records, wind speed and direction, relative humidity, ground and air temperature, rainfall and solar irradiation. Table 1 provides statistics of meteorological data collected for three sites over a very long period.

3. Cloud Monitors

The most important parameter for any astronomical sites is availability of large number of clear nights well distributed over all through the year. Therefore, monitoring local sky for the cloud is the first task before setting-up a new observatory. There are different methods adopted to monitor sky conditions. In the classical method, the whole sky is divided in eight equal parts and cloud is visually looked at in these sky regions. At IAO site, sky is being manually monitored over more than 18 years. In addition to this satellite imagery data can also be used to get qualitative estimates of clear sky. However, the best and more reliable method is to make use of all sky optical and/or IR camera, which captures whole sky on very regular basis.

3.1. CCD based All Sky Camera

CONCAM stands for CONtinuous CAMera, designed and built by Robert Nemiroff and his team at Michigan Technical University (Pereira 2003). CONCAM consists of a fish-eye lens that projects the night sky onto an SBIG CCD camera. Some time in 2004, one of CONCAM unit was installed at IAO Hanle and since then it is continuous working. The camera switches itself on during evening twilight and stops at the end of the night. Images are taken every four minutes of interval and they are immediately corrected for bias and flats. The reduced data is archived as well as made available on real time through Web. Over ten years of operation more than a million all sky night images are collected.

3.2. IR scanning cloud Monitor

A CCD camera based all sky monitor is an inexpensive tool, however, appearance of clouds in the optical images is very deceptive due to its strong dependency on apparent optical sky brightness which in turn depend on phases of the Moon, local light pollution and other atmospheric variables. Automated tools developed to derive a quantitative information of cloudiness from all sky images, usually fails to reliably extract cloud information and results are often found to be inconsistent with other methods. All Sky Scanning Cloud Monitor is one of the instrument what we have developed for the purpose of surveying potential sites. The instrument works on the principle of detection of the Infrared radiations from the clouds (Clay et al. 1998). A number of thermopile sensors are arranged in the form of a circular array and whole sensor array is rotated in azimuth to cover the full sky. The device has got high sensitivity, provides radiometric output, cover whole sky, works in fully automated mode and also very cost effective. The IR scanning cloud monitor has been calibrated and tested in IIA Bangalore then after recently it has been installed at IAO Hanle.

The analysis of cloud data indicates that as for as availability of number of clear nights is concern, IAO Hanle is not to different from Mauna Kea. Usable nights are more than 80%, whereas, number of spectroscopic nights are found to be about 70%. Even during peak Monsoon season, when rest of India is severely affected by cloud, one can expect about 35-50% clear nights at IAO Hanle.

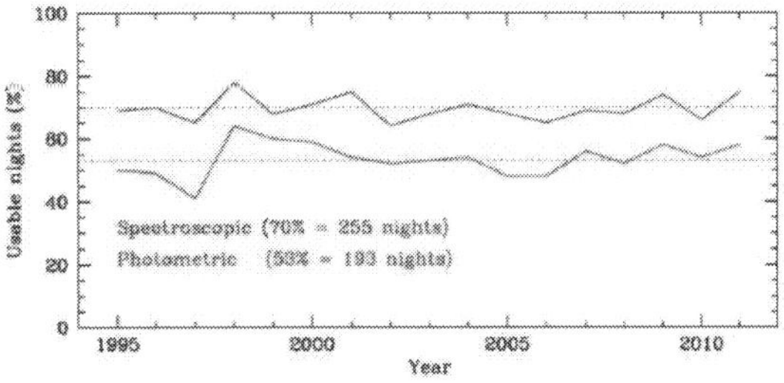

Figure 1. Automated extinction monitor

4. Extinction monitor

Quality of the nights whether it is photometric or spectroscopic is primarily judged by cloud coverage. Cloud free sky is usually considered to be photometric, whereas, partially cloudy sky is designated as spectroscopic. However, in practice, it has been noticed that even cloud free nights can have variable atmospheric transparency linked with change in atmospheric aerosol and absorption by water molecules. By doing photometric observations of stars, atmospheric extinction can be precisely measured. The best way to generate accurate statistics of photometric nights over longer run is to operate an automated extinction monitor (AEM). Our AEM is a standalone semi-robotic instrument designed to measure any small variation in the atmospheric transparency in one of optical band. The instrument comprises a wide field telephoto lens Nikon 300m F/4, an yoke mount equatorial tracking system, and a thermo-electrically cooled large format U32 Apogee CCD camera. The whole device is kept inside a tilted sliding roof enclosure which is again motorized and computer controlled. The key idea behind developing this instrument is to observe few hundred stars, brighter then 12th magnitude, over varying range of airmass. And make use of the brightness measurements of all stars to determine atmospheric extinction with very high accuracy. The same device can also provide sky brightness in one of photometric bands. The Extinction monitor was installed at Hanle during end of 2012, initial few months required to test the data quality of the device and also in improving its hardware . Then after device is being used almost on regular basis. We have collected more than 100 nights data which has not been completely analyzed. Initial test indicates that device works very precisely and extinction measured using AEM closely matches with 2m HCT observation.

Figure 2. Automated extinction monitor

5. DIMM Seeing Monitor

The most essential instrument for any site survey program is an automated seeing monitor. A new DIMM seeing monitor was developed using 14" Meade LX200 GPS telescope. It uses a U2000 CCD camera from Apogee which can have very short exposure and support fast read out, a prism as well as a mask with two holes. In order to make the device fully automated, we have incorporated, external limits switches into RA and Dec drives, day-night light and cloud detectors, a micro-controller driven power supply and a vertically sliding enclosure. Since the Meade telescope has got poor pointing accuracy, so we added a small finder attached with the web camera, which covers 1.5x2.0 degree FoV and automatically center the star. The instrument control and analysis software have been developed using C++ programming language. Images acquired by seeing monitor is archived and also processed in real time. The seeing monitor was installed at 4m high tower at IAO Hanle during September 2011. Since then it is in regular use and more than 180 nights seeing data have been collected. It was noticed that due to wind sake images acquired at higher wind speed were found to be elongated and such data points are excluded while computing the seeing values. The mean and median seeing value at Hanle is found to be 1.37±0.50 arc-sec as well as 1.27±0.31 arc-second respectively. Whereas, in the past seeing measured using old seeing monitor over less number of nights, shows median seeing value about 1.0 arc-sec. One way to interpret the differences, is the new seeing values which has been mostly collected during winter season. Whereas, it has been found from images collected using science camera of 2m HCT telescope that seeing is found to be usually very poor in winter season. Therefore, we expect better median seeing values, once we collect data equally spread all through year. Analysis of nightly seeing data show different behaviors/pattern. There are nights in which seeing values are considerably high all through night. Whereas, is in some night, seeing is found to be stable and exceptionally good. There are nights in which seeing may be large at the start of observations but later it improves or vice-versa.

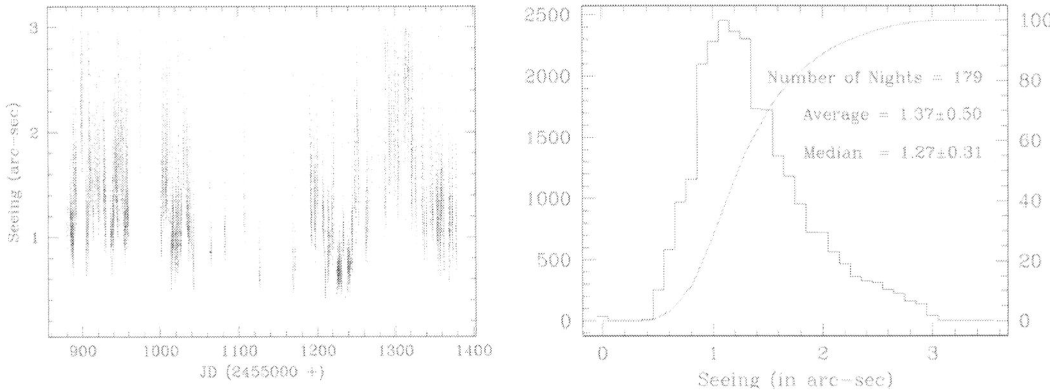

Figure 3. Distribution of seeing values collected during September 2011 to March 2013.

6. Lunar Scintillometer

Ground layer turbulence is one of the important site characterization parameters used for assessing the quality of an astronomical site. Lunar Scintillometer (LuSci) is a simple, yet an effective site-testing device for measuring the ground layer turbulence. The instrument consists of a linear array of photo-diodes which record the variations in moon's intensity caused by the lower layers of the Earth's atmosphere. The co-variance between all possible pairs of photo-diodes can be used to reconstruct the ground layer turbulence profile starting from the ground up-to a maximum height roughly determined by the distance between the furthest pair of detectors. Our 6-channel LuSci is close replica of turbulence profillometer recently built by the team led by Andrei Tokovinin of CTIO, Chile (Tokovinin et. al. 2010) A web-cam is used to ensure that the instrument is pointed to the moon. The whole assembly goes as piggyback on a small 12" Meade LX200 telescope which facilitate to point and track the Moon. Instrument control software which handles, Lusci, web camera and a Meade telescope has been developed using LabView. After testing the device in the laboratory as well as with the Moon, device was finally installed at IAO Hanle in September 2013. Then after Lusci data has been collected in more than 25 nights over several observing campaign. The average ground layer integrated seeing up to 784 meter high from telescope is found to be 0.38±0.28 arc-second, wheres as median values is 0.32±0.16 arc-sec. Most of the time it was found that the first two layers from device (3m and 12m) appears to contributing most. However there were instances when turbulence shifted from one layer to other during course of observations.

7. Other Survey Devices

7.1. 220Hz Radiometer

A 220-GHz (1.36 mm wavelength) Radiometer was jointly developed by Indian Institute of Astrophysics, Raman Research Institute and the University of Tokyo. The device was installed at IAO Hanle some time in 1999, since then it is continuously working. Radiometer works in scanning mode and measured opacity at 1.36mm wavelength. Opacity measurements carried out by this instrument over several years show that the absorption at this frequency is less than 0.1 for 70% of the time in winter months. This shows that at least during winter months IAO site is better than at Mauna Kea. However, annual average opacity has been found to be same for both the sites.

7.2. MASS-DIMM
The MASS-DIMM is one of very important site survey device which gives atmospheric turbulence profile, starting from 500m to 16km (Tokovinin & Kornilov 2007). We have got an MASS-DIMM from TMT USA as a loan for two years. MASS-DIMM instrument require a very sturdy telescope which is not affected by wind gust and can provide reliable seeing measurement at higher wind speed. Instead of using commercially available telescope which usually fails to meet this requirements, we have been developing a light weight, but stiff telescope driven by direct drive motors. The MASS-DIMM software originally developed by TMT also requires considerable changes before we can make use of it.

7.3. Micro-Thermal Seeing Measuring Device
In between ground and inversion layer there is strong convection which introduces micro-thermal fluctuations causing inhomogeneities in refractive index. By making use of accurate and fast temperature recording sensor, one can measure fluctuation in temperature and hence refractive index structure constant. A micro thermal measuring device is being developed at IIA.

Acknowledgment
We are thankful to night support staff at Indian Astronomical Observatory(IAO) Hanle, who tirelessly worked over several years in a very hostile climate to operate survey devices and collected very valuable data. We also gratefully acknowledge engineers and technical staff of IAO for their support in installation and maintenance of survey devices. We also like to thanks to Thimmaih N and Periyanayagam of IIA workshop who were instrumental in developing some of survey instruments. Finally, we like to acknowledge Andrei Tokovonin of CTIO for providing required code to analyze Lusci data.

References
[1] Clay R W, Wild N R, Bird D J, Dawson B R, Johnston M, Patrick R and Swell A 1998 *Publ. Astron. Soc. Aust.* **15** 332 – 5
[2] Cowsik R., Srinivasan R., Prabbu T.P., 2002, BASI, 30,105
[3] Pereira W. E., 2003, Ph.D. Thesis
[4] Tokovinin A., Kornilov V., 2007, MNRAS, 381, 1179
[5] Tokovinin A., Bustos E., Berdja A., 2010, MNRAS, 404, 1186

Validation of the WRF model for PWV forecasting at the Roque de los Muchachos Observatory

G. Pérez-Jordán,[1,*] J.A Castro-Almazán,[1,2] C. Muñoz-Tuñón,[1,2] B. Codina-Sanchez[3] and J. Vernin[4]

[1]Instituto de Astrofísica de Canarias, E-38200, La Laguna, Spain
[2]Dept. Astrofísica, Universidad de La Laguna, E-38200, La Laguna, Spain
[3]Dept. Astronomía y Meteorología, Universitat de Barcelona, Barcelona, Spain
[4]Dept. Astrophysique, Université de Nice-Sophia Antipolis, Nice, France
[*] *Visiting fellow*

E-mail: gpj@iac.es (GPJ); jcastro@iac.es (JACA); cmt@iac.es (CMT)

Abstract. The atmospheric Precipitable Water Vapour content (PWV) is crucial for astronomical observations in the infrared. In this work, we have validated the Weather Research and Forecasting (WRF) mesoscale numerical model as an operational tool for PWV at Roque de los Muchachos astronomical Observatory (ORM). We used radiosonde data obtained directly at the Observatory and at the sea level. An excellent agreement between PWV model forecasts and observations was found at both verification points with a correlation above 0.9.

1. Introduction

The atmospheric Precipitable Water Vapour content (PWV) is crucial for astronomical observations in the infrared [1]. We have validated the Weather Research and Forecasting (WRF) mesoscale Numerical Weather Prediction (NWP) model as an operational tool for PWV. The NWP models are increasingly being used as an additional and valuable tool for predicting weather variables affecting astronomical observations. The WRF model is a non-hydrostatic mesoscale meteorological model with hydrostatic option designed for both research and operational applications.

In the validation we used atmospheric radiosonde data. This allowed us to calibrate the model at the observatory and to validate it under different PWV values and atmospheric conditions. The forecasts will complement the PWV measured by a GPS monitoring system at the ORM[1]. The forecast of the instantaneous PWV value and its temporal stability may also improve the scheduled operation time at ORM.

An extension of the present work including the effect of different model resolutions is being published [2].

2. Datasets: atmospheric radiosoundings

In the present study we used atmospheric radiosoundings (the accepted standard used in atmospheric research) for model validation on the basis of its accuracy. Balloons were launched

[1] http://www.iac.es/site-testing/PWV_ORM

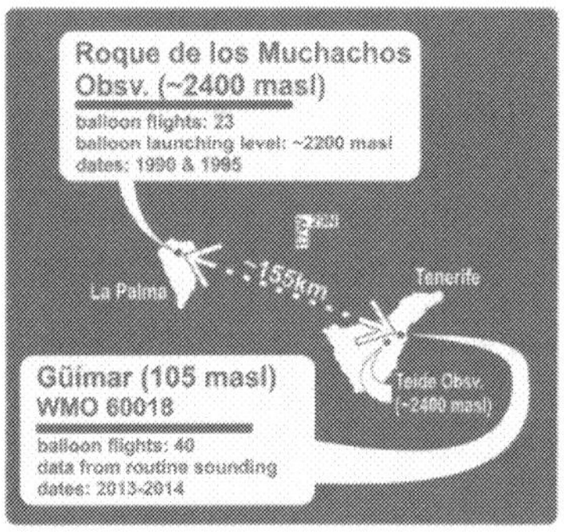

Figure 1. Location and characteristics of the two radiosoundings launching points (ORM and TFE), separated by ≈ 155 km.

Table 1. WRF model physics configuration and schemes selected for the physics modules.

LW Rad.	SW Rad.	Land Surface	Surface Layer	PBL	Cumulus	Microphysics
RRTM	Dudhia	Noah LSM	Monin-Obukhov	YSU	Kain-Fritsch	WSM6

at two separated locations (see figure 1), ORM at La Palma Island and Güímar WMO station 60018 at Tenerife Island (TFE, hereafter), close to the sea level.

A total of 23 balloons were launched directly from ORM during 3 intensive site testing assessments performed in April 1990, July 1990 and November 1995 [3], [4]. We have been used this unique dataset for the first time for Numerical Weather Prediction (NWP) model validation. As an additional verification point, we have used a sample of 40 operational soundings launched twice a day (00 and 12 UTC) near the sea level at TFE from May 2013 to April 2014 (1 year). These datasets allowed us to calibrate the model at the Observatory and to validate it under different atmosphere column heights and atmospheric conditions.

3. WRF model: configuration

We have configured the version 3.1. of WRF in a single model domain with a grid point separation of 54 km and dimensions 60×45 (thus covering a surface of 3240×2430 km). We have selected 32 vertical levels more concentrated near the ground in the case of ORM. The model physics configuration is shown in table 1. The ERA-Interim reanalysis produced by the European Centre for Medium Range Weather Forecasts (ECMWF) has been used to provide initial and boundary conditions (IBC) to WRF above ORM every 6 hours while the operational Global Forecasting Model (GFS) output has been used for IBC at TEN.

4. Results and discussion

An excellent agreement between PWV model forecasts and observations was found at both verification points (ORM and TFE) with correlations above 0.9 in both locations (figures 2 and 3). The WRF model performed quite well in different humidity conditions. The RMSE of the fit is ≈ 1 mm at ORM (1 mm $<$ PWV $<$ 13 mm) and ≈ 3 mm at TEN (12 mm $<$ PWV $<$ 40 mm).

In additional simulations we have also found that the WRF model is a very useful tool for reconstructing the history and evolution of the humidity field above the observatories as well as to assess the spatial structure of such field through 2D plots. The spatial structure allows to identify synoptic meteorological patterns in the PWV evolution. This circumstance may be used for an additional analysis based on the local expertise of the observer (figure 4).

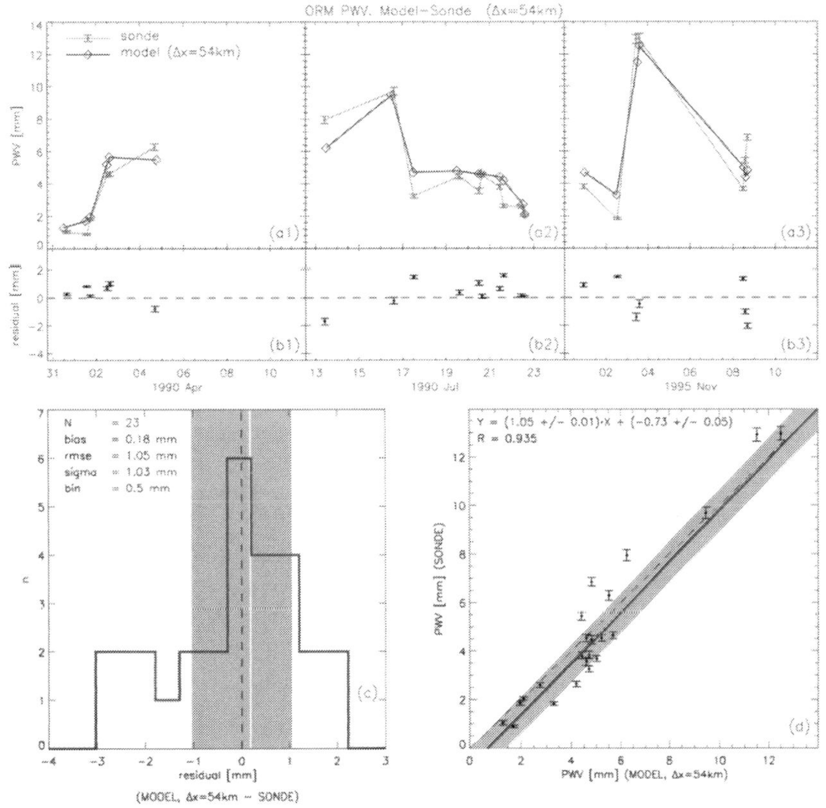

Figure 2. Time series of radiosonde measured (red) and forecasted (blue) PWV (a1 to a3), time series of residuals (b1 to b3), distribution of residuals (c) and regression analysis (d) at ORM. The indices 1, 2 and 3 refer to the 3 separated campaigns in April 1990, July 1990 and November 1995.

Acknowledgments

This work has been funded by the Instituto de Astrofísica de Canarias (IAC). The radiosonde balloons were launched at ORM by the IAC, the Centre National de la Recherche Scientifique (CNRS) and the University of Nice. Soundings at Güímar are launched by the Spanish Agencia Estatal de Meteorología and are available through the page of the Dept. of Atmos. Sci. at the University of Wyoming. We would like to acknowledge the scientific community behind the development of WRF model for their support, the European Center for Medium Range Weather Forecasts for access to their ERA-Interim Reanalysis dataset and the National Center for Enviromental Prediction (NCEP) for access to the operational GFS model output.

References

[1] Garcia-Lorenzo B, Eff-Darwich A, Castro-Almazán J, Pinilla-Alonso N, Muñoz-Tuñón C and Rodríguez-Espinosa JM 2010 *MNRAS* **405** 2683–2696
[2] Pérez-Jordán G, Castro-Almazán J, Muñoz-Tuñón C, Codina B and Vernin J 2015 *MNRAS* **(to be submitted)**
[3] Vernin J and Muñoz-Tuñón C 1992 *A&A* **257** 811–816
[4] Vernin J and Muñoz-Tuñón C 1994 *A&A* **284** 311–318

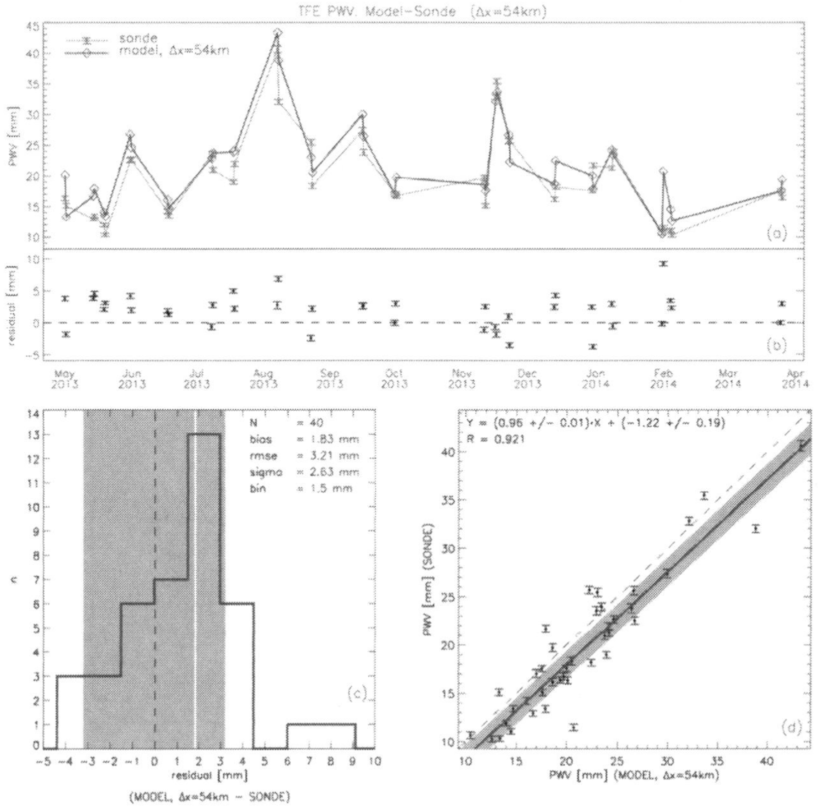

Figure 3. Time series of radiosonde measured (red) and forecasted (blue) PWV (a), time series of residuals (b), distribution of residuals (c) and regression analysis (d) at TFE. Radiosoundings are launched twice a day at 00 and 12 UTC as part of the WMO Global observing network.

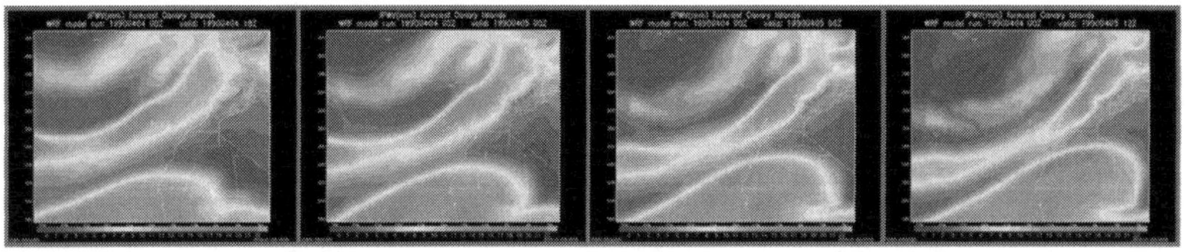

Figure 4. Passage of a moist air mass over the observatories (PWV integrated above 787 hPa) in the Canary Islands on the night of April 5, 1990, reproduced by the WRF model.

Adapting to the Atmosphere Conference 2014 IOP Publishing
Journal of Physics: Conference Series **595** (2015) 012027 doi:10.1088/1742-6596/595/1/012027

High-resolution ground layer turbulence from inside the CFHT dome using a lunar scintillometer

T Pfrommer[1] and P Hickson[2]

[1] European Southern Observatory, Karl-Schwarzschild-Str. 2, D-85748 Garching b. Muenchen, Germany

[2] University of British Columbia, Dept. of Physics and Astronomy, 6224 Agricultural Road, Vancouver, BC, V6T1Z1, Canada

E-mail: tpfromme@eso.org

Abstract. For ground layer adaptive optics systems, knowledge of the local height- and time-resolved ground layer (GL) turbulence is crucial to link local topography with optical turbulence. Such turbulence profiles have been obtained in the years 2009 and 2010 over 250 hours on Mauna Kea, Hawaii. Results from measurements inside the Canada-France-Hawaii Telescope (CFHT) dome indicate severe degradation of image quality due to a poorly vented dome and thus provide input for dome modifications and design aspects for a new ground layer adaptive optics system. The outside median GL seeing above 6 metres was determined to be 0.48±0.01".

1. Introduction

The lunar scintillometer *Portable Turbulence Profiler (PTP)* measures integrated moonlight with photo diodes to record intensity fluctuations due to atmospheric turbulence in the ground layer. We describe its design and measurement setup as well as first results from the measurement campaign at the CFHT. The primary goal of this study is to measure the turbulence contribution to the optical seeing from inside and outside the CFHT, and to compare with other existing seeing-instruments, such as the MKAM DIMM-MASS system and the CFHT dome DIMM instrument.

The principle of a lunar scintillometer is as follows: an array of photo diodes, separated by different baselines, is used to record intensity variations of moonlight, integrated over the lunar disk, at a sampling rate of order 1 kHz. Each diode has a slightly different line of sight to the Moon and probes a different atmospheric volume up to a certain height, above which two diodes share a common path. Thus, each diode pair with its unique separation baseline has its own response function defining the sensitivity range. By cross-correlating intensity fluctuations between different diode pairs, a height-resolved measurement of the turbulence strength in the ground layer up to about 1 km can be obtained. Above this height, the single diode pares share the same line of sight to the moon and the end of its sensitivity range is reached. One of the main advantages of such a device is that it measures dimensionless intensity variations that are directly related to the C_N^2-profile. No calibration of the instrument is needed.

The important atmospheric parameters seeing ε, Fried parameter r_0, atmospheric timescale τ_0, and the isoplanatic angle θ_0 define optical turbulence in the atmosphere. All these parameters can be determined if the height-dependent $C_N^2(h)$ (and the wind speed profile) is known. This structure constant defines the turbulence strength throughout the atmosphere and is the mean-square difference of index-of-refraction fluctuations of two horizontal locations, separated by 1 m. Several models of this turbulence profile have been suggested over the years [1–3], to note just a few. However, as stated by Beckers [4], the actual profiles vary from site to site and from time to time, and average profiles can only give a guideline. It is therefore important to characterize each site with respect to the C_N^2-profile, especially on the ground, where it is influenced by local topography, not just once but over years, in order to understand the long-term changes and fluctuations that could have local, diurnal, seasonal or other causes. The ground layer contributes a significant fraction (up to 80%) of the integrated C_N^2 [5]. Thus it is critical to vertically resolve $C_N^2(h)$ in the first km of the atmosphere to understand better the local effects of the topography and specifically the effect of the telescope dome and its micro climate on seeing on the optical turbulence, and hence the seeing. This requires measurements

Figure 1: PTP location inside the CFHT dome on the telescope floor. The picture is taken from the webcam on the Westside of the Mezzanine.

of the seeing from inside the dome to be carried out during normal telescope operations when the moon is suitably located so that moonlight enters through the slit in the dome. Fig. 1 shows the PTP acquiring data inside the CFHT dome, lit by moonlight. At other times, the seeing monitor was located outside and measured seeing concurrently with the DIMM-MASS monitor for comparison and cross-correlation.

2. Instrument Description

The PTP designed was designed and built at the University of British Columbia as an in-house project. The detector design is similar to the one used in the Arctic [6; 7]. Observations were made with the PTP from December 2009 until August 2010. A technician or telescope operator sets up the instrument at the beginning of the night near the DIMM-MASS weather station MKAM tower or, if the astronomical target at CFHT is favorable and moonlight enters the dome, the PTP is set up inside. As the slit moves during observation, the PTP is in this case repositioned accordingly.

2.1. Basic Principle

The basic principle is to measure the intensity covariance function $B(r)$ via intensity fluctuations over different baselines and relate it directly, with the appropriate response function, to the C_N^2-profile. This is done by placing diodes on a linear bar, separated such that each diode pair subtends a different separation distance (baseline). Light from the moon passes through the turbulent atmosphere and phase distortions occur as discussed previously. Such phase distortions create intensity fluctuations (scintillation) at the detector. Due to the overlapping fields of view above a certain height for each diode pair, which depends upon their baseline and orientation, the recorded intensity fluctuations are partly correlated, because part of the recorded light shares a common path. By calculating the covariances of the mean-subtracted and DC-normalized intensities, the amount of correlation is determined and with many baselines, a turbulence profile can be extracted. The electrical current, produced by each photodiode due to moonlight, is preamplified. This DC signal is then recorded with a high-resolution fast 24 bit analog to digital converter card. The intensity fluctuations we are interested in are of the order 0.01%, hence we need to record voltage changes down to μV-precision. A quarter moon produces DC values of about 400 mV, which means we need to resolve 40μV. The electrical current, produced by each photodiode from the incoming moonlight, is amplified by a low-noise FET operational amplifier employing a 100MΩ feedback resistor and subsequently digitized by a National Instruments data acquisition computer using an 8 channel 24 bit ADC card. DC-signals are digitized directly with no filtering. A control program written in the Labview SignalExpress environment ensures proper data storage.

2.2. Systematics and Noise Treatment

With an instrument capable of recording such high-precision fluctuations, it is important to understand the systematics. The advantage of a lunar scintillometer and its data analysis is that no calibration is required because normalized intensity fluctuations directly relate to a covariance function, which in turn directly enables us to estimate the turbulence profile C_N^2. In the following, possible noise sources that affect a lunar scintillometer are outlined.

All light sources that fluctuate at least on levels of 0.01% and whose light might find its way into the photo detectors will cause serious data contamination due to *optical interference* (not to be confused with coherent interference of light). Entrance aperture restrictions and a tracking device that ensures a perpendicular line of sight to the moon help to decrease the influence of this effect. Each photodiode was fitted with a circular baffle that limited the field of view to a $\pm 3°$ radius centred on the Moon and a low-cost amateur astronomical mount was chosen for the tracking. Even with such a large aperture, the sky background contribution is less than 10% of the DC value. *Electrical interference* may cause contamination of the diode signals on their way to the acquisition card. We reduce this by shielding and properly grounding the diodes with their directly-attached preamplifier board. These units are located inside an enclosed aluminum bar. Nevertheless, we do see a 60 Hz interference, which we filter out via software.

Such data contamination is usually readily identified by the analysis code and the one-minute data section is removed. With this list it becomes clear why the amount of useful data may be reduced, compared to the total raw data acquisition time. 20% of data taken outside near the MKAM tower, and 45% of data taken from inside the CFHT dome was discarded.

2.3. Theoretical Aspects

As already stated, intensity fluctuations due to turbulence can be used to determine the C_N^2-profile, from which all atmospheric parameters can be derived to characterize optical turbulence for astronomical observations. The intensity covariance function $B_{I,ik}$ is related to the turbulence profile, via the following equation [6; 8; 9].

$$B_{I,ik} = \int_0^\infty dh C_N^2(h) h \sec\zeta K_{ik}(\vec{r_{ik}}, h) \quad . \tag{1}$$

The response function $K_{ik}(\vec{r_{ik}}, h)$ is a function of baseline $\vec{r_{ik}} = \vec{r_i} - \vec{r_k}$. It is the product of several filter functions that can be determined prior to observing, as they don't depend on the data, but only on the response of the instrument to the current moon intensity distribution at the time of observation, a chosen turbulence model and instrumental design parameters such as baseline separations and sensitive photo detector area.

The intensity covariance function $B_{I,ik}$ for two diodes i, k is:

$$B_{I,ik} = \left\langle \left[\frac{I_i}{\langle I_i \rangle} - 1\right] \left[\frac{I_k}{\langle I_k \rangle} - 1\right] \right\rangle \quad . \tag{2}$$

$I_i = I(\vec{r_i})$ represents the intensity recorded by diode i and the inner ensemble average is the time average intensity over a 1 min time interval. This interval choice is justified because it is long compared to the atmospheric coherence time. For each 1-minute data set, the most probable C_N^2-profile is estimated by maximizing the posterior probability of the model [6; 7]. The insensitivity of the lunar scintillometer to high-altitude turbulence gives an additional constraint such that the last C_N^2 value is set to a very small value, close to zero. Interpolating between the reference heights gives C_N^2 values at other heights. The turbulence integral is defined by

$$J(h) = \int_h^\infty dx C_N^2(x) \quad . \tag{3}$$

This gives the contribution from turbulence at height h up to the outer range of the sensitivity of the lunar scintillometer at ~1 km.

3. Results of the CFHT GL Turbulence Test Campaign

The PTP was routinely deployed whenever the Moon was within one week of full Moon. If CFHT operations allowed, data were taken from inside the dome through the open dome slit if a direct line of sight to the Moon was available. If not, the PTP was set up outside and data were taken from near the MKAM tower. Regular data acquisition began in mid-May 2010 until end-August 2010, ~220 hours of outside data and ~37 hours of data from inside the dome were recorded.

For the analysis, ~80% (172 h) of the outside data and ~55% (21 h) from the inside data were used. The rest had shadowed diodes or showed other contamination that made the data unreliable. Data from inside the dome often showed shadow motion on some diodes as well as unusual spikes probably caused by electro magnetic interference. Absolutely photometric nights are required. Every minute of data acquisition is individually tested. In addition, for useful data with the PTP, the Moon should be at least 10° in altitude.

An example of the result of a lunar scintillometer can be seen in Fig. 2, where the ground layer turbulence is mapped as a time series of C_N^2-profiles, and the corresponding seeing values above the indicated heights are shown below. At the beginning of this specific night, a relatively strong boundary layer in the first 10 to 20 meters was observed. At around 6 UT, turbulence in the 100 m to 300 m region started to gain strength and stayed there for about an hour. The effect on the seeing was evident and bad conditions in the ground layer were detected. After the atmosphere calmed down, 3 hours of weak ground layer made for excellent conditions, while a slight increase in turbulent strength was still evident around 20 m height. Between 10:30 UT and 11 UT another incidence of turbulence was visible in the 100 m to 300 m region. This contamination disappeared from the line of sight to the Moon shortly after and beyond 11:30 UT. The distinct layer around 20 m gained in strength and extended to the ground after 12:30 UT, making for worse seeing conditions until the sun was rising. Over all, this night had a median GL seeing of 0.44" above 6 m, while the DIMM instrument, located right beside the PTP on a 6 m tower, measured 0.45". Besides the obvious uncertainties of local effects with the two instruments pointing to different positions in the sky, this shows that little free atmospheric turbulence was contributing during that night and most of the seeing was created in the first few hundred meters.

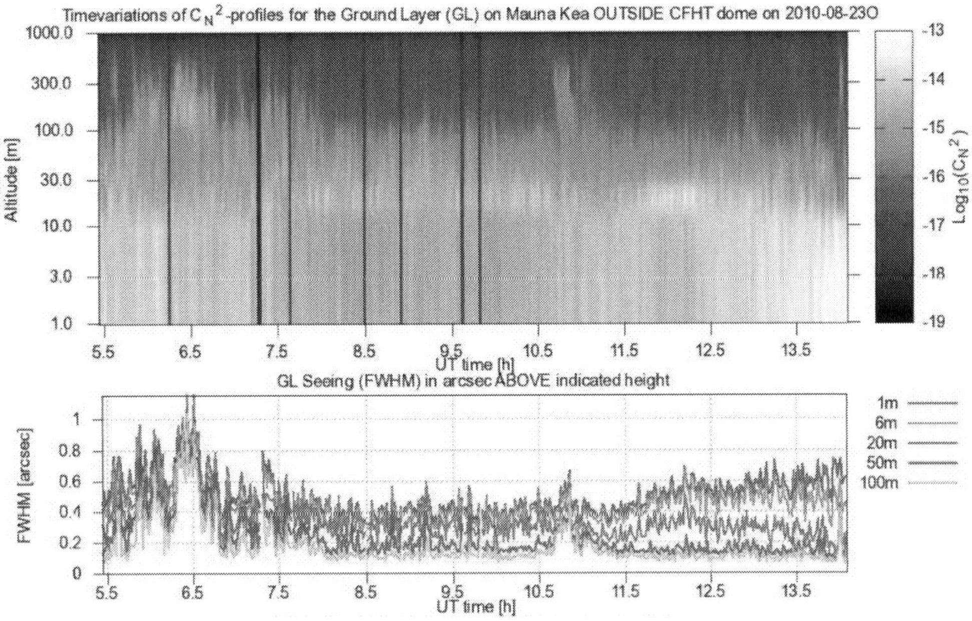

Figure 2: Example of a scintillometer result. The ground layer turbulence map shows many interesting features and the underlying seeing plots gives the information about what resolution performance a telescope at the indicated height, pointing at Zenith at 500 nm, would be able to achieve.

Fig. 3 shows a summary of GL seeing for all outdoor and indoor data. For comparison, the cumulative distribution of total seeing, as detected by the MKAM DIMM instrument, as well as by the CFHT dome DIMM, for the time ranges when PTP data are available, are shown. The CFHT dome DIMM is an instrument located inside the slit. Above 6 m from the instrument, the outside median GL seeing was 0.48"±0.01". This compares well with the results from Chun [10], who find 0.51" as median seeing value in the GL. Their data acquisition spanned over an 18 month period, while the PTP data was mostly recorded within 4 months. The PTP does not detect any free atmosphere turbulence, and hence while the data from the scintillometer only shows GL contribution, the other two instruments show integrated seeing. The strongest turbulence contributions arise from the region between 6 m and 50 m. In addition, the dome DIMM instrument usually detects turbulence through a very turbulent skin layer surrounding the CFHT dome and hence it is not expected that the PTP inside data should compare with its seeing values. With the PTP mostly located on the telescope floor, the line of sight to the moon passes nearly perpendicular through this layer while the line of sight of the dome DIMM generally subtends a larger distance in the strong turbulent region (private communication with R. Racine, 2010).

4. Conclusions

The PTP in Hawaii has recorded ground layer turbulence starting in December 2009 and regularly since May 2010, with 260 hours obtained from inside and outside the CFHT dome. The high-resolution of the PTP in the ground layer revealed outside dome influence from flow detaching and causing turbulence in front of the dome slit in the case of wind conditions perpendicular to the observing direction. While these results may seem obvious in hind sight,

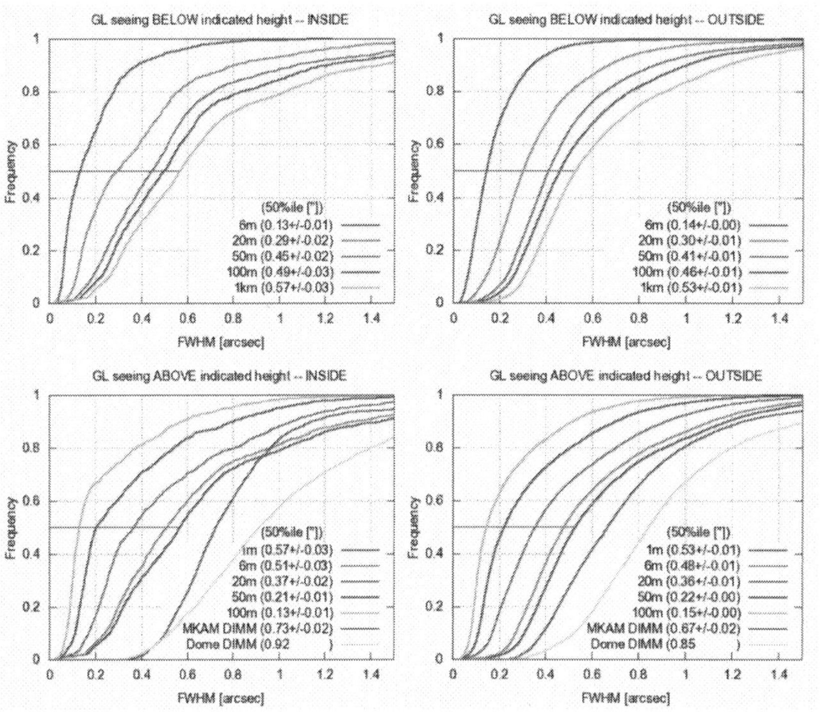

Figure 3: Cumulative distribution plots of ground layer (GL) seeing for all nights together, separated for INDOOR (left panels) and OUTDOOR (right panels) data. While the top two panels summarize GL seeing for heights BELOW the indicated values, the lower row is the summary for GL seeing ABOVE the indicated height. In the second row, the MKAM DIMM data as well as the CFHT dome DIMM data are shown in addition. Values are only taken when PTP data were taken, and separated for indoor and outdoor observation times.

---

the PTP is able to quantify these effects and reveal the spatial origin and contribution of each height to the optical turbulence in the ground layer.

The general shape of the optical turbulence strength in the boundary layer shows a sharp increase below 20 m to 30 m for the best nights, and below 50 m to 60 m for average nights. For bad nights, this increase is not as distinct anymore, instead a more general gradual increase in turbulence strength from a height of about 100 m to the ground is detected.

More detailed results will be published elsewhere. Those include a detailed analysis of different wind speeds and directions and their effect on dome-seeing. The outside ground layer turbulence results will be discussed with respect to their dependence on wind direction, speed and local topology.

Acknowledgments
I would like to thank *Dr. Christian Veillet* and the staff of the Canada-France-Hawaii Telescope for the possibility of operating the PTP on a regular basis on Mauna Kea. *Dr. Marc Baril* and *Dr. Kevin Ho* helped a lot to troubleshoot and improve the instrument for permanent usage.

References

[1] Hufnagel R E 1974 in Digest of Topical Meeting on Optical PropagationThrough Turbulence

optical Society of America, Washington, D.C.

[2] Valley G C 1980 *Appl. Opt.* **19** 574–577

[3] Tokovinin A and Travouillon T 2006 *MNRAS* **365** 1235–1242

[4] Beckers J M 1993 *Sol. Phys.* **145** 399–402

[5] Rajagopal J, Tokovinin A, Bustos E and Sebag J 2008 *Society of Photo-Optical Instrumentation Engineers (SPIE) Conference Series* (*Presented at the Society of Photo-Optical Instrumentation Engineers (SPIE) Conference* vol 7013)

[6] Hickson P, Carlberg R, Gagne R, Pfrommer T, Racine R, Schöck M, Steinbring E and Travouillon T 2010 *Society of Photo-Optical Instrumentation Engineers (SPIE) Conference Series* (*Society of Photo-Optical Instrumentation Engineers (SPIE) Conference Series* vol 7733)

[7] Hickson P, Gagné R, Pfrommer T and Steinbring E 2013 *MNRAS* **433** 307–312 (*Preprint* 1305.0615)

[8] Hickson P and Lanzetta K 2004 *PASP* **116** 1143–1152

[9] Hickson P, Pfrommer T and Crotts A P S 2008 *Conf. Proc. of the Imperial College Press World Scientific* Optical Turbulence Astronomy meets Meteorology, Sardinia, Italy

[10] Chun M, Wilson R, Avila R, Butterley T, Aviles J, Wier D and Benigni S 2009 *MNRAS* **394** 1121–1130

Numberical Calculations of Atmospheric Conditions over Tibetan Plateau by Using WRF Model

Xuan Qian, Yongqiang Yao, Hongshuai Wang, Liyong Liu, Junrong Li, Jia Yin

National Astronomical Observatories of China, Chinese Academy of Science, Datun Road A20, Chaoyang District, Beijing, China

E-mail: qianxuan@nao.cas.cn

Abstract. The wind field, precipitable water vapor are analyzed by using the mesoscale numerical model WRF over Tibetan Plateau, and the aerosol is analyzed by using WRF-CHEM model. The spatial and vertical distributions of the relevant atmospheric factors are summarized, providing truth evidence for selecting and further evaluating an astronomical site. It has been showed that this method could provide good evaluation of atmospheric conditions. This study serves as a further demonstration towards astro-climate regionalization, and provides with essential database for astronomical site survey over Tibetan Plateau.

1. Introduction

In view of the fact that the important meaning and physical demand of Chinese site survey, modeling of calculating atmospheric conditions over China is proposed, and the research mainly includes aerosol, precipitable water vapor and wind. Meteorological parameters of astronomical observatories are obtained by using the atmospheric numerical models, then precipitable water vapor is calculated with the method of water vapor mixture ratio parameterization model, and through the atmospheric numerical model computation output aerosol and wind parameters, and this method has been confirmed feasible for site survey assessment.

The atmospheric and terrain conditions are much complicated over Tibetan Plateau, and with its high altitude, it's much hard for people to measure the atmospheric conditions, so the meteorological model WRF has been employed. Different physical options and resolutions of WRF model have been investigated by comparing the simulated results with weather stations measurements, in order to obtain the most suitable options over Tibetan Plateau.

2. Experiment design

In this work, the NCAR's (National Center of Atmosphere Research) WRF model has been used to simulate the meteorological parameters, such as wind speed, wind direction, precipitable water vapor and aerosol (dust) at different levels from ground to the top of the atmosphere. The WRF domain is designed to locate over 25-40°N and 70-103°E, including all the Tibetan Plateau. Initial and boundary conditions for the large-scale atmospheric fields are given by the $1° \times 1°$NCEP Global Final Analysis Datasets. Five experiments with different model resolutions, microphysics and planetary boundary layer options, have been taken into account.

The wind field and precipitable water vapor over the plateau are analyzed by using WRF model with NCL reanalysis datasets of 2010, and the wind field includes the altitudes of 10m and 200hPa, and also by using WRF-CHEM model, the spatial distribution and temporal variation of aerosol is analyzed. For analyzing the vertical distributions of the factors, Ali site(32.5°N, 80.08°E) which lies in the southwest of the plateau has been chosen in this work.

3. Precipitable Water Vapor

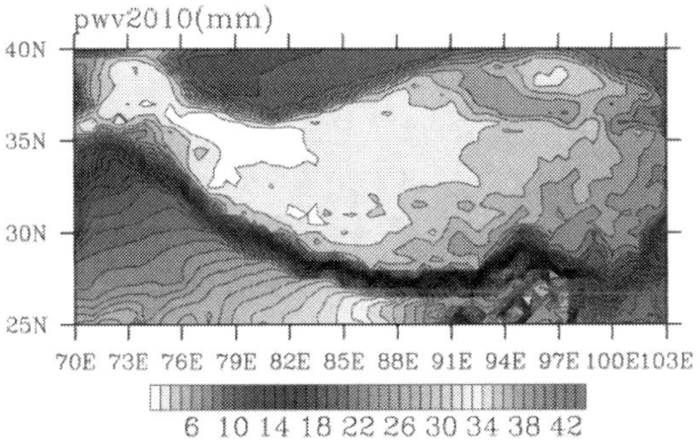

Figure 1. The spatial distribution of precipitable water vapor in 2010 over this domain, and the precipitable water vapor unit is mm.

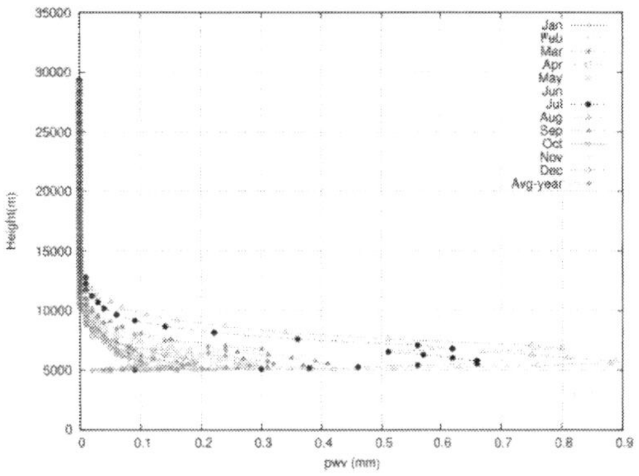

Figure 2. The vertical distribution of precipitable water vapor in 2010 in Ali site, and the precipitable water vapor unit is mm.

Fig.1 shows the spatial distribution of precipitable water vapor. The calculated results show that the mean yearly precipitable water vapor over most of the Tibetan Plateau is below 5mm, especially in the southwest of the plateau, the precipitable water vapor is the lowest, just around 3mm; and in the east of the plateau, the value is mainly between 5 and 10mm. To the southwest of the plateau in which belongs to India, the value of precipitable water vapor is the highest, more than 20mm. From the vertical distribution of precipitable water vapor in Ali site(32.5°N, 80.08°E) which is a new astronomical observatory station in southwest of Tibetan Plateau, according to Fig.2, precipitable water vapor mainly exists at 5000m to 12000m. And for the

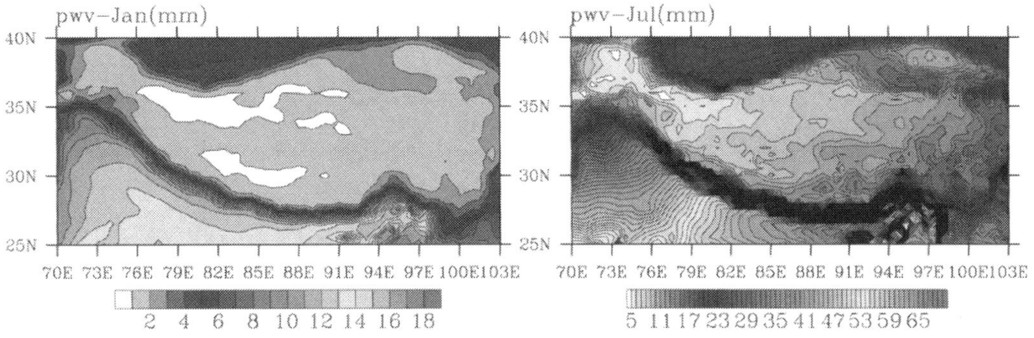

Figure 3. The spatial distribution of seasonal precipitable water vapor, and the precipitable water vapor unit is mm.

seasonal varieties, in summer its value is the highest, and in winter it's the lowest. In each month, the highest value exist at 6000m height.

Fig.3 gives the distribution of precipitable water vapor in January and July which represent winter and summer. In different seasons, precipitable water vapor in the west part is totally more than the east part. And in summer, precipitable water vapor is the most, the value is above 8mm over most of the plateau; and in winter, over most of the plateau the precipitable water vapor is 2-3mm; in the northwest and somewhere around Ali site, the precipitable water vapor is below 1mm.

4. Wind Field

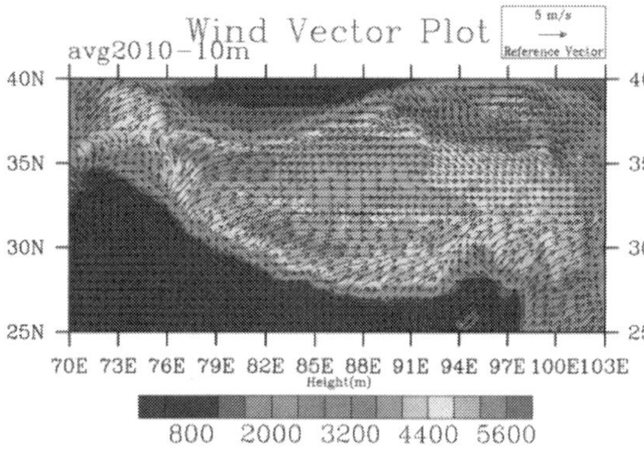

Figure 4. The spatial distribution of wind field at 10m in 2010 over this domain, and the wind speed unit is m/s.

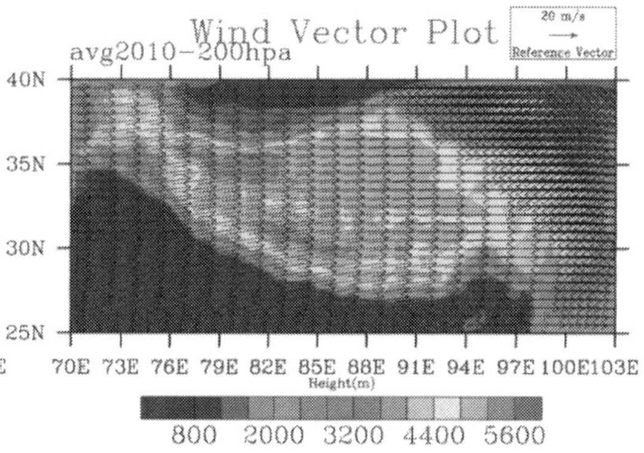

Figure 5. The spatial distribution of wind field at 200hPa in 2010 over this domain, and the wind speed unit is m/s.

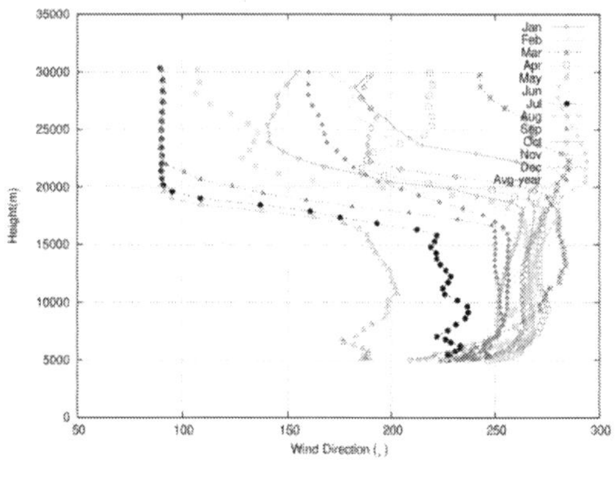

Figure 6. The vertical distribution of wind direction over this domain, and the wind direction unit is °.

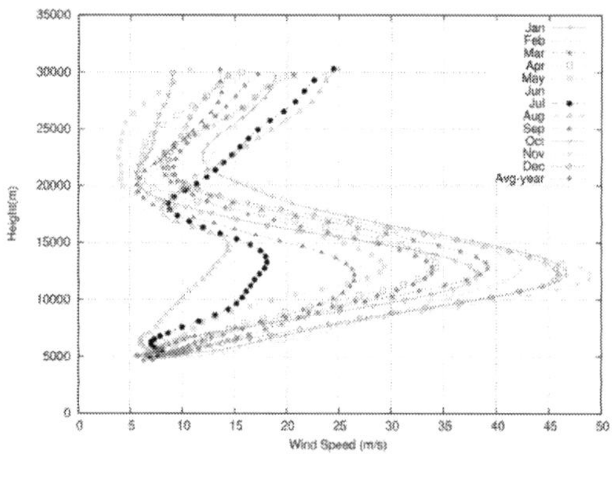

Figure 7. The vertical distribution of wind speed over this domain, and the wind speed unit is m/s.

Fig.4 and Fig.5 show the distributions of wind field at 10m and 200hPa. At 10m, there's always west wind in the middle of the plateau, the wind speed is nearly 5m/s; and in the west,

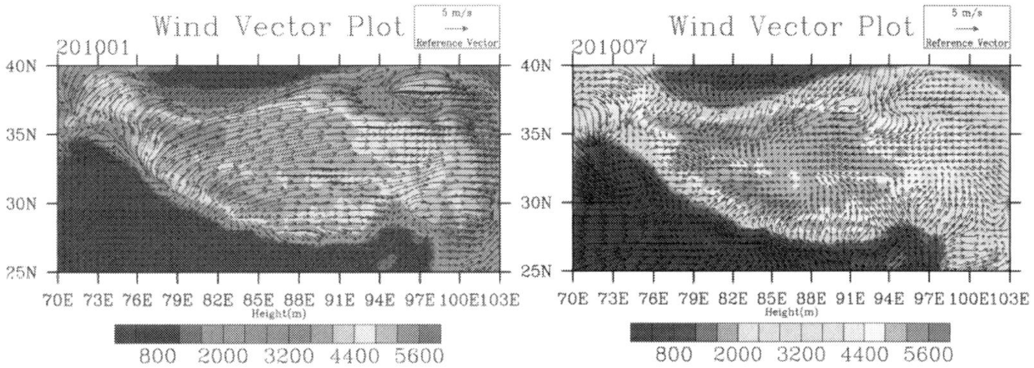

Figure 8. The spatial distribution of seasonal wind field at 10m, and the precipitable water vapor unit is mm.

southwest and southeast part, is mainly southwest wind, the wind speed is larger than 5m/s; in the northeast part of the plateau, there's always northwest wind, and the wind speed is below 5m/s. At 200hPa, the wind speed is high, almost around 20m/s, and is mainly west wind on the plateau.

According to Fig.6, below 15000m there's southwest wind over most of the Tibetan plateau in each month; between 15000m and 20000m is almost southeast wind; and above 20000m, in summer and autumn is west wind, and is southwest wind in other months.

Fig.7 shows the profile of wind speed of Ali site. At 5000m, the wind speed is the smallest in different months, and then it's getting stronger and stronger as the altitude grows, and until 13000-15000m the wind speed begins getting weaker and weaker with the altitude growing, and at 20000m level the wind speed is smaller, then it's still getting stronger and stronger with the altitude growing. Between 10000m and 15000m, the wind speed is the biggest in different seasons, especially in winter the wind speed is the biggest, and in summer the wind speed is the smallest.From the vertical distribution of wind speed in Ali site, in winter the wind speed is the highest, the value of the wind speed is higher than 4m/s; and next is in spring, in summer the wind speed is the lowest, the wind speed is lower than 2m/s. And at 10000m to 15000m height, the wind speed in each month is highest.

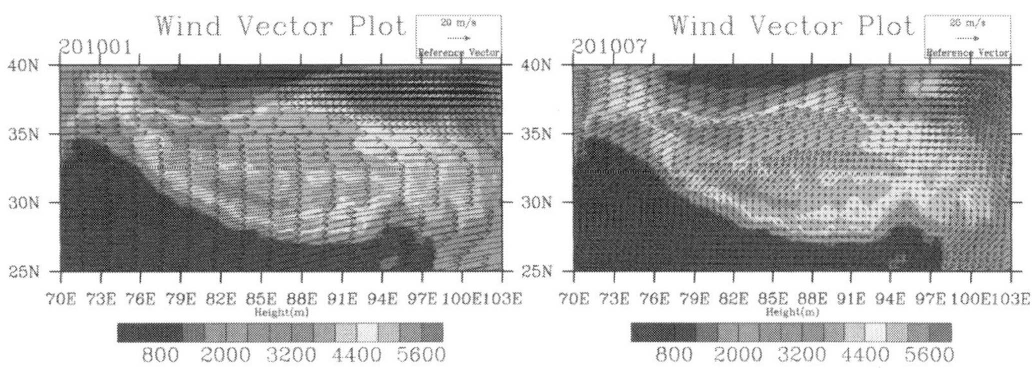

Figure 9. The spatial distribution of seasonal wind field at 200hPa, and the precipitable water vapor unit is mm.

According to Fig.8, for the seasonal varieties, wind speed is small in summer and large in winter and spring. The wind direction changes in different seasons, in winter, there's mainly southwest wind over the plateau, and in summer the wind direction forms a circle over the main plateau, in the middle part the wind speed is the smallest, and the wind flows towards the middle part in different directions.

As Fig.9 shows, for the seasonal varieties over the plateau at 200hPa, in winter and spring there's mainly west wind, and in summer there's mainly southwest wind in the middle and west of the plateau, and in the east of the plateau is mainly northwest wind. The wind speed is largest in winter and smallest in summer.

5. Aerosol

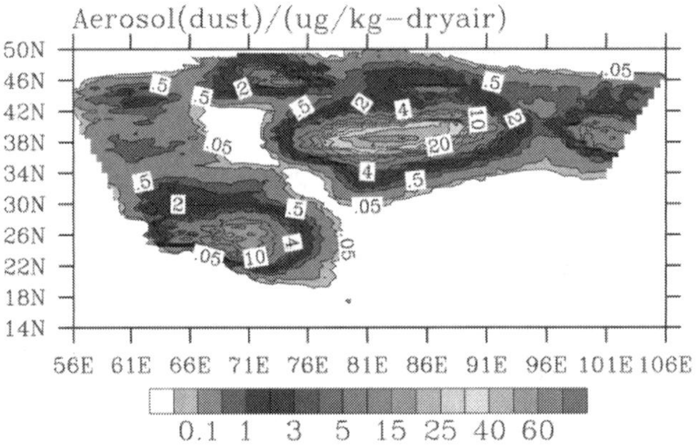

Figure 10. The spatial distribution of aerosol(dust)/(ug/kg-dryair) over this domain, and the aerosol unit is ug/kg-dryair.

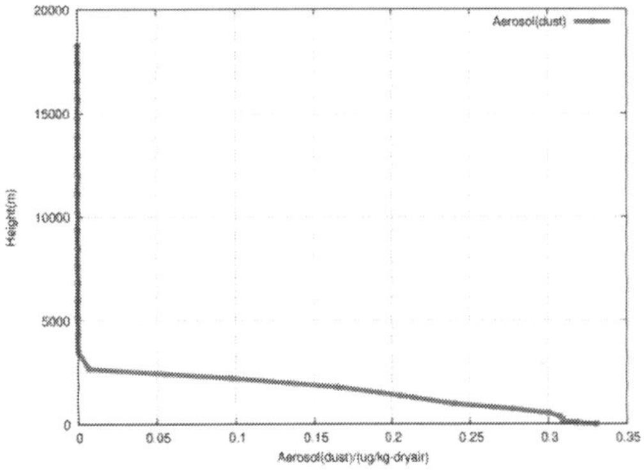

Figure 11. The vertical distribution of aerosol(dust) /(ug/kg-dryair) in the site (40°N, 101°E), and the aerosol unit is ug/kg-dryair.

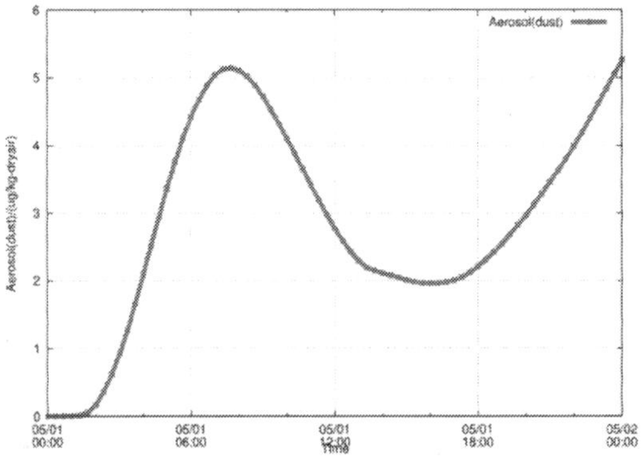

Figure 12. The temporal variation of aerosol(dust)/(ug/kg-dryair) in the site (40°N, 101°E), and the aerosol unit is ug/kg-dryair.

The aerosol(dust) is calculating by using WRF-CHEM model, Fig.10 gives the distribution of aerosol(dust), and the unit is ug/kg-dryair. The aerosol is below 0.5 ug/kg-dryair in the south of Tibetan plateau; to the north of the plateau where the Taklimakan Desert lies, the aerosol is above 25 ug/kg-dryair, for the sediments here can be readily uplifted into the atmosphere; and in the middle and nouth of the plateau , the aerosol is between 0.5 and 3 ug/kg-dryair.

Choosing the site(40°N, 101°E) of which the vertical distribution of aerosol(dust) is analyzing, as Fig.11 shows. Below 4000m, the aerosol decreases with the altitude increasing, and when it's next to the ground, the value is larger than 0.3 ug/kg-dryair. Nearly above 4000m, the aerosol is almost 0.

Fig.12 shows the temporal variation of aerosol(dust) of the site(40°N, 101°E) during a day. The aerosol is least after midnight, the value is 0 until 2am, and then it's increasing more and more, reaching the value of more than 5 ug/kg-dryair.

6. Conclusions

In this work, the new method of calculating atmospheric conditions by using WRF model is introduced, the distribution and variation of precipitable water vapor, wind speed, and aerosol are obtained, and its mature reliable application can bring the enormous convenience. The mean yearly precipitable water vapor over most of the Tibetan Plateau is below 5mm, and especially in the southwest of the plateau, the precipitable water vapor is the lowest, just around 3mm. At 10m level, the wind speed over the plateau is mainly around 5m/s, just on the north and east of the plateau, the wind speed is low. And at 200hPa level, the wind speed is high, almost around 20m/s, and is mainly west wind over the plateau. The aerosol is below 0.5 ug/kg-dryair in the south of Tibetan plateau, and in the middle and nouth of the plateau , the aerosol is between 0.5 and 3ug/kg-dryair.

7. Acknowledgment

This work was supported by the National Natural Science Foundation of China (Grant Nos. 11203044, 11373043, 11303055 and 11103042).

8. References

[1] Mian Chin, Paul Stefan Kinne, Omar Torres, Brent N.Holben 2002 *Journal of the atmospheric sciences.* **59** 461
[2] Xuan Qian and Yongqiang Yao 2013 it Chinese Astronomy and Astrophysics. **37** 212
[3] Yongjing Zhang, Yongqiang Yao, Xuan Qian 2012 *Sci Sin Phys Mech Astron.* **42** 1
[4] Liyong Liu, Yongqiang Yao, Jean Vernin, Xuan Qian 2002 it Research in Astronomy and Astrophysics. **15** 16
[5] Xuan Qian, Panxing Wang, Yongqiang Yao 2011 *Sci Sin Phys Mech Astron.* **41** 1

The Statistical Distributions and Evolutions of Seeing Values

R Racine

Association of Canadian Universities for Research in Astronomy & Département de physique, Université de Montréal, CP 6128, Succ. Centre-Ville, Montreal H3C-3J7, Canada

E-mail: racine@astro.umontreal.ca

Abstract. Extensive series of DIMM and MASS seeing values from thirteen astronomical sites are used to examine the shapes of their statistical distributions and their evolutions with time. At all sites, the distributions of seeing values can be satisfactorily reproduced over densities that span more than four orders of magnitude by random combinations of seeing values from two independent equal-population log-normal seeing components. The seeing varies typically by a factor of 4 throughout a night. At good sites, it reaches sub-half-arc second on 80% of the night. It is most stable when near its modal value where the delays for a 10% change in DIMM seeing average ~50 minutes. The average delays are ~25 minutes at the 10th and 90th percentiles of the distributions. These lifetimes differ by a factor of 4 between the fastest and the slowest seeing sites. The MASS seeing evolves ~7 times faster than the DIMM seeing. These characteristics make forecasting seeing a tall challenge.

1. Introduction

Much literature discusses the techniques for and the results of optical turbulence measurements above astronomical sites. Here we examines a relatively obscure sub-topic: the *shape* of the seeing values distributions, i.e. of the distributions of the full widths at half maximum intensity ε_0 imposed by atmospheric optical turbulence to stellar point spread functions. The evolutions with time of the seeing values are also examined, as this is of interest in planning seeing-critical astronomical observations and also in measuring the challenges of seeing forecasting using climate models. Seeing values are inferred from differential image motion monitor (DIMM) [1] and multi-aperture scintillation sensor (MASS) [2] observations assuming Kolmogorov turbulence of infinite outer scale L_0. MASS and DIMM are usually operated together on a same mount. Seeing is traditionally expressed in second of arc (") at a wavelength of 500 nm at the zenith.

2. Seeing distributions

The data used are years-long web-accessible archival series of ε_0 values obtained at thirteen sites. The measurements, averaged over 1 minute, were all made from elevations of about 7 m above local ground. For each site, the differential density distribution values were compiled in successive intervals in ε_0 that are multiples of 0.01" wide.

To model the distributions, numerical Monte-Carlo (MC) simulations adding in 5/3 power 140 000 pairs of ε_0 values randomly drawn from two independent distributions of same populations and adjustable medians (μ_1, μ_2) and dispersions (σ_1, σ_2) were run and the process was iterated until the least χ^2 fit of the resulting distribution to the data was obtained. In a first pass through the data the two model distributions had the form of log-normals but the powers n of the arguments of the exponentials, which are 2 for log-normals, were left as free fit parameters. This allowed testing if log-normals yield appropriate representations. This was confirmed, the values of n returned by the fits averaging 2.02 with a dispersion of 0.04. Final fits with log-normals were then adopted.

The results of the exercise are summarized in table 1. Column 1 identifies the sites and their altitudes. Columns 2, 3 and 4, 5 give the medians and the logarithmic dispersions of the DIMM and MASS ε_0 data. Columns 6, 7 and 8, 9 give the values of the medians and dispersions μ_1, σ_1 and μ_2, σ_2 respectively of the log-normal components whose combinations best fit the DIMM ε_0 distributions and column 10 gives the χ^2 per degrees of freedom of these fits. The two components have quite different dispersions and are accordingly termed the "narrow DIMM components" and "wide DIMM components". Columns 11 through 15 contain analogous results for fits to the MASS distributions which are improved by adding the effects of "weak MASS components" to those of "strong MASS components".

Table 1. Characteristics of Seeing Distributions and of their Components.

1		2	3	4	5	6	7	8	9	10	11	12	13	14	15
Site	Altitude (m)	$<\varepsilon_0>$ (asec.) / dispersion				narrow DIMM component		wide DIMM component		χ^2	weak MASS component		strong MASS component		χ
		DIMM	σ	MASS	σ	μ_1 (")	σ_1	μ_2 (")	σ_2		μ_3 (")	σ_3	μ_4 (")	σ_4	
Mauna Kea, Hawai'i															
13N	4040	0.74	0.46	0.33	0.68	0.36	0.30	0.56	0.66	4.8	0.15	0.64	0.23	0.93	4.2
MKAM	4200	0.64	0.42	0.38	0.59	0.30	0.30	0.48	0.62	5.3	0.04	0.68	0.37	0.74	8.9
UH 2.2-m[1]	4200	0.65	0.33	--	--	0.50	0.27	0.30	0.71	2.4	--	--	--	--	--
Chile, Norte Grande															
Tolar	2290	0.63	0.36	0.44	0.51	0.36	0.26	0.44	0.60	0.0	0.20	0.88	0.30	0.49	17.1
Paranal	2640	0.87	0.41	0.43	0.47	0.31	0.25	0.67	0.55	3.5	0.13	0.21	0.41	0.53	6.0
Armazones	3060	0.65	0.40	0.42	0.52	0.32	0.31	0.50	0.55	1.8	0.18	0.65	0.31	0.69	17.6
Tolonchar	4480	0.64	0.35	0.47	0.57	0.36	0.27	0.45	0.56	1.1	0.07	0.85	0.45	0.58	14.8
Chile, Norte Chico															
Tololo	2210	0.86	0.39	0.64	0.50	0.48	0.22	0.62	0.65	1.3	0.08	0.23	0.63	0.54	6.4
Vizcachas	2050	0.86	0.34	0.42	0.49	0.48	0.08	0.55	0.53	7.5	0.16	0.32	0.34	0.60	16.8
Pachon	2850	0.76	0.34	0.42	0.52	0.49	0.27	0.48	0.60	8.4	0.12	0.82	0.36	0.60	15.7
North America															
Palomar	1710	1.11	0.38	0.43	0.63	0.69	0.21	0.74	0.66	0.8	0.09	0.45	0.44	0.70	4.2
SPM	2830	0.79	0.46	0.38	0.66	0.40	0.24	0.60	0.68	0.5	0.08	0.37	0.35	0.72	4.4
Islas Canarias															
ORM	2396	0.78	0.44	0.31	0.56	0.45	0.22	0.55	0.75	3.2	0.11	0.33	0.27	0.67	11.2
averages		0.77	0.39	0.42	0.56	0.42	0.25	0.53	0.62	3.04	0.12	0.54	0.37	0.65	10.63
mean errors		0.04	0.01	0.02	0.02	0.03	0.02	0.03	0.02	0.79	0.01	0.07	0.03	0.04	1.67

[1] No MASS data are available for the UH site.

Figure 1 illustrates the results for the DIMM and MASS fits that have the smallest and the largest χ^2 per degree of freedom. Log-log displays allow visualizing the shapes of these distributions over large density ranges and trace an inverted parabola centred on the median log($<\varepsilon_0>$) for pure log-normal distributions. The figure shows that two-components models can reproduce the data quite well over a range of more than four orders of magnitudes in density. The DIMM plot for Las Vizcachas is a good example of a distribution compressed on its better seeing side by a strong narrow component.

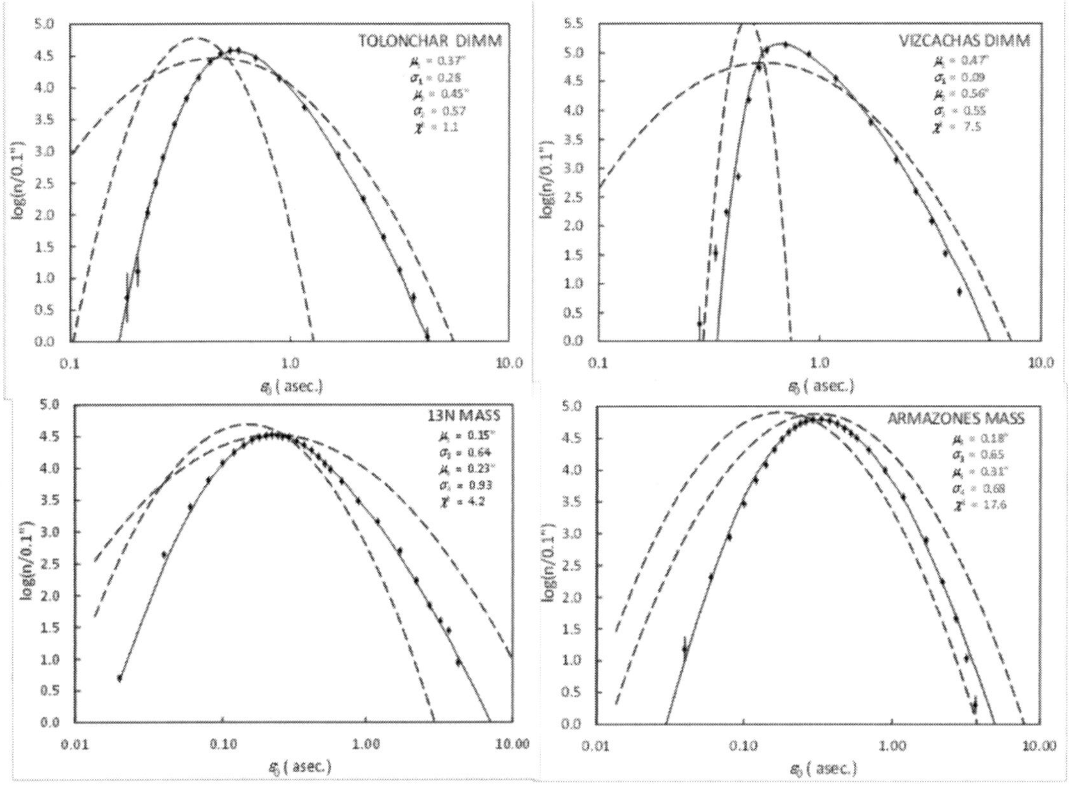

Fig 1 – These show the best (left) and the worst (right) two-component (dashed lines) fits to distributions of DIMM (above) and MASS (below) seeing values.

A detailed discussion of the relations between the DIMM and MASS components is beyond the scope allowed by this paper. Important clues are that, as can be seen in table 1, the strong MASS components and the wide DIMM components have comparable dispersions and must be related. But, figure 2 shows that the strength μ_2 of the wide DIMM component is generally stronger than that the MASS strong component, very much so for some sites. This indicates that the wide DIMM component includes, in addition to the MASS component, a large-dispersion contribution that arises below an altitude of 250 m and is therefore not sensed by the MASS. Those sites where this contribution is the strongest are known to have particularly strong low altitude turbulence [3] [4] [5].

All DIMM distribution models require the contribution of a narrow ($\sigma_1 \sim 0.25$) rather weak ($\mu_1 \sim 0.4$") component (figure 3). As figure 1 illustrates, these narrow component limits and steepens the better seeing side of the ε_0 distributions produced by turbulence at altitude and must arise by processes different from those that prevail in the upper atmosphere. In our model the components are additive. In contrast, the DIMM seeing above Dome C in Antarctica results from two or three *sporadic* turbulent components [6] [7], the surface layer being very dominant at times.

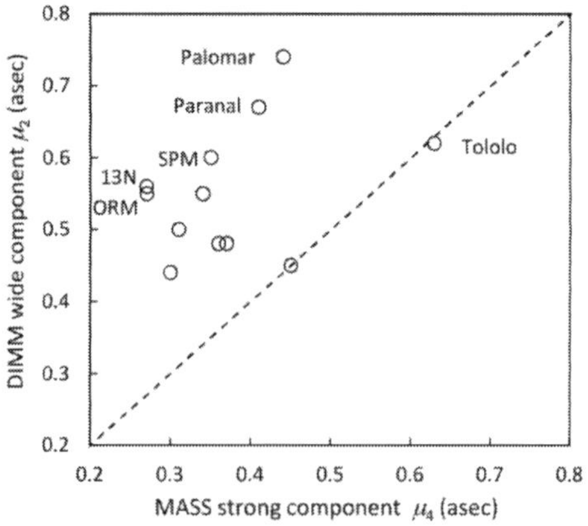

Fig 2 - The wide DIMM components are generally stronger than the strong MASS components.

It will have been noticed that in table 1 the χ^2 are generally much larger for the MASS fits than for the DIMM fits. This must be because at any site the DIMM always sample the full effect of the two components whereas the sensing of the high altitude, broader dispersion component by the MASS is variably contaminated by effect of the lower altitude turbulence. This variability generates additional noise in the shape of the distributions of MASS seeing values and lowers the χ^2 of the fits.

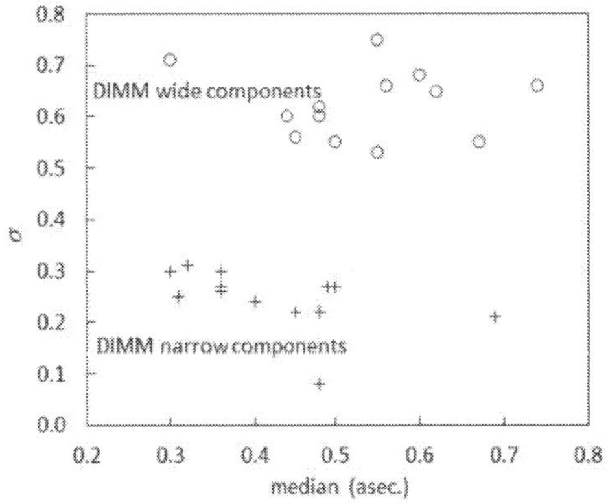

Fig. 3 – Models of DIMM seeing values distributions require the combined contributions of two components with notably different dispersions.

3. Seeing evolutions

Figure 4 shows DIMM seeing values sampled every minute over 10 consecutive summer nights at Mauna Kea 13N and Cerro Armazones. It says all there is to say about seeing evolution: "Seeing is a fickle friend: There is no such thing as a stable seeing night." More quantitative evidence will now be presented that forecasting seeing and successful "seeing sensitive" telescope scheduling are tall challenges.

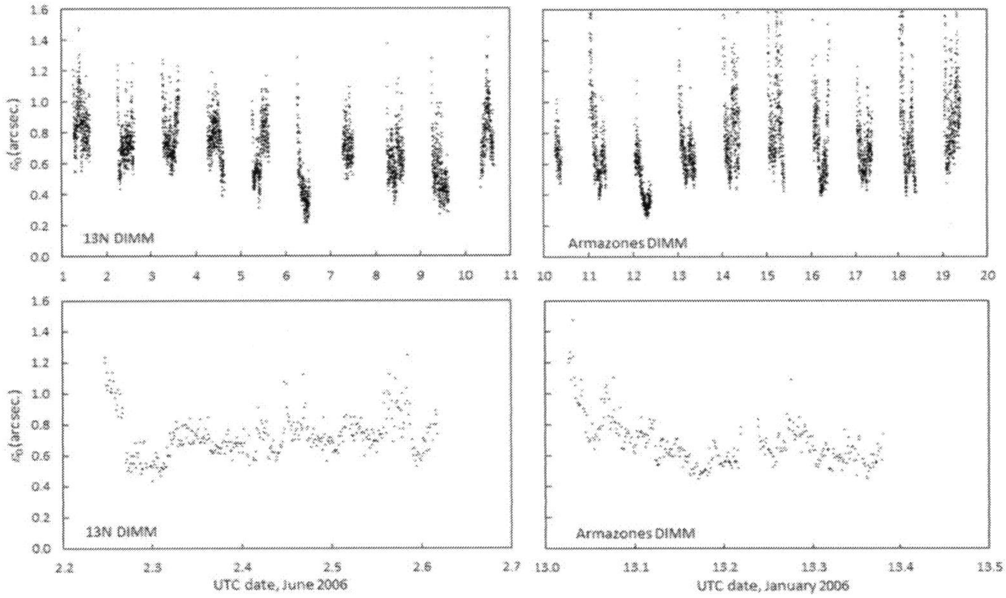

Fig. 4- Seeing is a fickle friend.

During a single night, seeing values typically span a range of a factor of 4 (figure 5). At Armazones, as at all good sites, DIMM seeing episodes better than 0.5" (telescopic image quality < 0.40" for $L_0 = 30$ m) occur on 80 percent of the nights.

Seeing evolution can be described by the time delay dependence of the seeing fractional difference *FD* as defined by Racine [8] and by Skidmore et al. [9]:

$$FD(\Delta t) \equiv |\varepsilon(t + \Delta t) - \varepsilon(t)|/(\varepsilon(t + \Delta t) + \varepsilon(t)). \quad (1)$$

The DIMM and MASS *FD* values for 2 sites are displayed against the delay Δt in figure 6. Such relations seeing were generated for all sites and for seeing values ranging from ¼ to 4 times the median values. The delays required for *FD* to reach 0.1 are shown as a function of the prevailing seeing in figure 7. The most stable seeing is the modal seeing. At the 10th and 90th percentiles of the distributions the seeing evolves two times faster. The maximum values for each site are given in table 2. The seeing speeds differ significantly between sites.

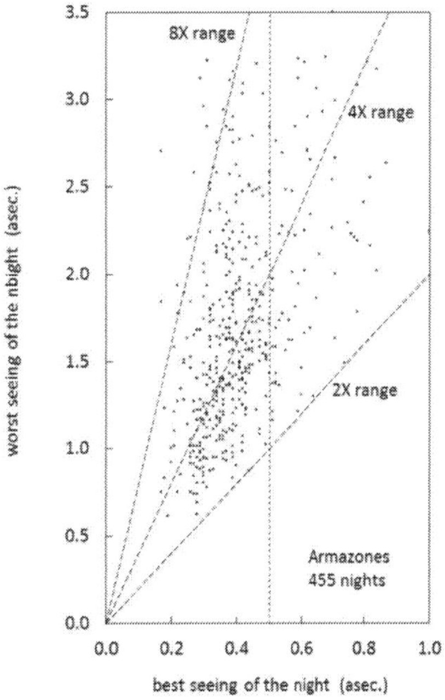

Fig. 5 – The best and worst seeing values for 455 Armazones nights are compared.

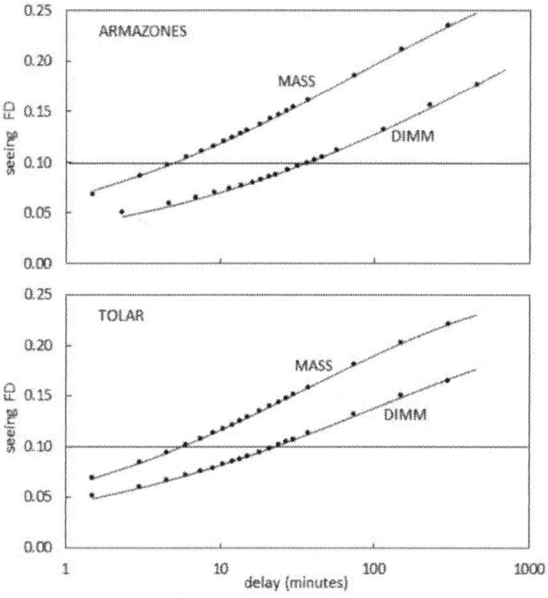

Fig. 6 – These show the evolutions of the fractional seeing differences with time delay for 2 sites.

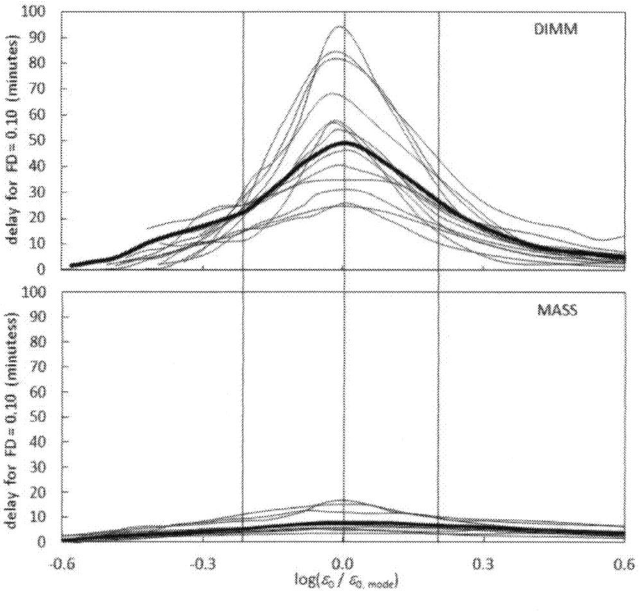

Fig. 7 – The delays for the fractional differences to reach 10 percent are plotted against the prevailing seeing values for each site. The bold curves are the average relations. The vertical lines are at the 10th, 50th and 90th percentiles of the seeing values distributions.

4. Conclusion

At all sites studied, the density distributions of DIMM and MASS seeing values can be satisfactorily reproduced over density ranges that reach four orders of magnitude by random combinations of the seeing values from two independent equal-population log-normal seeing components. The wider of these components is stronger than the integrated turbulence sensed by a MASS and includes the effects of optical turbulence below an altitude of 250 m. The narrower and weaker DIMM probably arises at low elevation.

The statistical evolutions of seeing values differ markedly between nights and between sites. There is no such thing as a stable seeing night. The seeing varies typically by a factor of 4 throughout a night. At good sites, the seeing from an elevation of 7-m above grade reaches sub-half-arc second on 80% of the night. It is most stable when near its modal value where the average time for a 10% change in DIMM seeing is ~50 minutes; it is ~25 minutes at the 10th and 90th percentiles of the distributions. The MASS seeing evolves ~7 times faster than the DIMM seeing on average. These lifetimes vary by a factor of 4 between the fastest and the slowest seeing sites.

Table 2 : Maximum Delays for 10% *FD*

Sites	DIMM (minutes)	MASS (minutes)
Pachon	94	7
Palomar	85	16
Vizcachas	81	17
Tololo	68	13
SPM	57	22
UH	57	--
Tolonchar	54	6
Armazones	46	5
Tolar	41	8
Paranal	35	15
13N	31	3
CFHT	26	4
ORM	25	5

5. References

[1] Sarazin M & Roddier F 1990 *A&A* **227** 294
[2] Kornilov V, Tokovinin A, Vozyakova O, Zaitsev A, Shatsky N, Potanin S & Sarazin M 2003 *SPIE* **4839** 837
[3] Sarazin M, Melnick J, Navarrete J. & Lombardi G 2008 *New Facts about the Evolution of Seeing at Paranal*, ESO Messenger **132** 11
[4] Tokovinin A, Bustos E & Badja . 2010 MNRAS **404** 1186
[5] Els S, Travoullion T, Schöck M, Riddle R, Skidmore W, Seguel J, Bustos E & Walker D 2009 *Pub. Astron. Soc. Pacific* **121** 527
[6] Fossat E, Aristidi E, Agabi K. Bondoux E, Challita Z, Jeanneaux F & Mekarnia D 2010 *A&A* **517** 69
[7] Aristidi E, Fossat E, Agabi A, Mékarnia D, Jeanneau F, Bondoux E, Challita Z, Ziad A, Vernin J & Trinquet H 2009 *A&A* **499** 955
[8] Racine R 2005 *Pub. Astron. Soc. Pacific* **117** 401
[9] Skidmore W, Els S, Travouillon T, Riddle R, Schöck M, Bustos E, Seguel J & Walker D 2009, *Pub. Astron. Soc. Pacific* **121** 115

Acknowledgements

This study was made possible thanks to colleagues who provided access to gigabytes of DIMM and MASS data: CFHT Billy Mahoney, ESO Marc Sarazin, NOAO Andrei Tokovinin, IAC Casiana Muños-Tuñón, TMT Mathias Schöck and Warren Skidmore and UH Mark Chun. Their help is gratefully acknowledged. Perceptive and encouraging comments on earlier versions of this work by Elena Masciadri, Andrei Tokovinin and Marc Sarazin help devising a more realistic and, hopefully, clearer analysis. Mark Sarazin kindly presented this paper at the Durham 2014 "Adapting to the Atmosphere" Conference when its author was unable to do so.

Near ground results of the CO-SLIDAR C_n^2 profiler

C Robert[1], **J-M Conan**[1], **L M Mugnier**[1] **and J-M Cohard**[2]

[1] ONERA - The French Aerospace Lab F-92322 Châtillon, France
[2] Laboratoire LTHE, 70 rue de la Physique 38400 Saint Martin d'Hères, France

E-mail: `clelia.robert@onera.fr`

Abstract. CO-SLIDAR jointly uses the slopes and scintillation indices from a double source recorded with a Shack-Hartmann (SH) wavefront sensor which leads to a robust C_n^2 profile restoration. This technique coupled to a 0.35-m telescope in the mid-IR has provided horizontal C_n^2 profiles over 2.7 km with a 300 m resolution. The SH wavefront sensor data revealed a very good fit of the spatial spectra of the phase to the Kolmogorov model, and of the scintillation distributions to the small perturbation regime as well. Quantitatively the inversion of the measurement covariances allows us to retrieve the C_n^2 profile with small error bars and stable structure. These C_n^2 distributed estimations are comparable with the averaged C_n^2 measurements from scintillometers set in parallel.

1. Introduction

The development of C_n^2 profile measurement is flourishing in the astronomy domain. In the past years, several techniques have been developed, based on the analysis of the beam coming from a single or binary star. These techniques consist in analysing the fluctuations of intensity (scintillation) or those of the wavefront angle of arrival. To measure the C_n^2 profile in the whole atmospheric column, ONERA developed the CO-SLIDAR technique based on a Shack-Hartmann (SH) wavefront sensor [1]. The joint use of correlations of slopes and of scintillation, both measured on a binary source with a Shack-Hartmann, allows CO-SLIDAR to combine sensitivity to both low (close) and high altitude (far) layers.

In astronomical instrumentation, the knowledge of the C_n^2 profile is a key element to design and to evaluate the performance of future wide-field adaptive optics systems (WFAO) [2]. In this perspective, recent CO-SLIDAR experiments in the visible at the Observatoire de la Côte d'Azur (France) have measured C_n^2 profiles (with error bars) over 17 km of atmosphere with a resolution of 600 m. These C_n^2 profiles were compared successfully to those calculated from the re-analysis of meteorological data [3]. The CO-SLIDAR technique is of course applicable to oblique or horizontal paths provided one stays in the Rytov regime. In practice, limitations appear for horizontal 100-m paths in the surface layer where turbulence intensity is the highest, which leads to a saturation effect of scintillation indices in the visible range and thus prevents the inversion of the C_n^2 distribution. Longer distances are possible using the CO-SLIDAR in the thermal infrared domain for km paths. Thus our IR Shack-Hartmann [4] was used to provide distributed C_n^2 values along a horizontal path near the Earth surface. These C_n^2 horizontal profiles can be further used to estimate sensible and latent heat surface fluxes at parcel scale which is relevant for climate, hydrological, and agronomics studies.

As the CO-SLIDAR principle and the on-sky results obtained at Observatoire de la Côte d'Azur have been presented in [3], the proceeding will focus on the near ground results.

2. Multi-instrument campaign for near ground C_n^2 profiles

A field experiment named AMOSC was conducted at the Centre de Recherche Atmosphérique (CRA) in Lannemezan (France) to first optically calibrate CO-SLIDAR horizontal C_n^2 profiles. This experiment was conducted in fall 2012 (ONERA with French labs LTHE, INRA, CESBIO). The SH ($\lambda = 3.8$ μm) was installed on the CRA roof, aiming at a vertical double halogen source placed 2.7 km apart in the church tower of the Campistrous village. In addition, 3 scintillometers (using a diode at $\lambda = 0.9$ μm as a source) were co-aligned with this line of sight, that was 22 meters in average above ground (see Fig. 1). The CO-SLIDAR profiles were to be compared with the C_n^2 values estimated from the scintillometers. Each one had been configured with a specific configuration of apertures - with their own weighting function - to highlight different parts of the path. We got finally 21 days of SH data synchronized with the scintillometers and micro-meteorological data. Data processing is still in progress and we present first encouraging results in this paper.

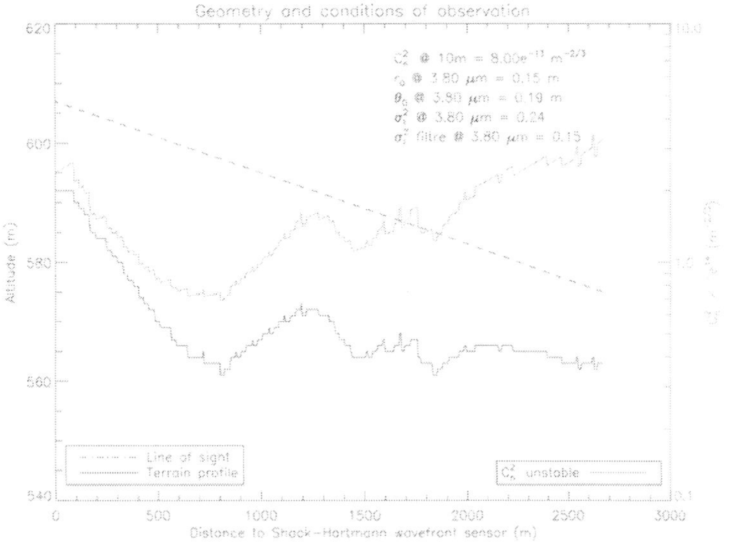

Figure 1. Terrain profile (black) and line of sight (dashed). Model of atmospheric boundary layer C_n^2 (h) profile (red) according to C_n^2 $(h_0 = 10m) \times h^{-4/3}$, corresponding to diurnal conditions (unstable), with h being the height of the line of sight. Derived turbulence parameters are given at $\lambda = 3.8$ μm.

3. CO-SLIDAR principle: direct problem, inversion and error bars

We currently only exploit correlations of x-slopes, of y-slopes and of scintillation to speed up the computing time since coupling between x and y-slopes are weaker. Correlations are averaged over all pairs of subapertures with given separation and represented as auto- or cross-correlation map as it relates to one or two sources. Then, one pixel of these six maps is stacked into a vector $\mathbf{C_{mes}}$. The latter is related to the discretized C_n^2 profile at different distances z, denoted $\mathbf{C_n^2}$, by the following linear relationship:

$$\mathbf{C_{mes}} = M\mathbf{C_n^2} + \mathbf{C_d} + \mathbf{u}. \tag{1}$$

M is the matrix of the weighting functions for slopes and for scintillation [3] derived here for a spherical wave since sources are at a finite distance. Because of detection noises, the pseudo-measurements $\mathbf{C_{mes}}$ are biased with the averaged correlations of these noises $\mathbf{C_d}$. As we estimate the correlations from a finite number of frames, $\mathbf{u}$ represents a Gaussian convergence noise.

The $\mathbf{C_n^2}$ profile is retrieved by minimizing the Maximum *A Posteriori* (MAP) metric composed of the Maximun Likelihood criterion with a spike preserving regularization $J_\mathrm{p}\left(\mathbf{C_n^2}\right)$ (white L1L2 [5]) assuming independent layers, which is relevant for horizontal profiles:

$$J_{\mathrm{MAP}}\left(\mathbf{C_n^2}\right) = \left(\mathbf{C_{mes}} - \mathbf{C_d} - M\mathbf{C_n^2}\right)^{\mathrm{T}} C_{\mathrm{conv}}^{-1}\left(\mathbf{C_{mes}} - \mathbf{C_d} - M\mathbf{C_n^2}\right) + J_\mathrm{p}\left(\mathbf{C_n^2}\right). \tag{2}$$

The covariance matrix of the convergence noise $C_{\text{conv}} = \langle \mathbf{u}\mathbf{u}^{\mathrm{T}} \rangle$, is approximated with the pseudo-measurements correlations. In addition, the diagonal values of matrix $\left(M^{\mathrm{T}} C_{\text{conv}}^{-1} M\right)^{-1}$ represent an upper bound of the variances of the $\mathbf{C_n^2}$ estimates [3], from which 3σ error bars on the reconstructed profile are readily computed.

4. Resolution in distance of the horizontal C_n^2 profiles

Since our IR Shack-Hartmann looks at a double source with an angular separation $\theta = 60$ arc second and has 5 subapertures in its pupil diameter D, we can retrieve at least 5 layers up to the maximal distance of triangulation $H_{max} = D/\theta = 1.3$ km thanks to the cross-correlations of slopes and of scintillation. Beyond the triangulation range H_{max} we will also estimate the profile from the auto-correlations of scintillation averaged on every single source and involving several subaperture baselines sampling the structure of the farthest weighting functions of scintillation (Cf. Fig. 2). Note the FWHM[1] of the 3 functions at $z = 1.7, 2.0$ and 2.3 km span over 2 subapertures allowing to exploit their mutual scintillation. In the following we will estimate the C_n^2 profiles on a grid of 8 layers regularly spaced along the path.

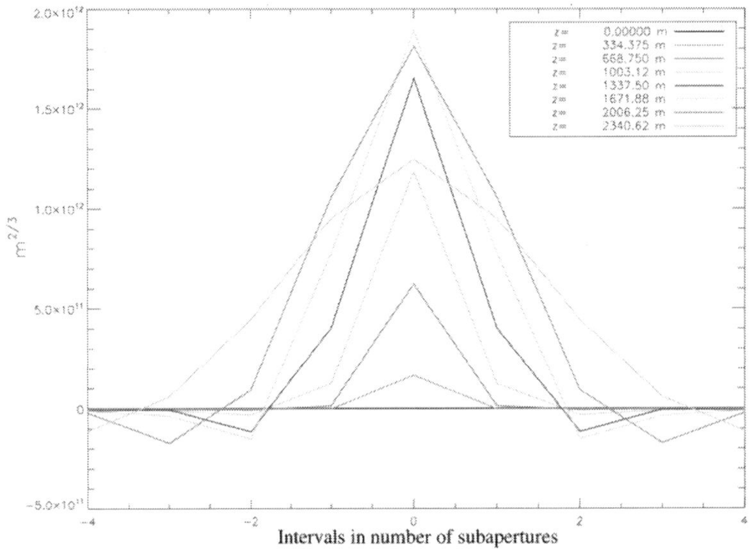

Figure 2. Cross sections of the scintillation weighting functions varying the distance z to the SH.

5. Shack-Hartmann data quality check

Before performing inversion of (1) to estimate the C_n^2 profile from the slopes and scintillations, we need to check if the data respect the model assumptions, namely Kolmogorov and the weak perturbation regime. To do so, we extract slopes and scintillation from the SH image spots (25×25 pixels) in moving windows of 7×8 pixels, and we check the turbulence statistics from the two kinds of data. From the slopes we compute the Zernike coefficient variances. It shows a Kolmogorov turbulence without noticeable outer scale effect (see Fig. 3). We validate the weak perturbation regime using intensity fluctuations, by fitting a log-normal distribution (see Fig 4) and estimating the Rytov variance. The latter was found to be far below 0.3, validating the hypothesis of weak perturbations at $\lambda = 3.8$ μm. Derived log-amplitude from intensity is in excellent agreement with the Gaussian law. The correlation maps in Fig. 5 present the expected patterns: The cross-correlation map of scintillation (Fig. 5a) and that of the slopes (Fig. 5d) shows peaks of correlation in the direction of the source alignment, representing the turbulent layers' signatures up to H_{max}. Last but not least the auto-correlation map of scintillation (Fig. 5b) covers 2 subapertures, and this signal is useful to estimate the $\mathbf{C_n^2}$ values beyond H_{max}. All these verifications confirm the data consistency with the model assumptions.

[1] Full Width at Half Maximum.

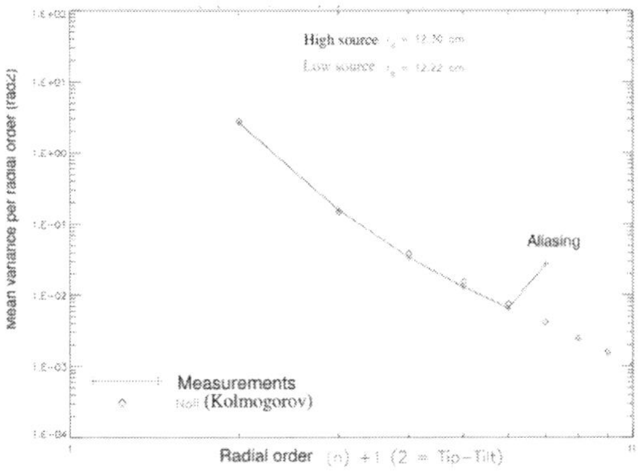

Figure 3. Mean Zernike variances over 1 min. (8520 wavefronts) on 14/09/12 at 14:19.

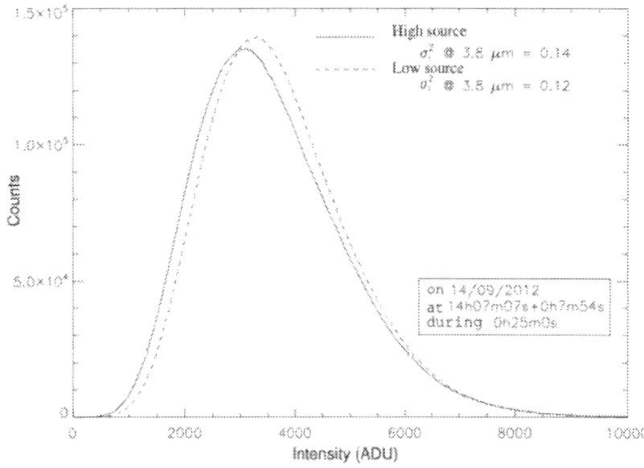

Figure 4. Log-Normal intensity distributions of the high and low sources.

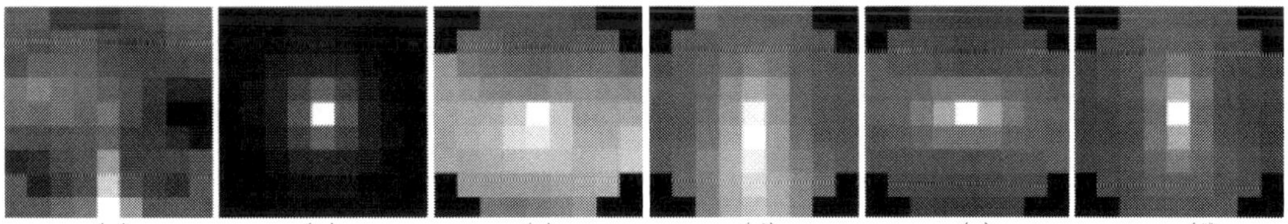

Figure 5. Correlation maps from scintillation and slopes measured with the mid-IR SH. Example of 1−min. records (8520 frames) at 14:15 on 14/09/12 . From left to right: (a) cross-correlation of scintillation between the sources, (b) auto-correlation of scintillation on each source, (c) cross-correlation (two sources) of y-slopes, (d) cross-correlation (two sources) of x-slopes, (e) auto-correlation (one source) of y-slopes, (f) auto-correlation (one source) of x-slopes.

6. Reconstruction of the horizontal C_n^2 profiles

We present on Fig. 6 one example of C_n^2 profile obtained from 1 minute SH data. The structure of the profile is clearly visible on the 8 reconstructed layers with small error bars (convergence noise). There are three significant layers with C_n^2 ranging $3-5\ 10^{-14}\ m^{-2/3}$. Two of them are placed at the sight line ends (CRA roof and the village) and one thick layer is located between 900 m and 1200 m. Very low C_n^2 values are observed elsewhere, especially the one almost null located at 1500 m. The horizontal lines show the integrated C_n^2 measured from the scintillometers (dashed) and synthesized (continuous) from the estimated C_n^2 profile. Their length illustrates the effective footprint of the scintillometers weighting functions.

CO-SLIDAR C_n^2 profiles are now compared with C_n^2 values integrated from the scintillometers. A good match is obtained in the first part of the path (magenta) but for the other two parts (black, blue) the trend is opposite between integrated C_n^2 synthesized with CO-SLIDAR and that measured from scintillometers. Even if the orders of magnitude are correct, these discrepancies require further investigations.

In any case, r_0 and scintillation indices from the CO-SLIDAR C_n^2 profiles fit well those measured directly for the sequence of a dozen minutes (see Figs. 7-8). A 4% error is obtained for r_0, while a 10% error is obtained for the scintillation. This means that unsteady conditions cannot be involved in the discrepancies mentioned. Finally, Fig. 9 shows a day and night turbulence variation as expected and a stable structure of the C_n^2 profiles.

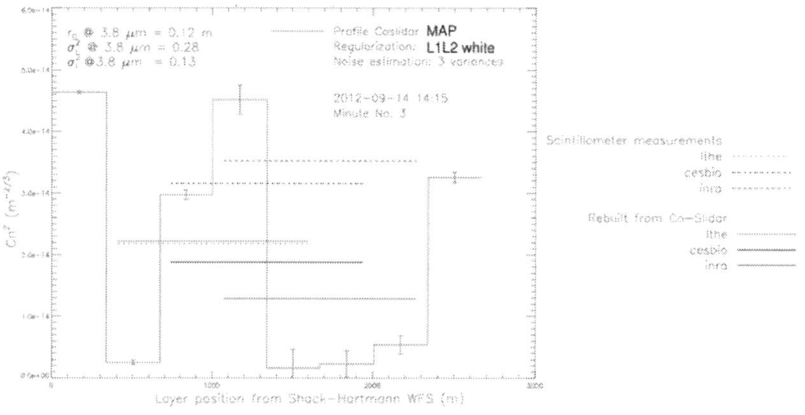

Figure 6. Comparison between C_n^2 profiles and integrated C_n^2 values from scintillometers on 14/09/12.

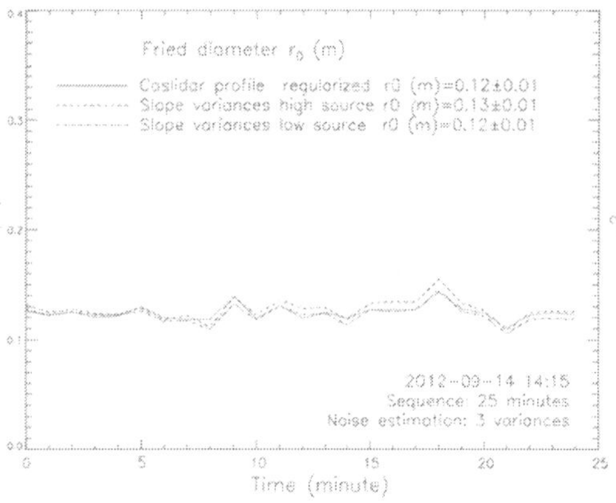

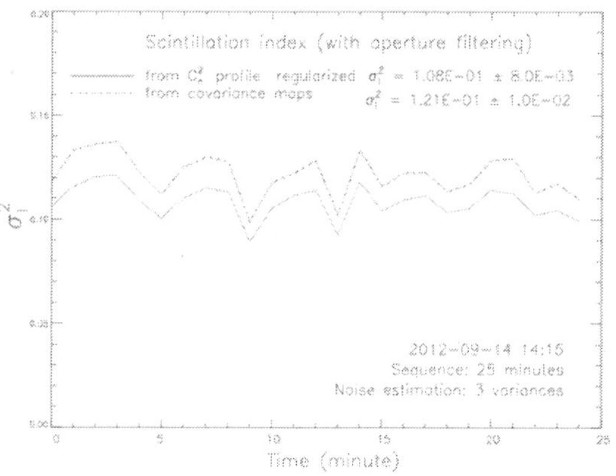

Figure 7. Fried parameter variation along time. **Figure 8.** Scintillation index variation along time.

7. Conclusion and perspectives

Near ground CO-SLIDAR in the mid-IR using a 0.35-m telescope provides C_n^2 profiles over 2.7 km with 300 m resolution. The joint use of slopes and scintillation leads to robust C_n^2 profile restoration with small error bars, even beyond the triangulation range thanks to scintillation. r_0 and scintillation indices from the CO-SLIDAR C_n^2 profiles fit well those measured directly from the slopes and scintillation variances. C_n^2 CO-SLIDAR profiles have been compared with C_n^2 values integrated from the scintillometers, both showing a day and night turbulence variation. But reverse order between both distributions require further investigations. To sort it out we have planned inter-comparison campaigns on homogeneous and restrained heterogeneous surfaces. CO-SLIDAR development is set out to build a lighter device to bring it outdoors easily and estimate in addition to the C_n^2 profile the transverse wind profile, perpendicular to the sight line. The motivation of this work is to obtain inputs for sensible and latent heat flux calculation.

CO-SLIDAR can also provide relevant inputs for site characterization, for design and performance evaluation of WFAO ELT instruments that will include multi-WFS. The need of an external vertical C_n^2 distribution to calibrate the internal low resolution C_n^2 profile is envisaged. Position with respect to other astronomical C_n^2 profilers is yet to be analyzed during inter-comparison campaigns [6]. Anyway whatever the application whether it be water management in environment or astronomic instrumentation, CO-SLIDAR C_n^2 profiles can help the forecast of atmospheric surface parameters and optical turbulence.

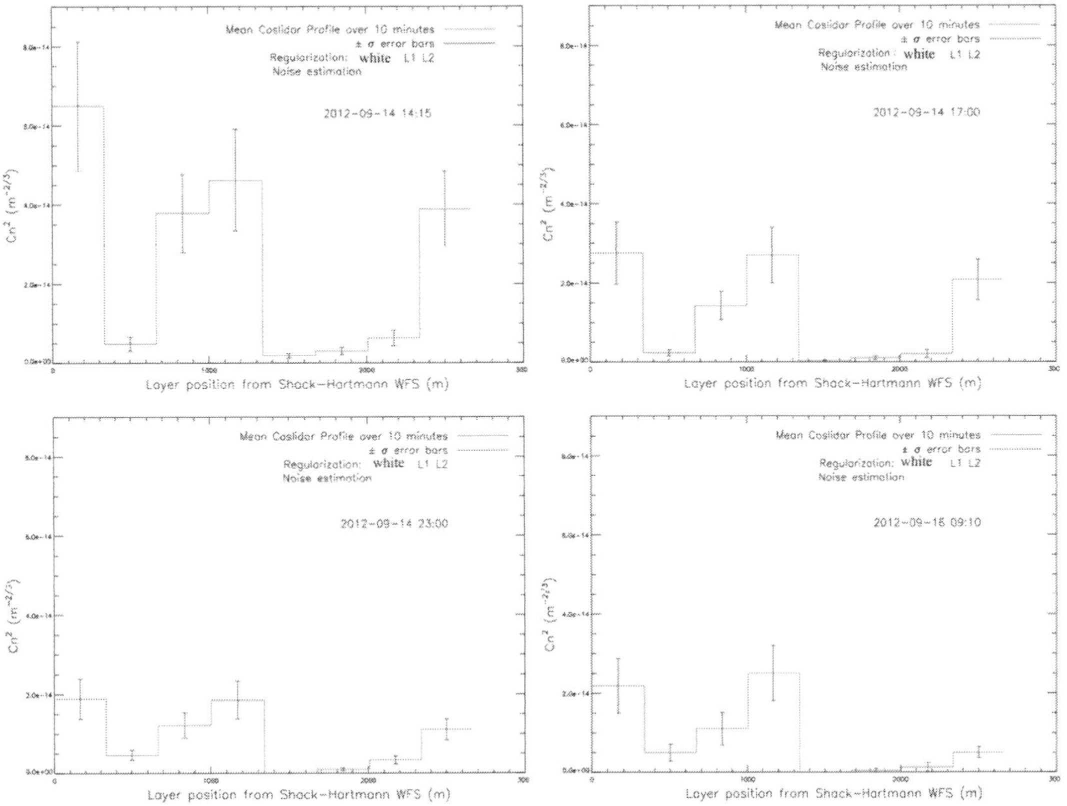

Figure 9. Mean C_n^2 profiles averaged over 10 minutes (red) and their standard deviations (blue). Variation during afternoon, evening (upper graphs), night, morning (lower graphs) on consecutive and sunny dates.

Acknowledgments
The authors wish to acknowledge assistance from PhD J Voyez and training student Y Tellier, technical work by F Mendez and B Fleury, fruitful discussions with M-T Velluet, V Michau and N Védrenne, then financial support from CNRS and ONERA. We would also like to thank J-P Lagouarde who initially proposed this study. We express our gratitude to M Irvine and A Brut for their dedicated assistance during the observations.

References
[1] Védrenne N, Michau V, Robert C and Conan J-M 2007 *Optics Letters* **32** 2659–2661
[2] Costille A and Fusco T 2011 *Second International Conference on Adaptive Optics for Extremely Large Telescopes*. vol 1 p 78 URL `http://ao4elt2.lesia.obspm.fr/spip.php?article603`
[3] Voyez J, Robert C, Conan J-M, Mugnier L M, Samain E and Ziad A 2014 *Opt. Express* **22** 10948–10967
[4] Robert C, Michau V, Fleury B, Magli S and Vial L 2012 *Opt. Express* **20** 15636–15653 URL `http://www.opticsexpress.org/abstract.cfm?URI=oe-20-14-15636`
[5] Meimon S, Mugnier L M and Le Besnerais G 2009 *J. Opt. Soc. Am. A* **26** 108–120
[6] Masciadri E, Rousset G, Fusco T, Basden A, Bonifacio P, Fuensalida J, Robert C, Sarazin M, Wilson R and Ziad A 2013 Conference date: May 2013, Florence

New generation LOLAS: Redesign of an Optical Turbulence Profiler with High Altitude-Resolution

L J Sánchez[1], R Avila[2], C A Zúñiga[2], I. Cruz-González[1], J J Tapia-Rodríguez[3], J L Avilés[4]

[1] Instituto de Astronomía, UNAM, México D.F., México
[2] Centro de Física Aplicada y Tecnología Avanzada, UNAM Campus Juriquilla, Qro., México
[3] Instituto Tecnológico de Morelia, Mich., México
[4] Instituto Nacional de Astrofísica, Óptica y Electrónica, Pue., México

E-mail: leonardo@astro.unam.mx

Abstract. We present the instrument and first results of a Low Layer SCIDAR (Scintillation Detection and Ranging), called new generation LOLAS (LOLAS 2), which consists of the optical coupling of a Ritchey-Chrétien telescope and an EMCCD camera, and allows the measurements of optical turbulence profiles in the atmosphere from ground level with high altitude-resolution. The system is designed for widely separated double star targets, which enables a high altitude resolution of $5 \leq \Delta h \leq 35$ meters above the first kilometer, measured from ground level. The first results were obtained in June and November 2013, and May 2014 at the Observatorio Astronómico Nacional de San Pedro Mártir (OAN-SPM), México.

1. Introduction

This second generation instrument design Low Layer SCIDAR 2 (LOLAS 2), which includes several improvements (Zúñiga et al. 2015) yielding an optimized performance from our prototype LOLAS (Avila et al. 2008; Chun et al. 2009; Avilés et al. 2012), allows the measurements of optical turbulence profiles in the atmosphere from ground level with high altitude-resolution.

The LOLAS concept consists of putting into practice a Generalized SCIDAR (G-SCIDAR) on a dedicated portable telescope, using widely separated double stars as light sources. Wider star separations yield better altitude resolution but shortens the maximum detection altitude.

LOLAS are designed to use a small telescope, an Electron Multiplied Charge Coupled Device (EMCCD) to improve sensitivity and a real-time computation of the scintillation autocovariance.

2. G-SCIDAR method description

The LOLAS instrument is based on the G-SCIDAR method to obtain the optical turbulence profiles $C_n^2(h)$ at an altitude h. This parameter quantifies the intensity of the turbulence fluctuations of the atmosphere refractive index n.

The G-SCIDAR is based on the detection of stellar scintillation, the variance σ_2 of which is proportional to $h^{5/6}$.

In order for the instrument to be sensitive for turbulence at $h = 0$, the detection plane is placed instead at a distance h_{gs} below the telescope conjugated pupil, which then implies that σ_2 is proportional to $|h - h_{gs}|^{5/6}$.

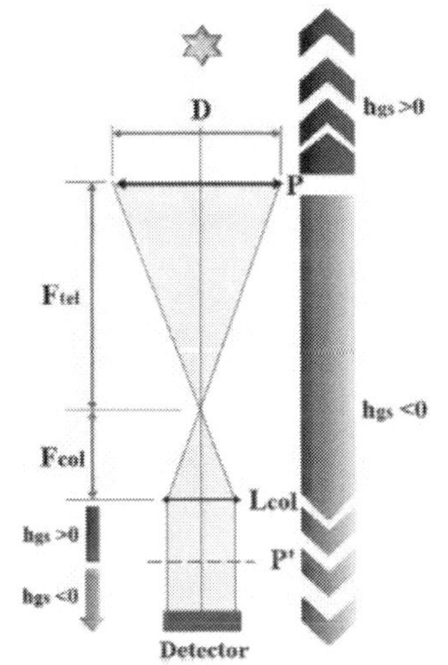

Figure 1. Optical scheme for the G-SCIDAR method.

Figure 1 shows a schematic G-SCIDAR description of the method. D and L_{col} are the telescope and the collimator diameters, and F_{tel} and F_{col} their focal distances. P' marks the telescope conjugated pupil plane. The CCD detector is located below this plane.

Figure 2. The instrument LOLAS second generation working at the Observatorio Astronómico Nacional at San Pedro Mártir, Baja California in México.

3. LOLAS prototype instrument and LOLAS 2

The components of the LOLAS prototype instrument were: a Schmidt–Cassegrain telescope 40.6 cm in diameter, an EMCCD camera, optics (two achromatic doublets), and real time data processing software (GUI).

The new generation LOLAS makes use of an RC Optical 40.6 cm $f/9$ Ritchey-Chrétien open telescope, a robust Astro-Physics German equatorial mount, an EMCCD Andor iXon camera, an SBIG guiding camera and an improved real-time data processing software (GUI).

The LOLAS 2 instrument improved software, focus stability and guiding allows to achieve an optimized performance.

4. LOLAS 2 first results

The instrument was tested at the Observatorio Astronómico Nacional at San Pedro Mártir, Baja California in México (OAN-SPM) during three observing runs June and November 2013, and May 2014. Figure 2 shows the instrument working at the OAN-SPM.

Scintillation images are captured on an active zone of 256×80 pixels of the CCD detector. Frames are binned 2 by 2, with a resultant pixel size of $32\,\mu$m.

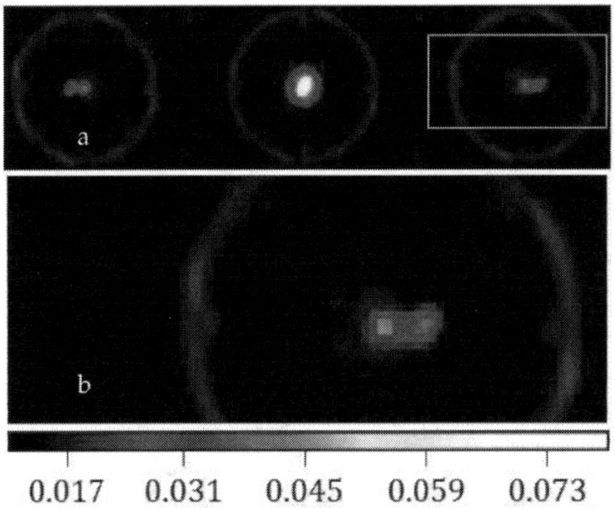

Figure 3. (a) Example of a LOLAS 2 measured autocovariance map. (b) Enlarged view showing the signature of two layers, one at the ground and one near the ground.

The autocovariance of this stochastic illumination is obtained by computing the spatial normalized autocorrelation of each image and averaging those autocorrelations over thousands of statistically independent image samples. Each layer of optical turbulence contributes to the resulting scintillation autocovariance with three covariance peaks.

The exposure time was set to 2 ms. The number of images to calculate one autocovariance was set to 30000. An example of one autocovariance map obtained with LOLAS 2 is shown in Figure 3.

The $C_n^2(h)$ profile shown in Figure 4 was obtained analyzing the autocovariance using a modified CLEAN algorithm. The source was a double star (12 Cam) with 181 arcsec separation, $h_{gs} = -1212$ m and $\lambda = 0.5\,\mu$m. For these parameters we obtained an altitude resolution of 14 m.

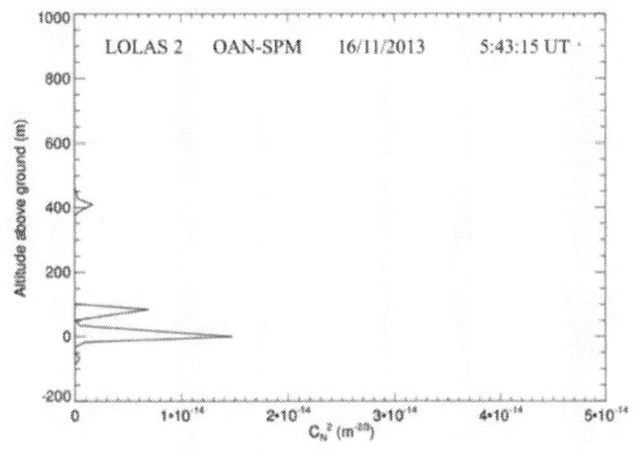

Figure 4. LOLAS 2 $C_n^2(h)$ profile corresponding to the autocovariance in Figure 3.

5. Conclusions

We designed and build a new generation LOLAS instrument with improved focus stability, guiding and software. The optimized performance of the instrument allows better measurements of optical turbulence profiles with high altitude-resolution above the first kilometer, measured from ground level.

Acknowledgments

We acknowledge the technical and administrative staff of the OAN-SPM for their support. Partial founding was provided by DGAPA (Universidad Nacional Autónoma de México) PAPIIT grants IN103913 and IN115013.

References

Avila R, Avilés J L, Wilson R W, Chun M, Butterley T and Carrasco E 2008 *MNRAS* **387-4** 1511

Avilés J L, Avila R, Butterley T, Wilson R, Chun M, Carrasco E, Farah A and Cuevas S 2012 *MNRAS* **423-1** 900

Chun M, Wilson R, Avila R, Butterley T, Avilés J L, Wier D and Begnini S 2009 *MNRAS* **394-3** 1121

Zúñiga C A, Avila R, Tapia-Rodríguez J J, Sánchez L J, Cruz-González I, Avilés J L, Valdés-Hernández O and Carrasco E 2015 *PASP* submitted

All sky scanning cloud monitor for NLOT site survey

T K Sharma*,†, Padmakar Parihar* and P M M Kemkar*

*Indian Institute of Astrophysics, Block II, Koramangala, Bangalore, Karnataka, India-560034

E-mail: †tarun@iiap.res.in

Abstract. Monitoring local sky for the cloud is one of important task before setting-up a new observatory. Here we present the design, implementation and initial results of a scanning type cloud monitor developed in Indian Institute of Astrophysics, Bangalore, India. The new cloud monitor is expected to be used in search for a potential site for India's National Large Optical Telescope project. The instrument works on the principle of detection of the Infrared radiations from the clouds. A number of thermopile sensors are arranged in the form of a circular array and are rotated in azimuth to cover the whole sky. An analog circuit was designed and fabricated to amplify the weak output of the thermopile. A customized data acquisition devise is developed for recording the output of the sensors on SD card. LabVIEW based data analysis software is developed to process raw data as well as to generate the cloud map of the sky.

1. Introduction

Availability of large number of clear nights is one of the prime requirement for any observatory and before setting-up a new observatory, this parameter needs to be reliably determined. For any existing observatory, monitoring of clouds on real time also helps to effectively use their automated telescopes. Furthermore, in case of manual operation of a telescope, observer can select from equally desirable targets based on local sky transparency. All these objectives can be met if observatory operate a very reliable automated all sky cloud monitor which regularly broadcasts local sky condition.

The traditional method for monitoring sky-condition by means of visual observation is extremely tedious and very subjective. Whereas, using a CCD camera attached with a fish-eye lens to capture whole sky in optical wavelengths is a great improvement over manual observation method. Several such devices usually called All-Sky Night Camera have been used by different groups [1, 2, 3]. A CCD camera based all sky monitor is an inexpensive tool, however, appearance of clouds in the optical images is very deceptive due to its strong dependency on apparent optical sky brightness which in turn depend on phases of the Moon, local light pollution and other atmospheric variables. Automated tools developed to derive a quantitative information of cloudiness from all sky images, usually fails to reliably extract cloud information and results are often found to be inconsistent with other methods[4].

One of the best methods of observing clouds from the ground is to make use of infrared radiation coming from cloud either due to thermal radiation of cloud itself or cloud can act like screen which reflects radiation originating from relatively warmer Earth surface. In the $8-14\mu$ mid-IR atmospheric window, clear sky appears cold and clouds appear warm. The effects of variable sky brightness becomes minimal and hence a mid-IR sensor provides the greatest cloud/sky discrimination. A mid-IR imaging camera can be best suited for this purpose,

Journal of Physics: Conference Series **595** (2015) 012032 doi:10.1088/1742-6596/595/1/012032

however it is too expensive and only few such camera have been used for the cloud monitoring [5, 6]. In attempts to provide an inexpensive cloud detector working in mid-IR range, Ashley & Jurcevic (1991) developed a cloud detector based on pyroelectric infrared element which requires mechanical chopping[7]. Subsequently Clay et.al. (1998), used infrared thermopile device for cloud detection[8]. Thermopile is very inexpensive infrared sensor, most suitable for cloud monitoring without cooling the systems and requires simple detector electronics. Both cloud monitors mentioned above uses single detector with limited field of view and can only detect cloud present in certain direction.

In Indian Institute of Astrophysics we have initiated a program to develop site survey instruments. The instruments are developed as a part of the site survey program for the NLOT (National Large Optical Telescope) project, which will be an 8−10 meter class state of the art telescope. The telescope will be installed at some suitable site in the country. All Sky Scanning Cloud Monitor is one of the instrument what we have developed for the purpose of surveying potential sites. The device has got high sensitivity, provides radiometric output, cover whole sky, works in fully automated mode and also very cost effective.

2. Operating principle

Measurements of the sky infrared radiation above 5 μm are sensitive to the presence of the clouds in particular between 8 to 14 μm[9]. Presence of clouds produces an enhanced signal compared to the clear sky. The signal also has an effect of atmospheric humidity, and gets enhanced at lower elevations because sensor has to look through increased air−mass or more water vapor. The device works on the principle of measurement of sky temperature for detection of clouds.

Thermopile is device which is composed of a large number of thermocouples connected in series. The device works on the principle of Seebeck effect and converts the thermal energy to electrical energy, when its two junctions are subjected to a temperature difference. The two junctions are called measuring and the reference junctions. A thermopile sensor generates a voltage proportional to the temperature difference between the two junctions, which means thermopile can be used only for differential temperature measurement. Absolute temperature measurement requires output of the thermopile sensor be corrected for changes in the reference junction temperature. This correction is called the ambient temperature correction.

TPS334L5.5 thermopile sensor from Perkin Elmer is used for measuring sky temperature. Sensor comes with an inbuilt IR lens which gives it a field of view of 7° and has a flat pass-band from 2 to 22 μm. The sensor has a sensitive area of 0.7×0.7 mm^2, and high sensitivity of 55 V/W, and a small noise voltage of 35 $nv/\sqrt{Hz}$.

Figure 1. Laboratory picture of the cloud monitor instrument.

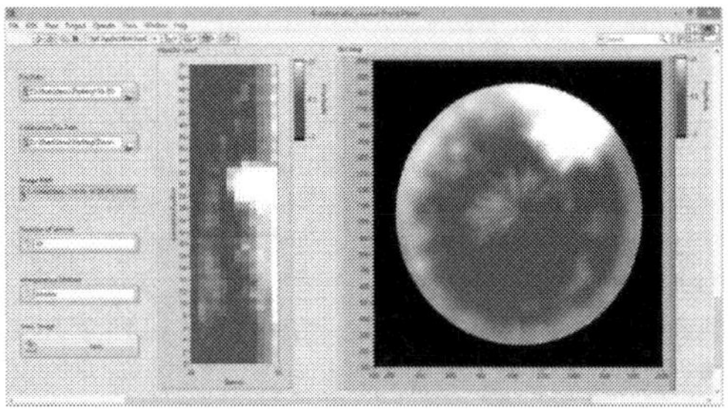

Figure 2. A screen captured image of the front panel of software developed using LabVIEW.

Figure 3. Sensor interface board.

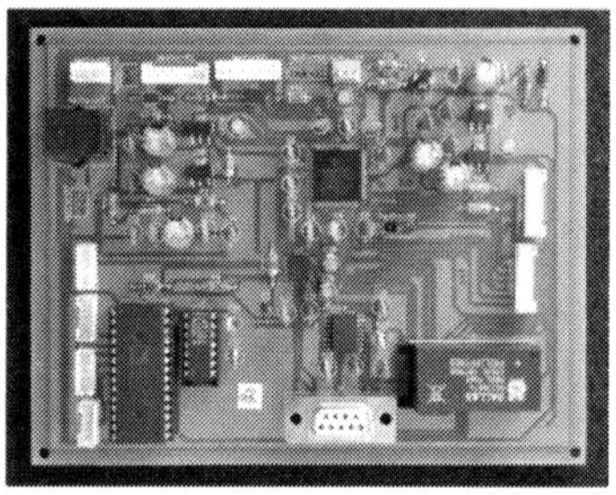

Figure 4. Data acquisition and logging board.

3. Instrument structure

Instrument structure is shown in figure 1 . Each Thermopile sensor has a field of view of 7°, which means single sensor can cover only a small part of the sky. 12 sensor units are mounted on the circular arc, which covers 84 ° of the sky. This scanning unit is rotated around azimuth in steps of 7° to cover the whole sky. The scanning unit is rotated by a stepper motor and timer belt–pulley based drive. All the electronics like data acquisition and logging board, SD card module, stepper motor driver and optical isolation boards are mounted inside the hexagonal box structure of the instrument. A Hall effect based home and limit switch is also provided to determine the initial position of the scanning unit.

4. Instrument electronics

Different electronic modules of the instrument like sensor interface board, data acquisition and logging board, optical isolation board, stepper motor driver and limit switch interface ware designed and fabricated locally. Two important boards are described in this section.

4.1. Sensor interface board

Thermopile generate a voltage output which is proportional to the temperature difference between the two junctions of the sensor. This makes it necessary to either operate the sensor at a constant ambient temperature or to compensate the sensor output for changes in ambient temperature. Figure 3 shows the sensor interface board. We have modified the circuit used by Clay et.al. (1998) for their single channel cloud monitor[8]. To minimize noise as well as EMI, sensor is mounted on the board itself. The board amplifies the weak output of the thermopile, it can also compensate the thermopile output for changes in ambient temperature.

4.2. Instrument control and Data acquisition board

A customized network based data acquisition and logging board is developed for the instrument[10]. Figure 4 shows the board with all the components mounted. The board also control all the functionalities of the instrument like initialization, homing, movement of scanning unit, data acquisition and logging. A high end micro-controller PIC18F97J60 from Microchip is used as the main processor. An external ADC MAX1144 converts analog output of the sensor to digital and this data is stored on an SD card. On board RTC unit provides the time information which is required for time-stamping the data. Data can also be transmitted to remote server using either serial or Ethernet connectivity.

5. Sensor calibration and ambient temperature compensation

Multiple sensors are used in the instrument to cover the whole sky. No two sensors give same response when subjected to the same excitation source. Also the sensor interface boards used to amplify the weak output of the sensors do not provide same gain characteristic. The main reason being the different offset voltages in each op–amp and the tolerances of the gain providing components. This makes it necessary to calibrate each sensor board.

The output voltage of the thermopile is a function of the object temperature and the ambient temperature i.e. the temperature of the reference junction. This relation can be expressed as

$$V = K.F(T_{obj}, T_{amb}) \tag{1}$$

Where V is output voltage of thermopile, K is a constant, T_{obj} and T_{amb} are object and ambient temperature respectively.

Equation can be expanded to include the calibration temperature T_{ref}, which can be any arbitrary temperature at which sensors are calibrated. Expanded equation can be written as

$$V = K.F(T_{obj}, T_{ref}) - K.F(T_{amb}, T_{ref}) \tag{2}$$

The function $F(T_1, T_2)$ is obtained by observing the output of the thermopile sensor which is kept at constant ambient temperature and is looking at an object whose temperature is changed in steps. Equation 2 establishes the relation between the output voltage of thermopile sensor, object temperature and the ambient temperature. For a fixed ambient temperature we can use an equation or a lookup table to deduce the object temperature from the output voltage of thermopile sensor. But if the ambient temperature itself varies, the output voltage will vary for the fixed object temperature. This means we have to correct the output of the thermopile sensor for changes in the ambient temperature. We have implemented the software based or digital ambient temperature compensation scheme. In this scheme uncompensated output of thermopile and the thermistor output are read, and all calculations are performed by the software to measure the actual object temperature. The actual procedure of this compensation scheme is discussed in section 5.3.

5.1. Thermistor calibration

Thermopile sensors are equipped with an inbuilt thermistor, which can be used for ambient temperature compensation. This thermistor is in contact with the reference junction and measures the ambient temperature. Two parameters $\beta - value$ and the resistance R_0 at temperature T_0 are used to characterize the thermistor. Thermistor used in our sensor has a $\beta - value$ of 3964 and value of resistance at 25 °C is 30 KΩ. Resistance of the thermistor at any other temperature can be expressed as

$$R = R_0.e^{\beta(1/T - 1/T_0)} \tag{3}$$

The thermistor is placed in series with another resistor of fixed value of 30 KΩ, and this combination is biased with a voltage of 1 V. The bias voltage is kept at minimum to avoid the self-heating effect in thermistor. Voltage output is taken at the other resistor and is fed to the op-amp with gain of 2. This voltage output is read by the data acquisition board, and is converted back to the equivalent ambient temperature by the high level software.

5.2. Thermopile calibration

Thermopile sensor is calibrated by exposing the thermopile to an object whose temperature can be changed, while ambient temperature remains constant. The output voltage is measured for different object temperatures. For small temperature range the thermopile response is linear with constant sensitivity. But this is not the case for wide range, also results obtained using constant sensitivity are less accurate.

Journal of Physics: Conference Series **595** (2015) 012032 doi:10.1088/1742-6596/595/1/012032

We have designed a setup for calibrating the sensors. The setup consists of a Peltier element, which acts as an object whose temperature can be varied as desired. This Peltier cooler is mounted on a heat sink. Sensor module is placed near to the Peltier element in such a way so that the whole sensor field of view is covered by the Peltier element, and no radiation falls on the sensor from the nearby objects. One thermistor is placed in close contact with the front surface of the Peltier element to measure its temperature. Heat sink compound was used to keep this thermistor in its place to have better thermal coupling between the thermistor and the Peltier element, so that the thermistor measures the correct object temperature. Using this setup, we were able to calibrate all the sensors from -8 degree to 50 degree centigrade. the response of the sensor is found to be quadratic in nature and can be expressed as

$$V = I + B_1 T + B_2 T^2 \tag{4}$$

Calibration procedure gives separate values of I, B_1, B_2 and calibration temperature for each sensor. All these values are supplied to the software in a calibration file.

5.3. Ambient Temperature Compensation

Followings are the steps, which are followed in software based ambient temperature compensation

(i) Sensor is calibrated at fixed ambient temperature.
(ii) A relation is obtained between the output voltage of thermopile and the object temperature.
(iii) The uncompensated thermopile output voltage is measured.
(iv) Thermistor output is measured and converted to the ambient temperature.
(v) The calibration temperature is put in the equation 4.
(vi) Present ambient temperature is put in the equation 4.
(vii) A difference is taken and this is added to the uncompensated thermopile output voltage.
(viii) Now the object temperature is calculated back by putting the compensated voltage in the equation 4 and solving for the temperature.

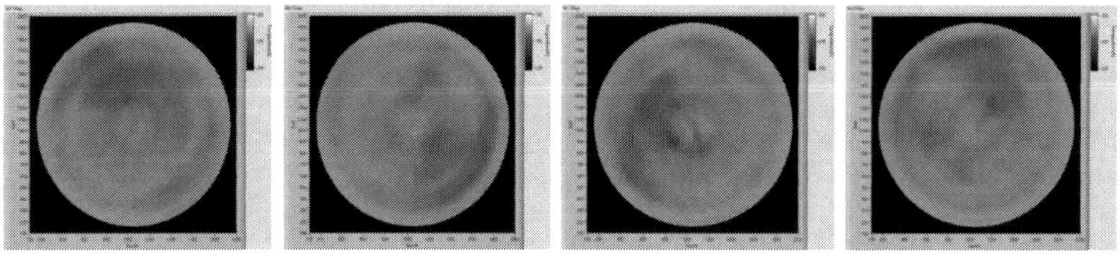

(a) 19-39-24 (b) 19-49-57 (c) 20-13-6 (d) 20-36-17

Figure 5. Sky maps generated by the instrument.

6. Instrument Software

Two different software's are developed for the instrument. These are:– instrument control and data logging software, and the data analysis software. Instrument control and data logging software is the low level software, which directly controls the hardware. The software sits inside the data acquisition and logging board and is developed using embedded C language from Microchip. The software perform tasks like controlling stepper motor for operating scanning unit, operation of analog multiplexer, analog to digital conversion of sensor outputs, storage of data onto SD card and finally providing serial and Ethernet connectivity to the instrument. A LabVIEW based data analysis software is developed. Figure 2 shows the snapshot of the software with sky map. This software reads the raw data from a file called Data File generated by the instrument, and converts it to the corresponding sky map. The software performs different operation like converting the ADC numbers to the corresponding voltages, compensating these voltages for ambient temperature variations, calculating the sky temperature from the compensated voltage, and finally generating the sky map. Other than data file, one more file called Calibration File is also used by the software. This file keeps all the calibration coefficients and the calibration temperature for each sensor.

7. Results and discussion

Before proceeding with the full scanning system, individual sensors are tested for sensitivity to the clouds. A setup was made consisting of two sensor modules and a webcam looking in the same direction. Data was acquired continuously using NI USB-6008 data acquisition module, and images were stored. Data was collected over the period of one month. A direct relationship was found between the voltage level of the sensor output and the percentage cloud cover in the sensor field of view. Also the outputs of the two sensors showed almost similar signal pattern, suggesting similar behavior for same kind of excitation. Hardware based ambient temperature compensation was used for this test.

Careful calibration procedure combined with the software based ambient temperature compensation scheme provides the temperature measurement accuracy of $\pm1°$. This kind of measurement accuracy is quite good considering the sky temperature for the clear sky ($-60°$ to -40°) and the clouds (-20° to 10°) at IAO Hanle, Ladakh. Before installation at Hanle, instrument was tested in IIA Bangalore. As evident in figure 2, The temperature of the cloudy and clear sky are found to be about 20° and $-5°$. The cloud temperature was found to depend on the thickness of the cloud, thicker the cloud higher the temperature is measured. The reason for the higher observed sky temperature at Bangalore is primarily due to increased airmass and higher water vapor content in the local atmosphere. Figure 5 shows few sky maps generated by instrument at Hanle along with time and temperature scale. The bottom and left side of each frame indicates the north and east directions.

While operating the device at IAO Hanle, we found that at higher wind-speed, scanning unit of device start shaking and also gets rotated, which smears out the sky map. The primary reason being the use of simple Timer belt−pulley based drive. We are working on improving the instrument structure and drive mechanism, while keeping the electronics same. The new design will also protect the instrument in case of rain and snowfall. Data analysis software will be improved to provide the percentage cloud cover. Before we start regular use of the device, a thorough comparison will be made between the results generated by the IR scanning cloud monitor and CCD based all sky cloud monitor.

References

[1] Pereira W E 2003 PhDT
[2] Pickering T E 2006 *SPIE* **6267** 62671*A*
[3] Skidmore W, Riddle R, Schöck M, Travouillon T, Els S, Walker D and Magnier E 2011 *RMxAC* **41** 70 – 3
[4] Shamir L and Nemiroff R J 2005 *PASP* **117** 972 – 977
[5] Suganuma M, Kobayashi Y, Okada N, Minezaki T, Aoki T, Enya K, Tomita H and Koshida S 2007 *PASP* **119** 567 – 82
[6] Sebag J, Andrew J, Klebe D and Blatherwick R D 2010 *SPIE* **7733** 773348
[7] Ashley M C B and Jurcevic J S 1991*Publ. Astron. Soc. Aust.* **9** 334 – 35
[8] Clay R W, Wild N R, Bird D J, Dawson B R, Johnston M, Patrick R and Swell A 1998 *Publ. Astron. Soc. Aust.* **15** 332 – 5
[9] Sloan R, Shaw J H and Williams D 1955 *J. Opt. Soc. Am.* **45** 455
[10] Sharma T K and Parihar P S 2014 *International Journal of Engineering* **8** 22 – 29

The Thirty Meter Telescope Site Conditions Monitoring System

Warren Skidmore & Tony Travouillon

Thirty Meter Telescope, 1111 South Arroyo Parkway, Suite 200, Pasadena, CA 91105, USA

E-mail: was@tmt.org

Abstract. We examine the experiences and ideas from operating observatories regarding the measurements of the characteristics of the atmosphere that must be gathered within the locality of the observatory in order to support safe, efficient and scientifically optimized observatory operations as well as commissioning, performance monitoring and support the scientific analysis of telescope observations. We describe the expected requirements for the measurement capabilities of the the TMT Site Conditions Monitoring System (SCMS) and discuss how these plans are being developed with input from staff at operating observatories and active observational astronomers.

1. Introduction

The Thirty Meter Telescope Project moved into the construction phase in May 2014. The nominal time for the end of construction and beginning of early science operations and full commissioning is 2024. At the end of construction the present plan calls for a multiple laser guide star system supporting a fully functioning MCAO system. The MCAO system will feed a dual channel diffraction limited NIR imager and IFU instrument and an AO enhanced NIR multi-object spectrograph. An optical seeing limited wide field optical spectrograph will be available. Both classical and queue scheduled observing modes will be provided. Commissioning will include optimizing the AO system, developing observing and data reduction procedures and optimizing the image quality through the use of the enclosure vents.

The present plan for the SCMS is based on the TMT site testing system [2] (see Section 2). However a system to evaluate a site for an observatory has different requirements to a system that supports observatory operations and commissioning. It is important to explore the requirements for the TMT SCMS.

The TMT SCMS will work in a complimentary way with the Engineering Sensor (ESEN) system. The ESEN consists of many air and structural temperature probes, sonic anemometers, precipitation and humidity sensors, accelerometers to measure vibrations and tilt meters to measure structural deformations on and around the telescope.

2. Present Requirements for the SCMS

The TMT SCMS is required to support all of the commissioning and operations activities as well as provide input for local conditions forecasting and provide astronomers with information on local conditions during their observations to enable treatment of systematic effects in their astronomical data that may limit interpretation.

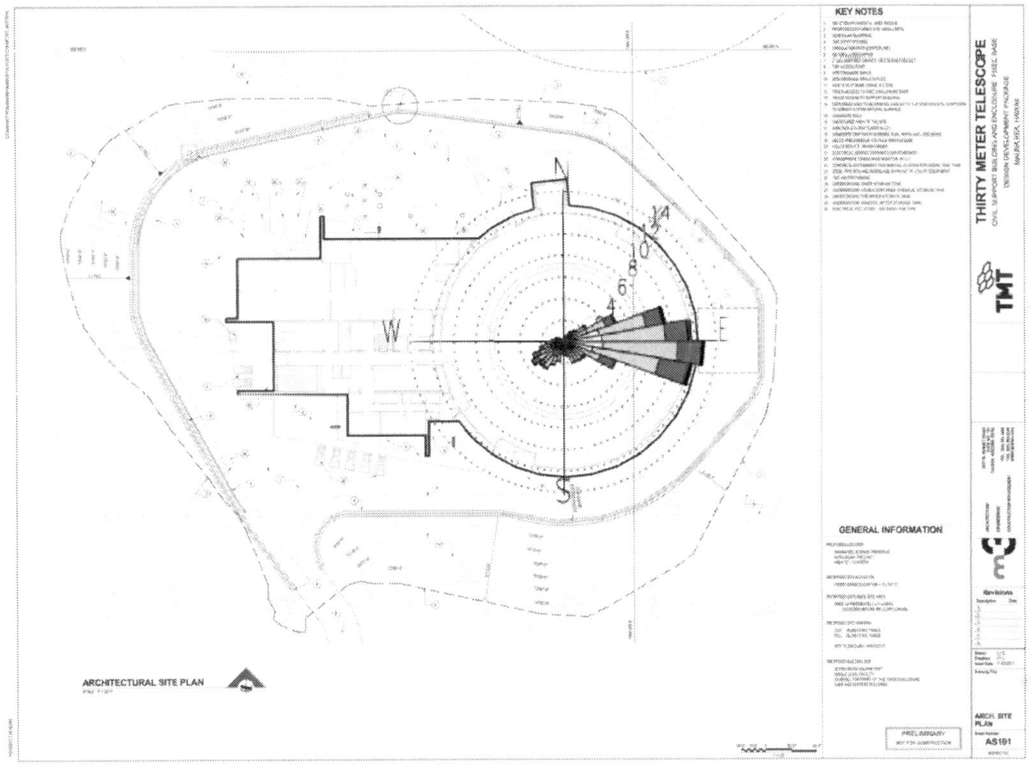

Figure 1. Plan of the TMT observatory at the its on Maunakea, showing the location of the SCMS (circled near top middle) with the measured wind rose super-imposed over the location of the telescope pintle bearing.

The location of the main part of the SCMS will be quite close to the observatory, see Figure 1 and the validity of local seeing measurements will be significantly reduced. The main components of the site testing system were a MASS/DIMM to monitor the atmospheric turbulence profile and integrated seeing, an All-Sky Camera (ASCA) for cloud and light pollution measurements [4], an Automated Weather Station (AWS) with a full suite of typical meteorological instrumentation, Sonic Anemometers at many locations around the observatory and an Infra-Red radiometer (IRMA) device to measure the PWV [1]. Sonic anemometers have been baselined as both wind and optical turbulence sensors [5] to be located around the telescope enclosure vents, working with sonic anemometers located on the telescope itself so that the optimal vent configurations can be derived to give the best image quality for a given set of external wind and temperature conditions and to help diagnose image quality problems that may or may not be related to wind flow through the enclosure.

3. Re-examining the SCMS Requirements

3.1. Information Gathering
A lot of communications were carried with operations staff at the Gemini North, Paranal, IRTF, CFHT and Keck observatories, with astronomers that use AO corrected images for extremely precise astrometric measurements, mid-IR spectroscopic and seeing limited optical observations and with meteorologists that make forecasts of the astronomical observing conditions for Maunakea. All were asked what types of atmospheric measurements are required for their specific needs. Where appropriate specific requirements were extrapolated or adjusted for the TMT.

3.2. Telescope Operations
In telescope operations the SCMS measurements would be used for optimizing queue scheduled observations, e.g to decide whether to follow or change the plan of scheduled observations. To do this would require the measurements listed in Table 1. Also listed in Table 1 are the measurements needed to ensure safe telescope operations and to support the segment alignment process because segment alignment works best during periods of good seeing.

Table 1. List of SCMS measurements needed for telescope operations, commissioning and performance monitoring and supporting scientific analysis.

Queue implementation and forecasting support	Clouds, transparency, sky brightness, PWV, seeing Turbulence profile if using AM2 Wind speed and direction
Optimized telescope performance (wind shake, vents, pointing)	Wind speed and direction
APS operations (M1 alignment)	R_0 and seeing
Telescope safety	Fog, dust, wind speed and direction, humidity Dew point (air pressure, temperature, humidity) Solar radiation
MCAO performance estimation and monitoring	R_0, seeing, turbulence profile Wind speed and direction
Seeing limited instrument performance evaluation	Transparency, cloud cover, seeing
Scientific interpretation	Wind speed and direction, air temperature, air pressure Sky brightness, PWV, seeing, transparency Turbulence profile

Another important aspect of SCMS support of operations is the supporting of conditions forecasting by providing information on observing and atmospheric conditions to the enable anchoring of forecast models and fine tuning. The Mauna Kea Weather Center carries out this work and continually seeks to improve the quality of forecasts and respond to the needs of the observatories on Maunakea.

3.3. Commissioning and Performance Monitoring
Observatory commissioning must be carried out whilst ensuring the safety of the telescope and other equipment. A good understanding of the atmospheric characteristics will greatly enhance

the ability to isolate and diagnose problems that all ultimately affect image quality. Understanding the fundamental expected image quality due to the atmosphere and being able to identify the cause of any degradation, be it due to incorrect AO correction, wind induced image jitter or primary mirror segment mis-alignment will be a powerful utility during the commissioning period. Table 1 lists the measurements needed during commissioning.

There was no identified need for high fidelity wind and optical turbulence measurements in the telescope enclosure. This is due to existing telescopes not needing such measurements, however there is uncertainty about whether this can scale up to an observatory the size of the TMT. The baseline plan for the TMT SCMS and Engineering Sensors (ESEN) proposes for sonic anemometers to be located in the enclosure vents and on the telescope structure and requires that the communications and power necessary to support such an array of sensors is in place.

3.4. Science Analysis

There are general types of measurements that are often considered by astronomers when interpreting their observations, these are listed in Table 1. However a common experience is that the scientific interpretation is greatly assisted with the availability of AO telemetry, telescope control system and Engineering Sensor (ESEN) measurements. It should be noted that the utility of measuring the turbulence profile is minimal because of the availability of AO telemetry and dominance of optical aberrations in the instrument over any smaller AO related effects.

Here we present two examples of sets of observations whose analysis and interpretation is limited by effects that can be better understood with appropriate measurements from an SCMS in the first case and from engineering data in both cases. This is done to illustrate that the SCMS must be considered part of a suite of sources for observatory related information that may be needed to maximize the scientific output of the observatory.

3.4.1. Uncorrected artifacts in diffraction limited images

The quality of correction of an AO system varies with time, on the most rapid timescales this is purely due to limitations in the wavefront sensing and correction abilities of the system. Some AO systems introduce long lived features that mimic diffraction patterns. When bright, these patterns are easy to identify but if two or more faint patterns below the detection limit coincide then they may create a false detectable 'source' in the image.

Identification of transient features in AO corrected images is a major limitation. Figure 2 shows the Milky Way Galactic Center and the G2 object that both moved and showed large variations in brightness. Discerning between variable artifacts and the combination of real motion and brightness variations was a very difficult task.

AO telemetry is extremely useful for discerning between real transient sources and variable artifacts. Also the number of artifacts has some dependence on seeing. Thus a combination of SCMS seeing measurements and AO telemetry is needed.

3.4.2. 1.27s oscillation in rapid Keck spectroscopy

Telescope control systems are not perfect and can introduce image motion. Figure 3 shows rapid t_{samp}=72ms optical spectroscopic observations of the Cataclysmic Variable AE Aqr that were folded on the 33 second spin period of the magnetic white dwarf [3]. During the analysis of the observations relatively large oscillations were found at a period of 1.27s in both the wavelength dispersion direction (seen in variations in wavelength of static atmospheric absorption lines) and the temporal direction (seen in variations in the flux levels). These 1.27s

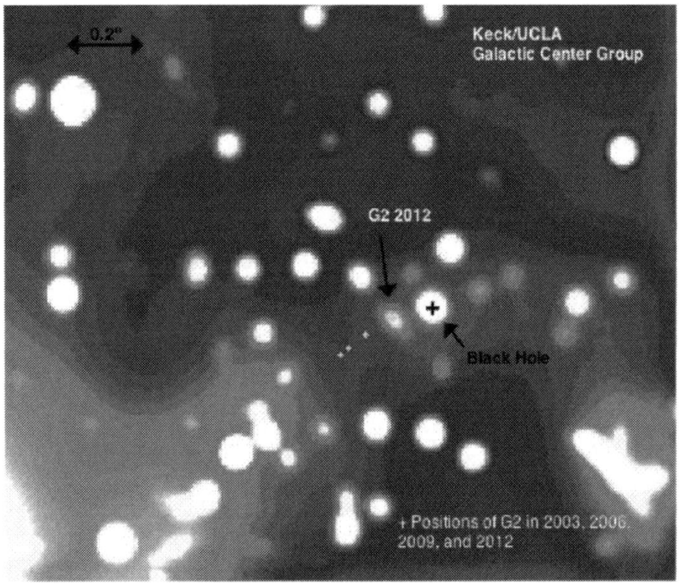

Figure 2. The G2 object and its path toward the SgrA* SMBH at the center of the Milky Way. Identifying the G2 object along its path was greatly complicated by the occurrence of false positive sources created by mis-corrections from the AO system and the G2's intrinsic variability.

variations corrupted and masked the very subtle intrinsic signals in the data. However the same 1.27s variation was seen in other objects observed during the same observing run and enquiries with the observatory engineers identified the 1.27s period as being related to an instability or resonance in the telescope drive system. Applying appropriate corrections to the data to cancel the 1.27s signal allowed the signature of the white dwarf rotation to be extracted.

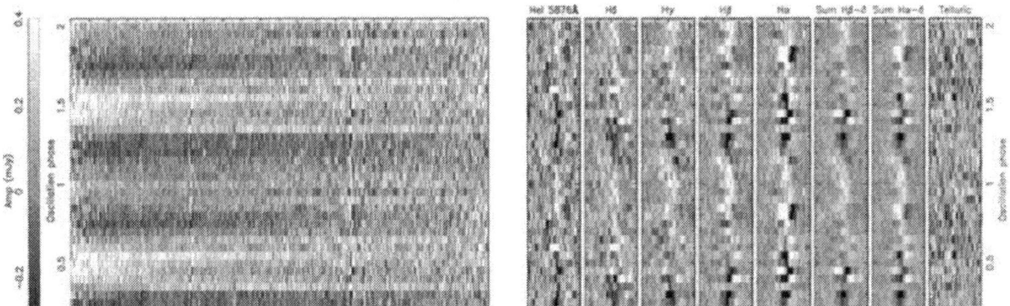

Figure 3. Trailed spectrograms of AE Aqr folded into 20 phase bins on 33 second white dwarf spin period and repeated for clarity. The left image shows the continuum emission with the pulsation amplitude rising toward the blue and the major peak occurring at phase 0.5, the minor at phase 1. The right panel zooms in on the emission lines and S curve type signatures can be clearly seen.

4. Conclusions

Clearly some form of an SCMS is a critical component for telescope operations. An important aspect that was identified was the need for engineering data that would be used in concert with site conditions measurements for all operations, commissioning and science analysis activities. Engineering data would include but isn't limited to AO telemetry, observatory engineering sensors and control system telemetry. Interestingly, the need for optical turbulence profiles was deemed 'interesting' but not of high importance, for example, for astrometric measurements of AO corrected images, instrumental distortion mapping will dominate over systematic AO correction limitations and also there is not a strong need for turbulence profiles during operations unless an adaptable secondary mirror is being used for ground layer AO corrections. Seeing and wind speed and direction measurements are used for many different aspects and are thus high priority measurements for an SCMS.

5. Acknowledgements

Many members of staff at operating observatories and active observational astronomers have provided input to help guide the development of technical requirements for the TMT SCMS.

The TMT Project gratefully acknowledges the support of the TMT partner institutions. They are the Association of Canadian Universities for Research in Astronomy (ACURA), the California Institute of Technology and the University of California. This work was supported as well by the Gordon and Betty Moore Foundation, the Canada Foundation for Innovation, the Ontario Ministry of Research and Innovation, the National Research Council of Canada, the Natural Sciences and Engineering Research Council of Canada, the British Columbia Knowledge Development Fund, the Association of Universities for Research in Astronomy (AURA) and the U.S. National Science Foundation.

References

[1] Otárola, A., et al., 2010, PASP,122, 470
[2] Schöck, M., et al., 2009, PASP, 121, 384
[3] Skidmore, W., et al., 2003, MNRAS, 338, 1057
[4] Skidmore, W., et al., 2011, RevMexAA, 41, 70
[5] Travouillon, T., et al., 2015, this proceedings

Ukaliq: Seeing Long-Term with Small, Precise Arctic Telescopes

Eric Steinbring, Brian Leckie & Rick Murowinski

National Research Council Canada, Herzberg Astronomy and Astrophysics, 5071 West Saanich Road, Victoria, British Columbia, Canada V9E 2E7

E-mail: `Eric.Steinbring@nrc-cnrc.gc.ca`

Abstract. Time-domain astrophysics benefits from extreme-latitude sites, which can combine intrinsically extended nighttime with good sky conditions. One such location is the Polar Environment Atmospheric Research Laboratory (PEARL), at 80° North latitude, on the northwestern edge of Ellesmere Island, Canada. Experience gained deploying seeing monitors there has been incorporated into an automated system called "Ukaliq" after the common arctic hare, which is also very well suited to its local environment. Even with modest aperture, high photometric reliability may be achieved using simple adaptive optics together with observing strategies that best fit the unique set of advantages available at PEARL: excellent image quality maintained during many clear, calm, dark periods of 100 hours or more. A potential multi-year search for gravitational microlensing of quasars with Ukaliq helps illustrate this niche in the era of large wide-field survey facilities.

1. Introduction

High-precision optical/near-infrared photometry uninterrupted for many hours to days, and reliably repeated over months to years is a special observing cadence for which polar sites can provide a distinct advantage. This paper presents a concept that could allow a cluster of identical optical telescopes of up to 0.5 m aperture, for both site testing and science, to be deployed in the High Arctic. The combined aperture would be equivalent to a 1.2 m telescope. We call the unit systems "Ukaliq", the Inuktitut word for the abundant arctic hare. Included are adaptations specialized to the atmospheric statistics, particularly cloud cover (ice crystals) and boundary-layer seeing, which depend on local wind conditions. To show the good fit, an instructive example of dedicating one unit to monitoring quasar microlensing events is provided.

2. Site Environment

Near the Eureka weatherstation at 80° North latitude on Ellesmere Island, Canada, is the Polar Environment Atmospheric Research Laboratory (PEARL). It was designed for atmospheric studies with optical instruments and has a purpose-built observing platform on its flat roof: 7 m wide by 18 m across, with an external stairwell. Over the last 5 years this has supported astronomical observations, with new ones planned [1], possible in part thanks to the Canadian Network for Detection of Atmospheric Change (CANDAC) which maintains other instruments there. The observatory is situated on a 600 m-high ridge, accessed by 4x4 trucks along a 15 km-long road from the sea-level base facility. That is operated by the federal weather service of Environment Canada (EC) and serviced by an all-weather airstrip and a yearly resupply ship.

All-sky camera images obtained from PEARL over several winters show no clouds overhead 48% of the time, 68% was clear (extinction in $V < 0.5$ mag) with a mean duration near 100 hours, and 84% was deemed usable ($V < 2$ mag) [2]. Air temperatures are typically −30 C with light winds and often subject to suspended ice crystals, sometimes referred to as "diamond dust". These periods are interrupted by brief, but sometimes intense, storms. Winds can gust to over 30 m s^{-1}, usually from the north or south. Free seeing measured during several roughly week-long winter campaigns with Multi-Aperture Scintillation Sensor (MASS) had a mode near 0.23 arcsec and a median of 0.50 arcsec. It is possibly poorer when winds are completely calm. The total seeing measured from 8 m elevation with MASS combined with Differential Image Motion Monitors (MASS/DIMM) on the roof had a median of 0.76 arcsec. Seeing is best near the median ground windspeed of 4 m s^{-1}, degrading with higher windspeeds [3].

3. Instrument Heritage and Design

The Ukaliq approach is based on experience gained deploying MASS and DIMMs at PEARL using Meade 25 cm and 35 cm LX200 ACF optical tube assemblies (OTAs). A larger, 50 cm version is also available from Meade. These are relatively inexpensive, lightweight OTAs easily handled by a crew of one to three people. All are f/8 Schmidt-Cassegrain (or Ritchey-Chretien) OTAs with a field corrector lens at the aperture into which is mounted a central secondary (with approximately 20% obscuration fraction). A sealed OTA has the nice property of keeping ice crystals (snow) from entering the tube. A downside is that this sometimes "sticky" and optically thick contaminant can then build up along with a thin layer of frost.

In MASS and DIMM operation, the OTA was almost always pointed at Polaris, that is, within 10° of zenith from PEARL, which may exacerbate the problem in two ways. First, the aperture corrector plate is pointed almost straight up. Even a long dew shield or insulated baffle would not be useful in completely avoiding radiation of that surface directly to the cold sky. And second, precipitation of diamond dust is then falling perpendicular to the surface. Both situations might be improved somewhat for telescope positions at lower elevations, e.g. a grazing incidence of falling crystals, but neither would be entirely avoided.

Although not as critical for DIMM, measurements with MASS require strict control over the clarity of the aperture window, and image quality (IQ). These parameters were well monitored, in part because Polaris is a multiple system. In the simultaneous DIMM data, the fainter component Polaris B showed stable IQ over 10 arcsec scales with estimates of Strehl ratio (difference between a stellar image from that of a perfect diffraction pattern) over 60% with the 25 cm OTA. If similar for the 0.5 m OTA, the point-spread function (PSF) under best seeing could be maintained at 0.8 arcsec full-width at half maximum (FWHM).

3.1. Aperture Clearing and Enclosure

One method for clearing the aperture might be to continuously heat the corrector lens by sending a low current through either loops of conductive wire or a transparent metallic surface coating, similar to the method employed by the three 50 cm Antarctic Survey Telescopes (AST3) at Dome A, Antarctica. This is undesirable for DIMM operation, as even very small temperature differences within the OTA are well known to disturb the seeing near a telescope. For most of the PEARL seeing measurements, methanol and wipes were used periodically to remove frost and ice crystals, a highly manual process not easily automated.

Experimentation was made with radiative heating and blowers, "heat guns", employed at intervals of about an hour or two. Melting was effective, as was using slight heat to encourage sublimation. Results were satisfactory for DIMM operations, that is, only a short period was needed to clear the aperture (~ 2 min) with no detectable, lingering effect on seeing. The aperture window and OTA are low mass and equilibrate quickly in the open air. Best results

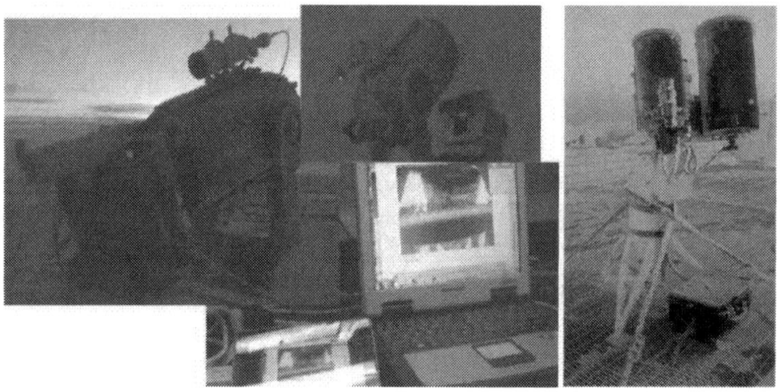

Figure 1. The Ukaliq parking box (left and top-middle) includes an electrical radiator to sublimate or melt frost and ice crystals, a soft brush which can be used to remove larger buildup mechanically, a blower, and a webcamera for inspecting the aperture - all controlled remotely (bottom). Discussion of operation with an improved equatorial drive system (right) follows.

came from using an electrical radiative coil element, basically an infrared heatlamp pointing out of an enclosure, over which the OTA could be pointed when needed.

This "parking box" (pictured in Figure 1) also doubled as the enclosure for the telescope aperture during high winds. A choice of orientation (into the prevailing wind) presents a minimal profile. With suitable anchorage to the rooftop platform this has provided sufficient protection for the OTA and instruments even during the worst winter storm conditions, and so it was found that Ukaliq does not require a dome. That avoids complexity and cost, and by presenting no extraneous obstacles also preserves the best seeing, which is known to occur during light winds (near the median windspeed) over the PEARL roof.

The reason that sublimation and melting of crystals and frost by radiative heating may work better at PEARL than by other methods is possibly that deposition of ice crystals there is mainly associated with periods of windblown dry snow, not from precipitation when calm. Frost tends to build up slowly under clear skies, which are known to be associated with calm winds. It also grows very uniformly over the aperture, which allows correction by employing good observing technique with careful relative stellar photometry. Thus, much of the time (when it is clear) aperture clearing is not needed. Otherwise, it is likely to be cloudy anyway. This is also found to be correlated with seeing poorer than 1 arcsec, that is, winds over 8 m s^{-1}.

3.2. Autonomous Weatherstation

Experience with deploying robotic instruments in the Arctic has taught that knowledge of changing weather and a quick response to it are key to preventing damage. The primary issue is wind, not the cold; it is always extremely dry, with relative humidity near saturation. The PEARL facility autonomous weatherstation (AWS) employs a sonic anemometer and is installed at 10 m elevation on a mast in the centre of the roof. It reports averaged conditions via the PEARL network on 10-minute intervals via a webpage. That system is reliable, but for redundancy a separate, dedicated AWS is also integral to Ukaliq. This uses a Young hydrophobic-plastic propeller anemometer and windvane with a Campbell Scientific CR10X datalogger; a common unit used in the High Arctic and previously employed with the "Inuksuit" devices.

3.3. Mount and Observing Platform

Replacing the original Meade LX200 mounts, the Ukaliq drive is an Astro-Physics GTO 3600 German equatorial mount. This was the same drive used by the Antarctic Search for Transiting

Exoplanets (ASTEP) 40 cm telescope at Dome C and later, the Dunlap Institute Arctic Telescope (DIAT) 50 cm telescope intended for PEARL. Like those, the Ukaliq drive is lubricated with low-temperature grease. Software from the manufacturer can account for periodic drive errors and drift, easily yielding blind pointing accuracies within a few arcseconds. Several parking positions can be pre-programmed, one of which is set to point the telescope at the parking box. For Ukaliq, an improvement to the "stock" drive is to replace its mechanical limit switches with those of the magnetic proximity type, avoiding a failure mode caused by packed ice crystals.

A transversely mounted plate, the largest available from Astro-Physics, is used to allow attaching various OTAs on the drive. This always includes a course guiding telescope: an Orion 8 cm ShortTube with a StarShooter USB-based CCD tracker. The mount can accommodate up to two more small OTAs at the same time; both the 25 cm and 35 cm have been installed together, as shown in Figure 1. These OTAs are each attached to pre-balanced dovetail plates, which mate to matching clamps, so all are interchangeable on the mount. Alternatively, by rotating the plate, a single 50 cm OTA could be mounted with the guider in tandem.

The pier is a squat portable tripod, custom built to fit the mount for the latitude of PEARL. The observing platform is subject to some vibration, and the most stable footing is achieved by placing the tripod over one of three I-beam members running under the floor. The pier is anchored to the members with high-tension cargo straps. The total weight of OTA plus mount and pier is approximately 250 kg and there is potentially floorspace for 5 more of these mounts on the roof. An alternative location (allowing only one unit) is to install on a 6-m tall tower further to the north along the ridge, already prepared for but not installed.

3.4. Adaptive Optics, Camera and Filter Wheel

The DIMM data show that mount vibration is minimized when winds are calm, under 4 m s^{-1}, and crew are not on the roof. Under these circumstances, image quality is dominated by seeing, which is still typically subarcsecond. To maintain image quality for windspeeds up to 8 m s^{-1}, when rooftop vibration dominates, each camera would be equipped with its own tip-tilt guiding system. A device could be readily built to meet the need: under 10 arcsec of throw at 10 Hz.

A commercial system already meeting similar requirements is the Santa Barbara Instruments Group (SBIG) AO-L combined with an ST-10 camera. We have not tested that, but have used this camera successfully in the "Ukpik" system. A scientific instrument might have standard V, R, and I filters and a 1k X 1k detector. This gives a field of view of about 40 arcsec on a side at 0.4 arcsec/pixel, or larger via a field lens. The built-in guiding detector in the focal plane would then be aligned with a bright ($R < 8$ mag) star within 16 arcmin by rotating the field, allowing fast guiding entirely internal to the camera, operated with only a start/stop command.

There is usually no need to protect cameras and electronics under the weather conditions at PEARL; waste heat has been sufficient to keep the SBIG and DIMM electronics operating, along with the MASS. Continuously cycling a filter wheel on a duty cycle of minutes can help ensure it remains "unstuck." Motors such as those in fans can fail, and are not needed, and so are to be shut off. On occasion, if instruments have been cold-soaked, an application of some heat to achieve normal operating temperature again was enough to restart them. The parking box can be positioned in such a way as to allow it to be used for this procedure as well.

3.5. Communications and Electronics

Communication between cameras, mount, and AWS is via RS232 or USB-to-ethernet converters, although direct communication with USB extender cables has also been successful. The bandwidth needed is modest: DIMM observations at 5 Hz rates generated 1 Tbyte of data in 10 days, although an imaging system with a similar number of pixels taking exposures once per minute would be less than that over a full winter.

Telescope electronics (other than instruments) are co-located at the base of the pier, protected inside a separate insulated "warmbox," somewhat larger than that shown in Figure 1. For Ukaliq, this box is also the shipping crate for the drive, which is small enough when disassembled to fit inside. When deployed, it houses a supervisory Linux-based single-board computer and USB to ethernet converters. Power is supplied from outlets on the roof. Inside the warmbox are power converters to supply 5 V, 12 V and 18 V power to the camera electronics, communications and the telescope drive. The data-storage unit is kept inside the warm building and communication to the outside world is via the PEARL network and its satellite link.

3.6. Logistics, Power Consumption and Operations

Each Ukaliq unit is a clone: OTA, instrument, mount, drive, pier, electronics, warmbox and parking box. Based on previous experience, it would take two to three days per unit to commission on sky. Once deployed in the early winter, the crew would leave the site, and the facility would be autonomous until sunup. Because the cost of crewing PEARL is high, it can be more cost effective to produce and ship duplicate systems (particularly by sea) rather than to maintain a single unit which may fail. During winter, there is limited technical support for fixing things. There has been success with repairs that require no technical expertise. For example, a simple swap of an electronics crate without need to open or adjust it could be feasible.

An attractive feature of Ukaliq is that it employs completely independent, swappable OTAs. Each has its own instrument pre-focused and ready to be deployed on the mount. An identical spare guider allows replacement within minutes. An experiment undertaken during MASS/DIMM observations at PEARL was to focus a duplicate 25 cm OTA for a colder temperature than typical, and later exchange it (when that temperature was reached again) without the need to refocus. The 25 cm OTA is relatively light, and can even be handled by one person, although larger versions would require a crew of just two or three to service.

Even for a full system of 6 units, operations are simple, and power consumption modest. The majority of the time, each unit is continuously viewing a single field, which for the equatorial mount requires only tracking in right ascension. The guiders are sufficient for acquisition of the target and guidestar for the tip-tilt system. Camera exposures would likely be in a uniform sequence, cycling through the common filter set, and sustained whenever on sky. A warning of winds over about 8 m s^{-1} or deviating from the prevailing northerlies from the AWS autonomously puts the telescopes into a safe mode; utilizing the parking boxes. A remote observer could then make a decision about resuming operations when safe to do so, first inspecting the apertures to ensure they are clear, or utilizing the parking boxes to make it so. Each radiant element can consume up to 100 W. The drive and electronics are under 10 W, so the maximum total power requirement would be 660 W, but typically below 60 W.

4. Example Survey Science

Each unit could independently track a single field for either site testing or science. One well-known gravitationally lensed system which illustrates the good fit of Ukaliq is Q0957+561. It is a doubly imaged quasar at $z = 1.405$ lensed by a foreground, $z = 0.355$ galaxy. Images A and B are relatively bright, $V = 16$, and separated by 6 arcsec on the sky; IQ of 1 arcsec PSF FWHM ensures no blending between them. Their time delay is 417 ± 2 days, that is, a path difference causes any change in the brightness of quasar image B to lag A, which has some intrinsic color dependence inherent to the source [4 and references therein]. It is remarkable that this is not known more accurately after roughly 30 years of monitoring, but that is in part because of uncertainty due to microlensing by substructure in the lensing galaxy. Since one such 0.25 mag event in 1981 to 1986 there have been several (disputed) detections in the last 20 years with amplitudes below 20 millimag and time periods shorter than about 5 days. A possibility is that variability over various timescales are superimposed. Ukaliq is an ideal tool to search for

Journal of Physics: Conference Series **595** (2015) 012034 doi:10.1088/1742-6596/595/1/012034

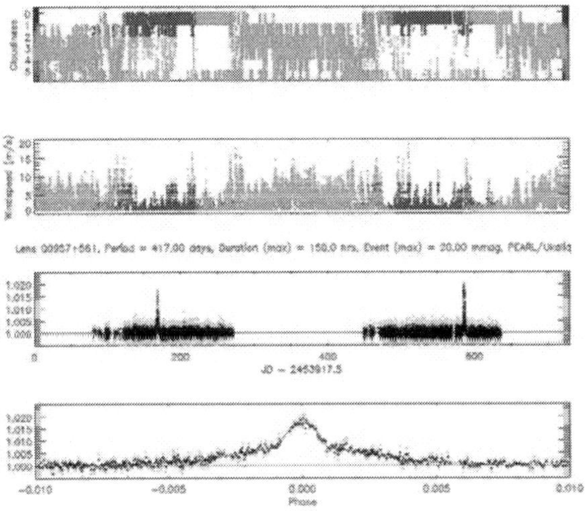

Figure 2. A simulated microlensing event in the lensed quasar Q0957+561 on Julian Day 2449706, lasting 150 hours. Based on the actual recorded PEARL sky conditions at the time, Ukaliq could have seen this and at least half of the time-lagged photometric fluctuation of the (non-microlensed) image the following winter. The short peak lasts 19 hours, requiring efficient round-the-clock observations to sample fully.

that due to its high observing efficiency. It can allow reliable, uninterrupted temporal coverage during successive dark periods, in roughly 100 hour blocks with clear skies and good seeing.

A simple simulation was developed to investigate potential detections. The results are shown in Figure 2: an artificial signal consistent with a microlensing event for QSO 0957+561 occurring in 2007, the A-B difference lightcurve. It is composed of three co-added symmetric sinusoidal components of 10, 5, and 2 millimags over time periods of 19, 75, and 150 hours. Real data for sky clarity and windspeed at PEARL were applied, using those for the years 2006-2008. Samples which were too cloudy (> 2.5) and windy (> 8 m s^{-1}) were excluded (shaded light brown) as have times with civil twilight or brighter (light blue). A realistic photometric error is roughly 1 millimag per combined hourly sample per filter, primarily from Poisson noise. Under these circumstances Ukaliq would detect the peak of the pulse and even be sensitive to the longest timescale fluctuation, because about half of the total duration is always viewed. Shifting the epoch of the event (until it occurred in daylight, not shown) does not change results; the same could be expected in subsequent winters. The full complement of 6 units, equivalent to 1.2 m aperture, could track 5 similar systems as well. Results on one of those for a network of mid-latitude sites might be more like the lightly-shaded (yellow) points, assuming the same average sky conditions, but incurring a penalty of 1 millimag of irreducible systematic error between sites. Even if monitored over 7 years, a much larger aperture at a single mid-latitude site would not help with the shortest-period component, i.e. night/day windowing of the Large Synoptic Survey Telescope (LSST) cannot allow sampling the full lightcurve (if this target was visible).

We thank Mark Halman for expertise in warmbox construction and that of Ivan Wevers during cold testing in the NRC-Herzberg environmental chamber. Helpful discussions with Nicholas Law and Suresh Sivanandam related their knowledge of the mount used with DIAT, as did Jerome Maire through operation of our prototype. Pierre Fogal and Jim Drummond of CANDAC kindly assisted in deploying instruments at PEARL, and we greatly appreciate the support of EC and the hospitality of the weatherstation staff during our Eureka observing runs.

References

[1] Law, N., Sivanandam, S., Murowinski, R. et al., 2012, *SPIE Conf. Series* **8444**, 5
[2] Steinbring, E., Ward, W., & Drummond, J.R., 2012, *Pub. Astro. Soc. Pac.* **124** 185
[3] Steinbring, E., Millar-Blanchaer, M., Ngan, W. et al., 2013, *Pub. Astro. Soc. Pac.* **125** 866
[4] Gil-Merino, R., Goicoechea, L.J., Shalyapin, V.N., Braga, V.F., 2012, *Astrophys. J.* **744** 47

Characterising daytime atmospheric conditions on La Palma

Matthew J. Townson[1], **Aglaé Kellerer**[2], **James Osborn**[1], **Timothy Butterley**[1], **Timothy Morris**[1], **Richard W. Wilson**[1]

[1] Department of Physics, Durham University, South Road, DH1 3LE, UK
[2] Cavendish Laboratory, University of Cambridge, JJ Thomson Avenue, Cambridge CB30HE, UK

E-mail: matthew.townson@durham.ac.uk

Abstract. Day-time turbulence profiles are important for the next generation of Solar Telescopes, which will be equipped with multi conjugate adaptive optics systems. We measured daytime turbulence profiles above La Palma from 17th to 19th September 2013 using a wide-field wave-front sensor. The measured profiles are consistent with previous measurements and suggest that the majority of the turbulence is located close to the ground. Our method does not yet permit precise measurements of high altitude turbulence due to limited instrument and analysis sensitivity.

1. Introduction

The next generation of Solar Telescopes, like the Daniel K. Inoue Solar Telescope (DKIST)[1] and the European Solar Telescope (EST)[2] will be equipped with multi-conjugate adaptive optics (MCAO). The optimal operation of these systems requires a precise knowledge of the atmospheric profiles at the telescope site, as well as the integrated seeing[3]. The profiles will influence the design of the MCAO systems: the deformable mirror (DM)s should be optically conjugated to the altitudes of the strongest turbulent layers in the atmosphere.

There is currently little information about the distribution of turbulence in the day-time atmosphere, most information has been extrapolated from night-time conditions, or from surface layer profiles[4].

We present here results from data gathered on La Palma from 17th to 19th September 2013 with the Swedish Solar Telescope (SST)[5] using an S-DIMM+[6], i.e. a wide-field wave-front sensor (WF-WFS). The WF-WFS data was reduced using a SLODAR style analysis[7] to generate turbulence profiles. The location of the instrument is especially relevant as the EST will be situated on Canary Islands[8], and will be equipped with an MCAO system to attain diffraction limited resolution over a $(2 \times 2)\text{arcmin}^2$ field of view.

2. Wide-Field Wave-Front Sensor

The WF-WFS used is similar to the S-DIMM+ previously installed at the SST[6]. The optical components remained unchanged, however the camera was replaced with a larger CMOS chip from Teledyne Dalsa. The camera was run at 100Hz, reading a single row of sub-apertures across the center of the pupil, as shown in fig 1. Typical data packets span 1 minute to ensure proper sampling of the atmospheric turbulence.

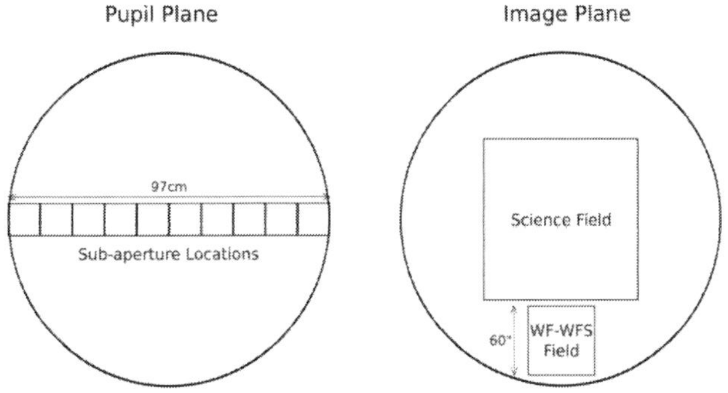

Figure 1. The left panel shows the layout of the sub-apertures across the pupil. This gives the maximal number of sub-aperture baselines in one dimension, but also keeps a small footprint on the detector, allowing for faster readout and reduced data sizes. The right panel shows the image plane of the SST: a field stop relays the science field, while a small pick-off relays the the WF-WFS field.

The WF-WFS measures local wave-front gradients as a function of field direction and pupil position. Small regions of the field, that cover a few solar granules, are cross-correlated to obtain the wavefront gradients[9]. The wave-front tilts induced by the turbulence change the position of the granulation on the camera. This displacement is then measured and converted into a local wave-front gradient. The wave-front gradients for different on-sky directions and pupil positions are used to tomographically reconstruct atmospheric profiles via the SLODAR technique[7], as illustrated in fig 2.

Figure 2. Two regions observed by two different sup-apertures. The correlation between the motion of the two sources behind their respective sub-apertures informs about the turbulence over the red surface. The strength of the correlation estimates the strength of the turbulence in that region where the two lines of sight overlap. A full turbulence profile is deduced from measurements behind several sub-apertures and along different field directions.

The cross-correlation of the solar granules is only successful if the seeing is below around 2":

for larger seeing values the contrast of the granules is too poor to ensure a proper wavefront gradient estimate. There are thus times when the WF-WFS cannot track the solar granulation, and no turbulence profiles can be measured. Typically the seeing deteriorates throughout the day, so in most cases there is a cut-off time after which it is not possible to measure the atmospheric profile due to the deterioration of the seeing. A typical frame from the WF-WFS is shown in fig 3: on this image it is indeed possible to see the solar granulation in the sub-apertures. We measure local wavefront gradients over 7×8 small 6" × 6" fields that are regularly distributed over the 53" × 60" sensor field.

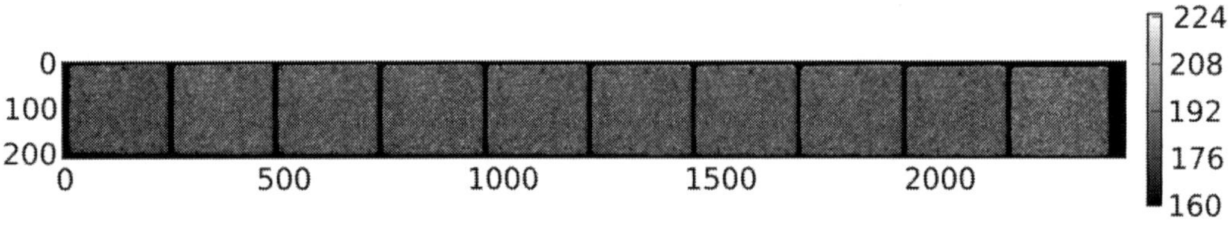

Figure 3. A raw frame from the WF-WFS. The ten sub-apertures across the pupil of the SST image the same region of the solar surface. A 610nm filter was used in all observations to optimize the contrast of the solar granulation. The granules are tracked to estimate local wave-front gradients as a function of field direction and sub-aperture. Each sub-aperture images a large field, so many different field directions can be used to estimate the atmospheric profile.

The estimates of the local wavefront gradients are affected by several sources of noise, most importantly tracking errors and wind shake on the telescope. Both shift the solar image uniformly in the focal plane, this shift is identical on all sub-apertures. To remove these effects we subtract the global tip/tilt from every frame, for all field directions. This also removes some signal from the layers of turbulence in the atmosphere: the SLODAR reduction takes this into account by appropriately scaling the global tip/tilt term in the response functions[10].

The resolution of the WF-WFS profiles is not as high as might be expected from the large number of baselines and field directions available. This is because our target is extended: the effective diameter of the sub-apertures therefore increases with altitude. This effect of increasing the size of the sub-apertures decreases the size of the local gradients across the sub-aperture and increases the overlap of the sub-apertures. The final profiles have 10 resolution elements, up to 20km in altitude.

3. Results

The profiles taken from the WF-WFS are shown in the left column of fig 4, where the plots run vertically from the 17th - 19th September 2013. The right column of fig 4 shows the average profile on each day.

The profiles all show a similar structure, with most of the power at the ground and some turbulence up to 5km. The lack of detected turbulence above this altitude is likely due to the extended nature of the source. The effective size of the sub-apertures increases with altitude, this not only reduces the resolution of the WF-WFS but also decreases the strength of the correlation signal for turbulence at high altitudes. This makes the profile reconstruction extremely insensitive to high altitude turbulence: the signal from these layers is typically below the noise level. This effect can be mitigated by estimating the wavefront gradients over smaller fields, but then the quality of the correlation - *i.e.* the precision of the gradient estimates - decreases. So although it is in principle possible to sense turbulent layers up to 20km, in practice the strength of high altitude signals is greatly reduced and turbulence at higher altitudes can thus barely be distinguished from noise.

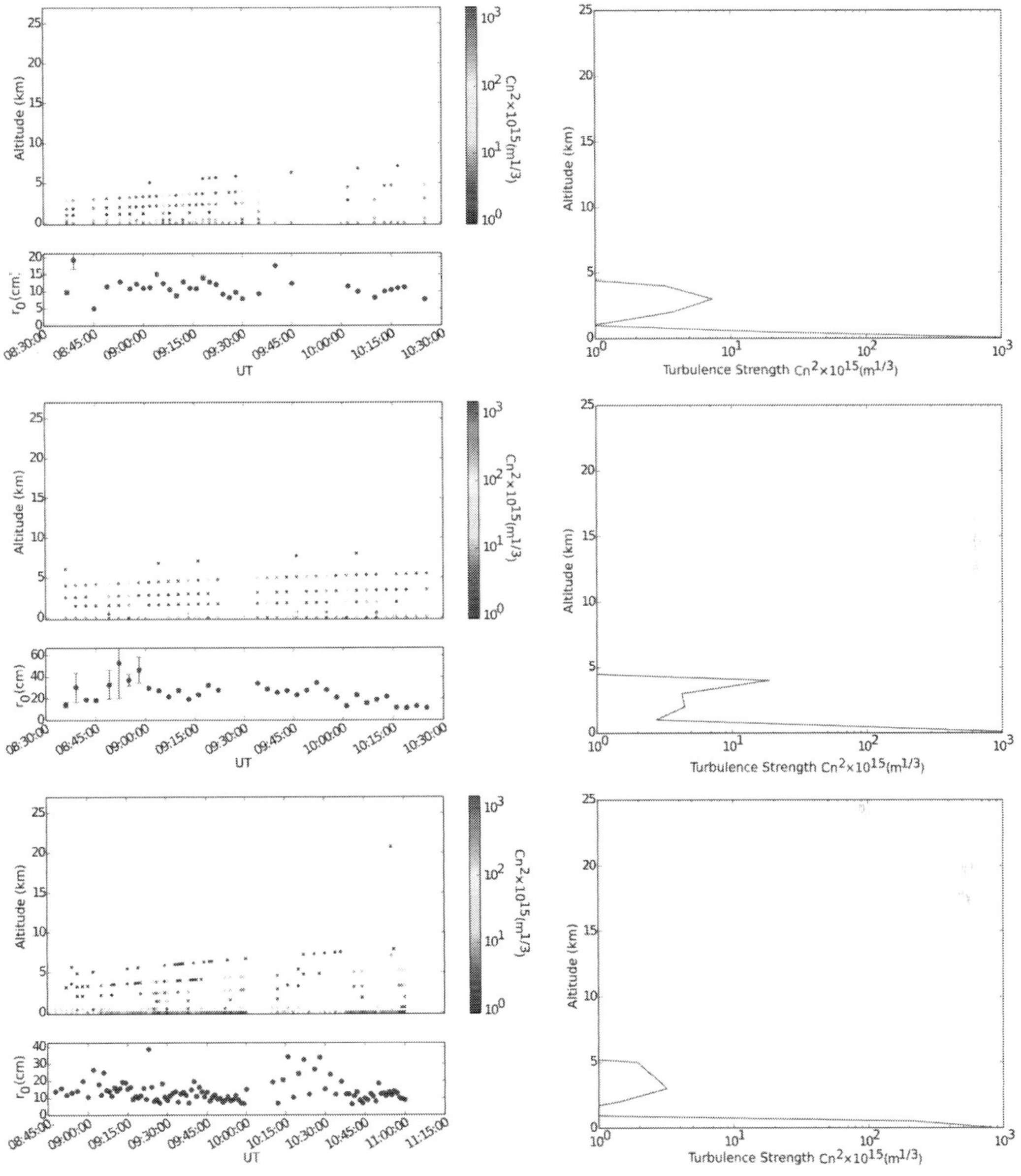

Figure 4. The profiles shown in the left column from top to bottom represent the 17th, 18th and 19th September 2013 respectively. The profiles shown on the right show the average turbulence profiles for each day. The profiles all show a strong ground layer, and an extended region of turbulence which peaks in strength around 2.5 to 5km. The estimates of integrated seeing with each of the profiles is shown underneath the profiles.

The average profiles in fig 4 give a clearer impression of the general structure of the turbulence, although all information about the temporal variability of the atmosphere is lost. The average profiles for each of the days show similar features, with the vast majority of the turbulence located at the ground and a weaker feature between 2.5km and 5km.

4. Conclusions and future work

The profiles presented here are consistent with previous measurements of the daytime atmosphere, which suggested that the majority of the turbulence is located at the ground. We detect no turbulence above 5km, even though the turbulence is expected from the jet stream. The WF-WFS is currently extremely insensitive to high altitude turbulence: this turbulence can only be sensed if it has a similar strength to to the ground layer.

We are currently working on the data analysis to improve the sensitivity to high-altitude turbulence. The WF-WFS will be used in 2015, with a plan to eventually automate the data acquisition and obtain profile estimates in real time.

References

[1] Keil S, Rimmele T and Keller C 2003 *Astronomische Nachrichten* **324** 303–304 ISSN 00046337 URL `http://doi.wiley.com/10.1002/asna.200310103`
[2] Collados M, Bettonvil F, Cavaller L, Ermolli I, Gelly B, Pérez a, Socas-Navarro H, Soltau D and Volkmer R 2010 *Astronomische Nachrichten* **331** 615–619 ISSN 00046337 URL `http://doi.wiley.com/10.1002/asna.201011386`
[3] Kellerer A, Gorceix N, Marino J, Cao W and Goode P R 2012 *Astronomy and Astrophysics* **542** A2 ISSN 0004-6361 URL `http://www.aanda.org/10.1051/0004-6361/201218844`
[4] Berkefeld T, Bettonvil F, Collados M, López R, Martín Y, Peñate J, Pérez A, Scharmer G B, Sliepen G, Soltau D, Waldmann T a and van Werkhoven T 2010 *Proc. SPIE* **7733** 77334I–77334I–10 URL `http://proceedings.spiedigitallibrary.org/proceeding.aspx?articleid=750430`
[5] Scharmera G B, Bjelksjo K, Korhonen T, Lindberg B and Petterson B 2003 *Proc. SPIE* **4853** 341–350
[6] Scharmer G B and van Werkhoven T I M 2010 *A & A* **513** A25 ISSN 0004-6361 URL `http://www.aanda.org/10.1051/0004-6361/200913791`
[7] Wilson R W 2002 *Monthly Notices of the Royal Astronomical Society* **337** 103–108 ISSN 00358711 URL `http://doi.wiley.com/10.1046/j.1365-8711.2002.05847.x`
[8] Berkefeld T, Bettonvil F, Collados M, López R, Martín Y, Peñate J, Pérez A, Scharmer G B, Sliepen G, Soltau D, Waldmann T a and van Werkhoven T 2010 *Proc. SPIE* **7733** 77334I–77334I–10 URL `http://proceedings.spiedigitallibrary.org/proceeding.aspx?articleid=750430`
[9] Löfdahl M G 2010 *Astronomy & Astrophysics* **524** A90 ISSN 0004-6361 URL `http://www.aanda.org/10.1051/0004-6361/201015331`
[10] Butterley T, Wilson R W and Sarazin M 2006 *Monthly Notices of the Royal Astronomical Society* **369** 835–845 ISSN 00358711 URL `http://doi.wiley.com/10.1111/j.1365-2966.2006.10337.x`

Accurate measurements of Optical Turbulence with Sonic-anemometers

T Travouillon[1], A Otarola[1], S Els[2], R Riddle[3], M Schöck[1], W Skidmore[1], D Bibb

[1]Thirty Meter Telescope, 111 South Arroyo Parkway, Pasadena, 91105 CA, USA
[2]Gaia DPAC, c/o ESAC
[3]California Institute of Technology, 1200 E. California blvd, Pasadena, 91125 CA, USA

E-mail: tonyt@caltech.edu

Abstract. The minimization of optical turbulence in and around the dome is key to reach optimum performance on large telescopes equipped with adaptive optics. We present the method and preliminary results of in-situ measurements of optical measurements made using sonic-anemometers. We show the impact of correcting the raw data for aliasing, path averaging, pulse sequence delays and Taylors' hypothesis. Finally, we highlight the occurrence of non-Kolmogorov turbulence which complicates the quantitative impact of the measurements on the telescope's resolution.

1. Introduction

The site testing campaign for the Thirty Meter Telescope (TMT) took place between 2003 and 2008 [1]. A minimum of 2 years of data were recorded at 5 sites: Cerros Armazones, Tolar and Tolonchar in Chile, San Pedro Martír in Mexico and finally Maunakea in Hawaii which ended up being selected for the construction of TMT. The resulting data set includes an array of key parameters including meteorological parameters, turbulence, all-sky images, atmospheric transparency and sky brightness. These data have been made publically available and accessible online (www.sitedata.tmt.org).

Several instruments used during this campaign are seen as having a useful purpose during the operation of the telescope as well. One of them, a CSAT3 model sonic-anemometer was used at every site gathering temperature and wind velocity data from the 7m tower of the DIMM telescope. Thanks to the high cadence of the measurements (60Hz), it is in principle possible to use the wind and temperature to calculate the turbulence parameter C_n^2. Such capability makes this instrument ideal for *in-situ* measurements of turbulence inside an observatory. In such an environment, remote sensing is impractical and inadequate for highly localized measurements. Scintillometers for example, measure the turbulence over a given path instead of at a specific location. The only instrument commonly used for making pinpointed measurements of optical turbulence is the thin-wire microthermal sensor. This instrument is commonly used balloon-born to measure turbulence profiles of the atmosphere, but when used at a stationary location, its fragile construct means that it breaks under tough meteorological conditions. A sonic-anemometer, on the other hand is built to handle any conditions due to its metal frame and its lack of moving parts.

The goal is to use the sonic-anemometers operated during the TMT site testing campaign around key locations around the dome of the telescope to measure the turbulence field and minimize this turbulence to obtain the best possible on-sky resolution. In this paper, we present the initial work made to transform the sonic-anemometer data into accurate turbulence measurements. We summarize how the raw temperature and wind velocity data were combined and corrected to obtain time series of the turbulence figure of merit: C_T^2. We also open a discussion on an observation made during the analysis of these data: non-Kolmogorov turbulence.

2. Sonic-Anemometer data

While the completion of the construction of TMT is still a few years away, testing the CSAT3 suitability can be done on existing data. A very large sample of raw temperature and wind velocity data was sampled at 60Hz, totaling to approximately 35GB per site, and is available for the five sites studied in the campaign. The sonic-anemometers were located at a height of 7m above the ground, attached to the structure of the DIMM telescope tower. They were installed upwind on an extended arm to keep the head of the sensor 1.5m away from the dome structure. In addition to this setup, 30m towers were installed on Cerro Armazones and Cerro Tolonchar and three anemometers were installed at 10, 20 and 30m to obtain a vertical profile of the site's weather conditions. In this paper however, we will focus on a small subset of the data using the first 3 months of the Maunakea dataset. As the next section will show, the data reduction is computer intensive and this smaller dataset allows us to more easily test the analysis routines and draw initial conclusions on how to proceed with the full analysis.

To understand the processing of the raw data described in the next section, it is important to know the basics of the instrument in question. The CSAT3 uses three pairs of acoustic transducers with the geometrical characteristics shown in Figure 1. A sonic pulse is sent from one transducer to its opposing counterpart. The temperature θ, often referred to as "sonic temperature", is deduced from its simplified relationship to the speed of sound:

$$c_{sound} = (331.3 + 0.606\,\theta) \tag{1}$$

This temperature is therefore an average of the temperature inside the cylindrical beam the pulse travels through. Each pair of transducer takes turn to go through a cycle of sonic emission/reception to give for each a wind speed along the line of the transducers pair. From these and using simple geometry, the x, y and z component of the wind velocity can also be calculated.

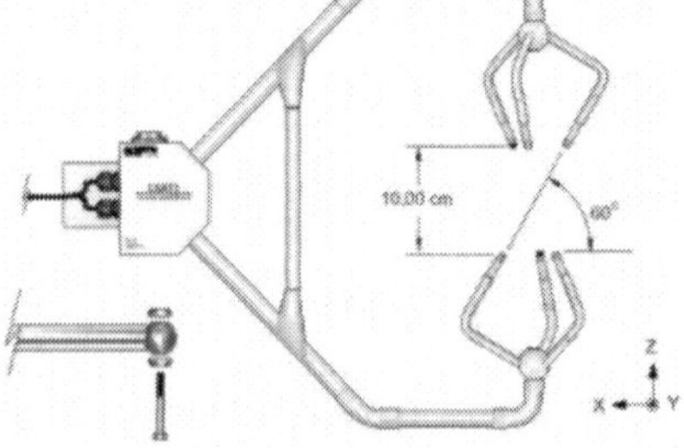

Figure 1. Visual representation of the CSAT3 sonic-anemometer.

Sixty times per second a sonic temperature and three wind speed components are measured and recorded by the instrument along with a quality flag. This quality flag is used to identify times when the sonic beam was blocked (by ice for example) or simply of poor accuracy due to a low signal to noise ratio. Such data were discarded all together from our analysis. Finally the data are divided into 2mn blocks from which the temperature power spectrum $S_T(f)$ and the mean wind speed U are calculated. It is from these parameters that the turbulence can be calculated:

$$C_T^2 = 13.67 S_T(f) f^{5/3} U^{-2/3} \tag{2}$$

In reality however, it is important to carry out some corrections to the temperature power spectrum before applying it to the above formula. In the section below, we describe and estimate the different types of correction that are required to transform the sonic-anemometer's date into real single point measurements of temperature and wind velocity. It is with this corrections that we can aim for the most accurate turbulence measurements possible.

3. Data correction

3.1. Aliasing and white noise

With a limited sampling speed of 60Hz, the range of measurement is limited to a Nyquist frequency of half this value and subjected to aliasing effects like any other digital sampling. The energy above this frequency is however folded back into the observed range which is visible in a raw power spectrum in the form of an "uptail" in the high frequency end of the spectrum. The corrected power spectrum is deduced from the aliased power spectrum $S^a(f)$ through the relation:

$$S(f) = S^a(f) / \sum_{m=-\infty}^{\infty} H(f_m) (\frac{f_m}{m})^{-5/3} \tag{3}$$

where $f_m = |f + 2mf_N|$ and $H(f)$ is the attenuation caused by the processes described in the subsections below. Figure 2 shows the visual effect of de-aliasing the power spectrum, removing the folded energy and linearizing the power spectrum in the log space.

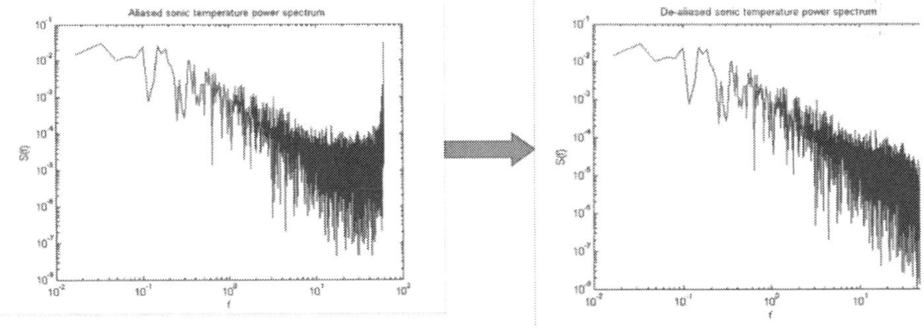

Figure 2. Example taken from our analysis of power spectrum de-aliasing.

From this de-aliased spectrum, white noise can easily be removed since the CSAT3's noise is independent of sampling rate and approximately equal to $\sigma = 0.004^{\circ}C$. A constant value of σ^2/f_N is therefore subtracted from the power spectrum.

3.2. Sonic path-averaging

As visible in Figure 1, the sonic pulse has to traverse a volume of air to go from the emitting transducer to its receiving counterpart. The size of this path is not negligible with a vertical distance of

10cm and therefore has the effect of averaging over that volume. This averaging, which affects both temperature and wind speed measurements, is aggravated by the fact that the anemometer needs to combine the data from the three paths. The temperature averaged over a path p is related to the punctual temperature at the centre of the path,x_0, following:

$$\tilde{T}(x_0, p) = \frac{1}{p}\int_{-p/2}^{p/2} T(x_0 + s)ds \tag{3}$$

The mathematical leap to the Fourier space, which is described in [2], leads to a transfer function H^p which must be applied to both temperature and wind speed power spectrum.

3.3. *Pulse sequence delays*

A full cycle of measurements with the CSAT3 requires 6 pulses: one in each direction for each pair of transducers. This sequence takes a total of 13.4msec during which the turbulence has not been stationary. This effect studied initially by Larsen et al. [3] and updated by Nielsen and Larsen [4] constitutes an attenuation and a cross contamination of the temperature and wind velocity spectra. For example, the temperature measurement made by the CSAT3, T_S, is an average of the measurements of the three paths and is related to the true sonic temperature, T_{sv}, via:

$$T_S = \frac{1}{6}\sum_{j=1}^{6} T_{sv}(t_j) + \frac{a}{6}[w_a(t_1) - w_a(t_2) + w_b(t_3) - w_b(t_4) + w_c(t_5) - w_c(t_6)] \tag{4}$$

where w is the wind speed along the three paths a, b and c of the instrument. Its Fourier transform leads to a correction factor dependent on spectra and co-spectra of both temperature and wind velocity. The temperature power spectrum is therefore attenuated by a factor H^τ which is used in the net spectral correction.

After the temperature power spectrum has been de-aliased and white noise was removed, a net correction that takes consideration of both pulse delay and path averaging:

$$S_T(f) = S_{Tcorrected}(f) \sum_{m=-\infty}^{\infty} H_T^p(\omega_m/U) H_T^\tau(\omega_m) \left(\frac{f_m}{f}\right)^{-\frac{5}{3}}$$
$$+S_{wa} \sum_{m=-\infty}^{\infty} H_{wa}^p(\omega_m/U) H_{wa}^\tau(\omega_m) \left(\frac{f_m}{f}\right)^{-\frac{5}{3}} + \cdots$$
$$+Co_{waT} \textstyle\sum_{m=-\infty}^{\infty} [H_{wa}^p\left(\frac{\omega_m}{U}\right) H_T^p\left(\frac{\omega_m}{U}\right)]^{1/2} H_{waT}^\tau(\omega_m) \left(\frac{f_m}{f}\right)^{-\frac{57}{3}} + \cdots$$

(5)

where $\omega = 2\pi f$ and the derivation of the attenuations factors can be found in the references above.

3.4. *Taylor's hypothesis*

The final spectrum is corrected for Taylor's hypothesis which is assumed above. This is done out of completeness only and not necessary for our data, since we found it to only contribute a correction of about 1%, Wyngaard and Clifford [5] show that the ratio of the measured to real power spectrum is:

$$\frac{S^m}{S} = 1 - \frac{1}{9}\frac{\overline{u'u'}}{\overline{U}^2} + \frac{1}{3}\frac{\overline{v'v'}}{\overline{U}^2} + \frac{1}{3}\frac{\overline{w'w'}}{\overline{U}^2} \tag{6}$$

where u, v and w are the wind speed along the x, y and z as labelled in Figure 1.

4. Results

As mentioned earlier, applying all the corrections to a power spectrum is quite computer intensive and the purpose of testing it on a subset of the data is to test the software and get an idea for the size of the correction. If the effect is found to be very small, the correction can simply be ignored and if the effect is systematic, it can at least be simplified. With the wide variations of C_T^2 that can be encountered throughout a night of measurement, a random variation of 5% could, for example, easily be considered acceptable. So the first obvious comparison is that of the corrected C_T^2 versus its raw counterpart.

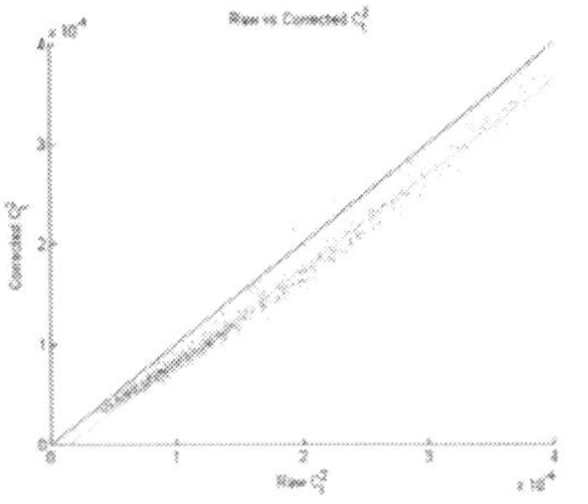

Figure 3. Scatter representation of the corrected and raw C_T^2.

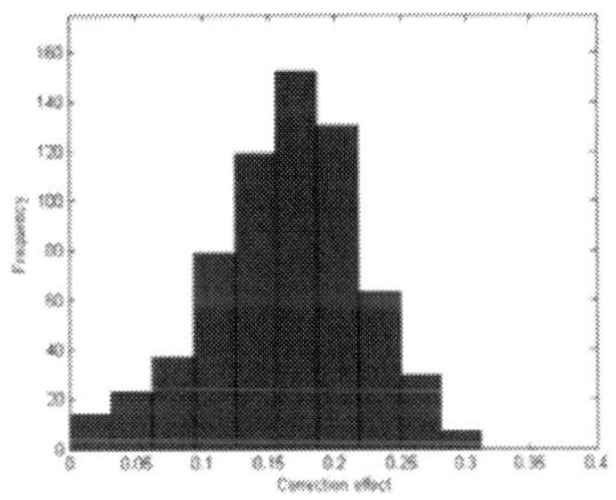

Figure 4. Statistical distribution of the correction factor in percentage.

Figures 3 and 4 show the effect of the correction on the data subset. The correction systematically decreases the value of the measured turbulence with a mean correction corresponding to a 16% decrease. The scatter in the correction is also small as it never exceeds 30%. The line of best fit, shown in grey on Figure 3 as a slope of 0.95, meaning that the intensity of the turbulence has little effect on the size of the correction. Looking at the data a bit deeper, we can see that the parameter affecting the correction is, not surprisingly, the wind speed since it impacts both the path averaging and pulse delay elements of the correction (i.e. a higher wind speed moves an air volume further between each pulse). Figure 5 shows an almost systematic increase of the correction factor with wind velocity. This relationship is even linear enough to be used as a crude correction method (the dip between 2 and $4m.s^{-1}$ is most likely due to low number statistics) and shows that inside an observatory, where the wind speeds will be low, the correction may even have a small effect

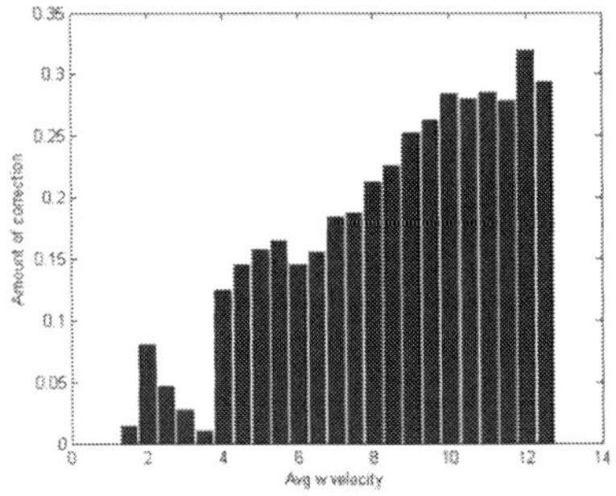

Figure 5. Wind velocity dependence of the correction factor.

5. The problem of non-Kolmogorov turbulence and conclusion

The analysis of these data had the unintended consequence of attracting our attention to a more fundamental problem. Equation 2 assumes a Kolmogorov turbulence spectrum, which has a frequency dependence with a -5/3 power law. In reality, however, the power spectrum does not always have such a frequency dependence. Leaving aside the practical difficulties and choices that are necessary to fit a power law to a noisy spectrum, we found that not only was the slope of the power spectrum in log space not always equal to the expected -5/3 but had some systematic variations with the time of the day (see figure 6). The nature of these variations are likely linked to the stability of the turbulence and its interaction with the ground and physical features. With an instrument only 7m from the ground, and in even closer proximity with a metallic tower, it is not surprisingly to find turbulence conditions that are not fully developped. But the question persists: as the intended purpose is to use this instrument inside and around an observatory, how does one measure non-Kolmogorov turbulence and translate the data to a quantity that can then be used to estimate the telescope loss of resolution contributed by this dome turbulence?

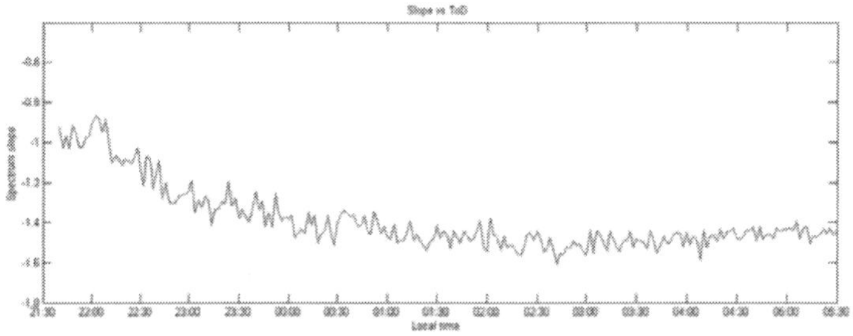

Figure 6. Slope of the power law (in log-space) as function of the time of the day. The instrument was only operated in night time but the data suggest that the turbulence conditions only approach Kolmogorov a few hours after sunset.

Acknowledgment

We would like to thank Tom Horst and Steve Oncley for establishing the theoretical framework around which this work is based. The TMT Project gratefully acknowledges the support of the TMT collaborating institutions. They are the Association of Canadian Universities for Research in Astronomy (ACURA), the California Institute of Technology, the University of California, the National Astronomical Observatory of Japan, the National Astronomical Observatories of China and their consortium partners, and the Department of Science and Technology of India and their supported institutes. This work was supported as well by the Gordon and Betty Moore Foundation, the Canada Foundation for Innovation, the Ontario Ministry of Research and Innovation, the National Research Council of Canada, the Natural Sciences and Engineering Research Council of Canada, the British Columbia Knowledge Development Fund, the Association of Universities for Research in Astronomy (AURA), the U.S. National Science Foundation and the National Institutes of Natural Sciences of Japan.

References

[1] Schöck M et al. 2009 *Thirty Meter Telescope Site Testing I: Overview* Publ. Astr. Soc. Pac. 121 (878) pp 384-395
[2] Horst T W, and Oncley S P 2006 *Corrections to inertial-range power spectra measured by*

CSAT3 and Solent sonic anemometers, 1. Path-average errors. Boundary Layer Meteorology 119 pp 375--395.

[3] Larsen S E, Edson J B, Fairall C W and Mestayer P G 1993 *Measurement of temperature spectra by a sonic anemometer* J. Atmos. Oceanic Tech. 10 pp 345–354.

[4] Nielsen M and Larsen S E 2002 *The influence of pulse–firing delays on sonic anemometer response characteristics* 15th AMS Conference on Boundary Layers July 15-19, 2002, Waginingen, The Netherlands, pp 436-439.

[5] Wyngaard J C, Clifford S F 1977 Journal of Atmospheric Sciences vol. 34 Issue 6 pp 922-929

Optical Turbulence Characterization by WRF model above Ali, Tibet

Hongshuai Wang, Yongqiang Yao, Liyong Liu, Xuan Qian, Jia Yin
National Astronomical Observatories, Chinese Academy of Sciences, 20A Datun Road, Chaoyang District, Beijing, China
E-mail: wanghs@nao.cas.cn

Abstract. Atmospheric optical turbulence modeling and forecast for astronomy is a relatively recent discipline, but has played important roles in site survey, optimization of large telescope observing tables, and in the applications of adaptive optics technique. The numerical approach, by using of meteorological parameters and parameterization of optical turbulence, can provide all the optical turbulence parameters related, such as C_n^2 profile, coherent length, wavefront coherent time, seeing, isoplanatic angle, and so on. This is particularly interesting for searching new sites without the long and expensive site testing campaigns with instruments. Earlier site survey results by the site survey team of National Astronomical Observatories of China imply that the south-west Tibet, Ali, is one of the world best IR and sub-mm site. For searching the best site in Ali area, numerical approach by Weather and Research Forecasting (WRF) model had been used to evaluate the climatology of the optical turbulence. The WRF model is configured over a domain 200km×200km with 1km horizontal resolution and 65 vertical levels from ground to the model top(10millibars) in 2010. The initial and boundary conditions for the model are provided by the $1° \times 1°$ Global Final Analysis data from NCEP. The distribution and seasonal variation of optical turbulence parameters over this area are presented.

1. Introduction

In April of 2012, EACOA(East Asian Core Observatories Association) reviewed the site survey in west china. The results implied that the south-west Tibet is one of the world best IR and sub-mm site. However, the present Ali site which located at E80.03,N32.32 with 5050m high might not be the final best site, and the nearby summits should be also further studied. For searching the best site in Ali area within hundreds of kilometers, numerical approach by WRF(Weather and Research Forecasting) model had been used to evaluate the climatology of the optical turbulence.

The WRF model was configured with 3 nested domains with increased horizontal grid resolutions 25km, 5km and 1km. The surface respectively covered 2000km×2000km, 500km×500km and 200km×200km(Fig 1). 65 vertical levels from ground to the model top(10millibars) were used. The initial and boundary conditions for the model were from $1° \times 1°$ Global Final Analysis data of NCEP. A simulation of 30 hours was run every 5 days on 2010.

2. Optical Turbulence Parameters

It has been proposed by H.Trinquet and J.Vernin in 2007[1][2], employing over 145 profiles of meteorological balloons, equipped with microthermal sensors and launched from ground to the

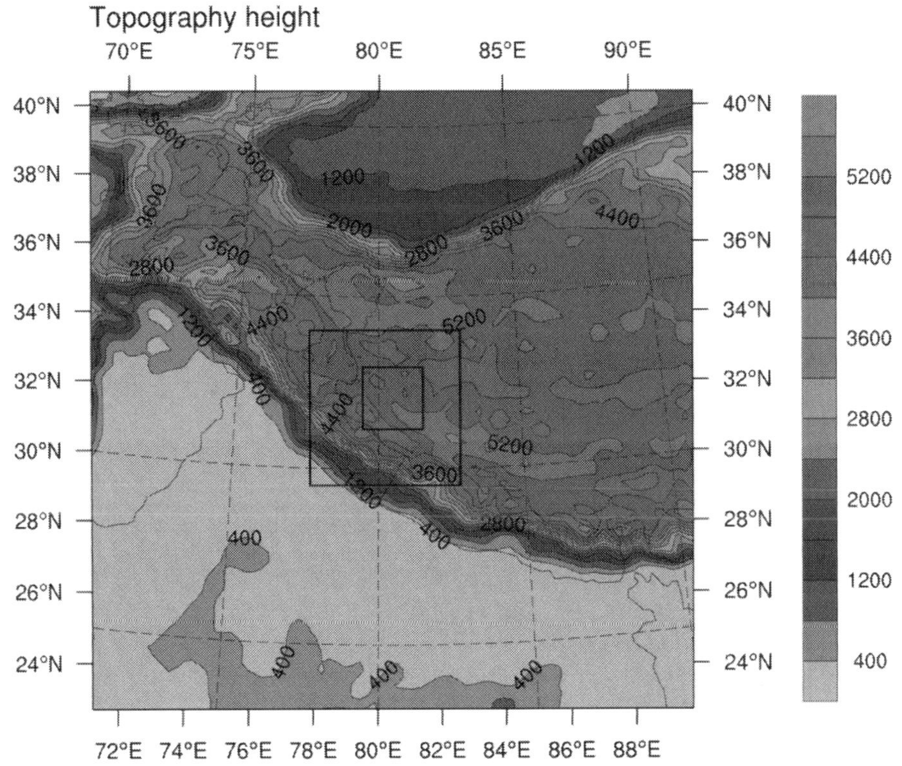

Figure 1. The configuration of the WRF model at Ali area.

mid- stratosphere. The relationship among temperature structure constant, buoyancy force, and vertical shear of the horizontal wind speed are

$$C_T^2 = < C_T^2 >_m \frac{\chi}{< \chi >_m} (\frac{S}{< S >_m})^{1/2} \tag{1}$$

$$\chi = \frac{d\bar{\theta}}{dz} \tag{2}$$

$$S = \sqrt{(dV_x/dz)^2 + (dV_y/dz)^2} \tag{3}$$

Here χ is potential temperature gradient, S is wind speed shear, $\langle\,\rangle_m$ means median value. then the C_n^2 can be obtained

$$C_n^2 = \left(\frac{80 * 10^{-6} P}{T^2}\right)^2 C_T^2 \tag{4}$$

The parameters related to optical turbulence, such as seeing, isoplanatic angle, and wavefront coherent time, are then obtained with the integrations of C_n^2, altitude and wind velocity.

3. Results

By averaged the 72 nights results, we obtain seeing distribution over 200km×200km of Ali area (Fig 2). The southern and south-east summit of current ali site have shown better seeing, will be further tested.

Fig 3 shows the variations of seeing on 2010 every 5 days. Seeing in summer is better than in winter.

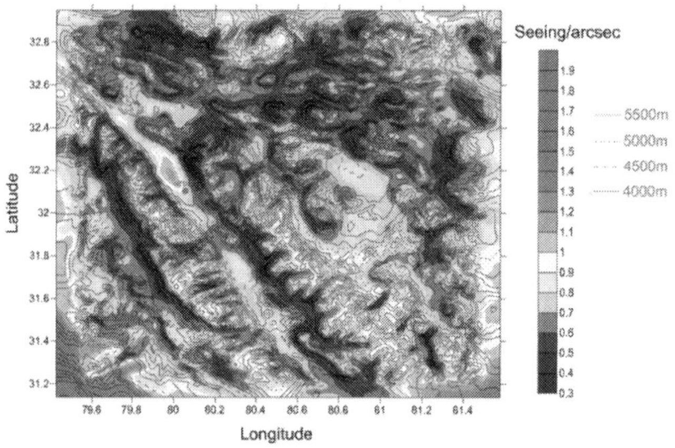

Figure 2. Averaged Seeing distribution over Ali area 200km×200km.

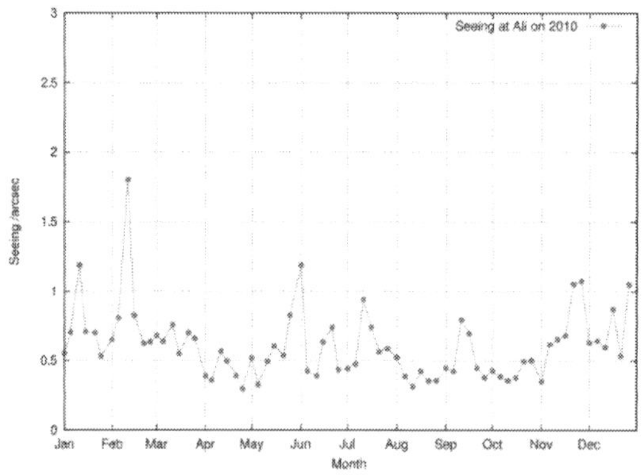

Figure 3. Averaged seeing(00:00-06:00) at Ali site on 2010.

4. Acknowledgment

This work was supported by the National Natural Science Foundation of China (Grant Nos. 11203044, 11373043, 11303055 and 11103042).

References

[1] Trinquet H, Vernin J. 2006, PASP, 118, 756
[2] Trinquet H, Vernin J. 2007, Environ. Fluid Mech, 7, 397

Atmospheric monitoring strategy for the Ali site, Tibet

Y Yao, Y Zhou, L Liu, H Wang, J Yin, X You, X Fu
National Astronomical Observatories, Chinese Academy of Sciences, China
E-mail: yqyao@nao.cas.cn

Abstract. The astronomical site survey in China has been carried out since 2003. Remote studies and local surveys are performed over the high plateaus, and candidate sites have been selected and performed site testing measurements. The monitoring results show that Ali area in western Tibet can be the best choice for astronomical observations over East Asian regions. Ali site, near the central town of Ali area, has been further identified for small telescope projects and simultaneously for detailed site characterization, and begun construction in 2010. This paper presents the site monitoring strategy and site development plan of the new Ali observatory.

1. Introduction

The site survey in western China has been carried out since 2003, in order to identify appropriate sites for constructing medium-size and large telescopes for East Asian communities. Remote studies and local surveys are performed over the vast land of plateau in the western part of China[1]. Two candidate sites, called Karasu, on Pamirs, and Oma, on Ali plateau in Tibet, are selected in 2005 to make the first phase site testing measurements. Both the sites keep away from cold snap in winter, warm current in summer, and seasonal dust storm, and possess good conditions of low temperature, dry air, less cloudy, less population and delayed development in industry. The topography of the sites shows wide flat summit with enough relative height and good accessibility.

Figure 1. The site testing stations at Karasu on Pamirs and at Oma on Ali plateau in Tibet.

The first phase site testing observations have been deployed in the period of 2005-2006. Cloud cover, seeing, precipitable water vapor, and wind speed are taken as the most important parameters to evaluate the sites, and other meteorological parameters and sky brightness are also measured. The monitoring results for the 1.5 years have shown that Oma site appears to be better than Karasu, and can be really superior and worth proceeding with further facilities for more reliable evaluation. The second phase site testing has concentrated on Oma site with improved facilities in the period of 2008-2010. The cloud monitoring with SBIG A340 at Oma site shows 60.7% photometric nights and 71.5% spectroscopic nights. The seeing measurements with DIMM and SBIG Polaris seeing monitors indicate a rather good seeing at Oma site of median 0.75"-0.8" [2]. Figure 1 shows the site testing stations at Karasu in Oct. 2007 and Oma in Nov. 2008.

In the period of 2008-2010, more candidate sites are explored and inspected following the remote studies to further examine potentially better sites. Several existing observatories and good candidate sites over China are selected for more than one year monitoring to further identify the qualities of Ali area. This site comparison research, together with the remote studies, proves that Ali area in west Tibet can be the best selection for constructing astronomical bases in wide wavelength bands.

2. The Ali observatory

For the requirements of detailed site characterization and small telescope projects, a new site in Ali area has been identified at the end of 2009 and begun construction in Aug. 2010. One of the key advantages of this site is near the central town, Shiquanhe, so that the observatory can be easily sustained. Shiquanhe town, the capital of Ali area, is the political, economic and cultural center in west Tibet, and also the intersection of main lines from Xinjiang, Qinghai, and from Lhasa.

Figure 2. The layout of Ali site in Dec. 2013.

The Ali site is located at N32°19' & E80°01', with an altitude of 5100m. The altitude of vast flat land nearby is around 4300m. The whole course from Lhasa to Ali is 1500km, with all-the-way paved road. The Ali airport is open in 2010, making it possible to fly from Ali to Beijing within one day, or from Beijing to Ali by two days, affording one day acclimation in Lhasa.

The Ali site comes into site testing work in 2011. There are two domes constructed in Oct. 2010, and 25KW solar power supply and satellite communication antenna are completed in May 2011. There is also electrical power and fiber network lines from Shiquanhe passing the way on summit to the Ali airport. In Sept. 2011, the site construction has pass through the tests of electric power and communication, and the site testing instruments have been installed on

Figure 3. The EACOA site survey workshop in April 2012 and Norio Kaifu visiting Ali site in June 2013.

site, including meteorological station, cloud monitor, and DIMM seeing monitor. A SCIDAR system for turbulence profile has made the first on-site run during the campaign in Nov. 2011. Following the site testing strategy, the site construction has continued to make progress in 2012-2013. Figure 2 presents the site layout by the end of 2013.

The construction of Ali site has attracted a lot of attention and support. In April 2012, the East Asian Core Observatories Association(EACOA) has organized a review committee to evaluate the site survey activities and results. The committee recognizes that the site survey team has made great efforts for the past years and agrees that the team has localized the candidate area qualitatively. The Ali site is very promising, although longer time monitoring and more international comparisons with standard instruments are required. The west Tibet is unique for astronomy because of longitudinal location and of high altitude, and the current results imply it is one of the world best infrared and sub-mm sites. The Science journal has reported the Ali site in Sept. 2012 as a world-class observatory rising on 'Roof of the World'[3]. Norio Kaifu, the IAU president, has visited Ali site with a team in June 2013, providing very good comments and instructions for the site development. Jun Yan, the general director of NAOC, has visited Ali site in June 2014, signing cooperation agreement with the local government and promoting to formally establish the Ali observatory.

Table 1. The site testing instruments for Ali site.

Item	Instrument
weather condition	Vaisala:WXT510, Huantron:CAWS620
sky condition	All-sky camera, MIR cloud monitor
dust counter	TSI:DustTrak 8520
water vapor	Sun photometer, RPG:TAU 225-340
turbulence in atmosphere	DIMM, MASS
turbulence profiling	Singe Star SCIDAR
turbulence in surface layer	C_T^2 sensors on tower
turbulence in upper layer	UNSW:SNODAR
opacity and background	50cm scope and instruments

3. The monitoring strategy

Following the EACOA recommendations in 2012, the atmospheric monitoring strategy has been made and implemented. More site testing instruments have been and will be installed at Ali site, with emphasis on complete turbulence measurements and with some facilities in the infrared and sub-mm wavelengths. Table 1 shows a list of site testing instruments for Ali site.

The first sets of site testing instruments have been installed in 2011 on the 10m tower. A 30m tower for microthermal C_T^2 sensors has been set up, and three new domes constructed with observing rooms in 2012; one 4.5m dome for SCIDAR telescope to measure turbulence profile, one 6m dome for a 50cm telescope to make observations and to provide some qualified data for the site, and one 8m dome for a newly planned wide field 1.2m telescope of Beijing Planetarium. The real observations will be much helpful to verify the site conditions, including practical operation, and provide with more parameters for practical application, such as sky background and transparency. Additional three 3m domes for the DIMM, MASS instruments, and for a 30cm monitoring telescope have been completed in 2013. The joint effort has been made to complete the connection of local electric power and network, and we can finally make remote observations in Beijing office.

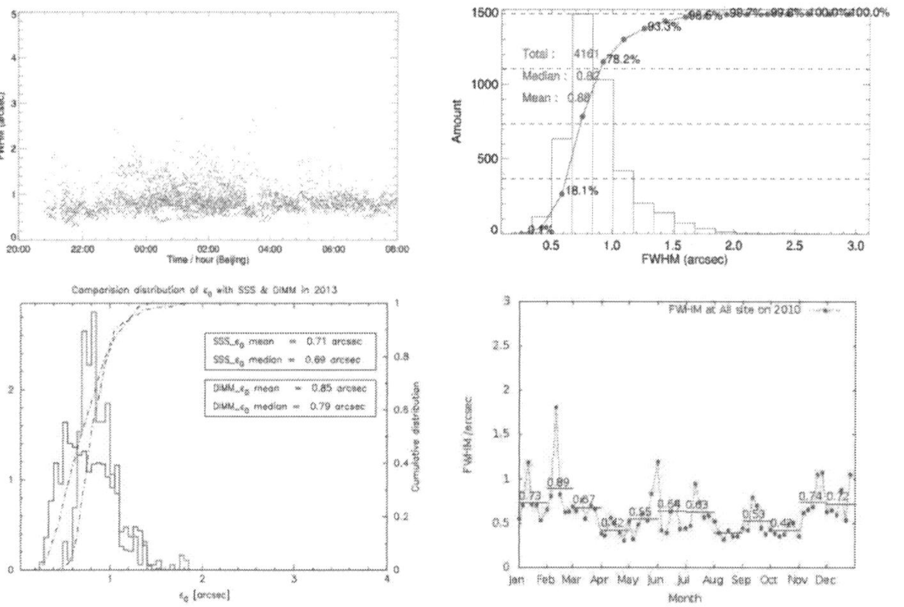

Figure 4. The DIMM seeing measured at Ali site in 2013-2014(top left) and the statistical distribution(top right). The comparison of seeing distributions measured in June 2013 by SCIDAR and by DIMM (bottom left). The monthly averaged seeing in 2010 calculated with WRF model for the Ali site (bottom right).

The recent cloud monitoring at Ali site shows percentages of 72.3% clear nights and 80.4% usable nights. The statistics of DIMM seeing measurements shows a median seeing of 0.82", and the quartile seeing of 0.72", as shown in Figure 4. The seeing measured by the single star SCIDAR is consistent in variations with the DIMM seeing, but tends to be lower in values than the latter. Figure 4 presents statistical distributions of the two instrument seeing during the measurements in June 2013; the median SCIDAR seeing is 0.1" lower than the median

DIMM seeing, and the seeing <0.6" measurements by SCIDAR seem to be undetectable for DIMM devices. It is probably understandable that the current DIMM seeing monitor with a sub-aperture separation of 20cm could not detect the better seeing conditions. Figure 4 further presents the monthly averaged seeing calculated with WRF model, which are also generally lower than the measured DIMM seeing of each month, with a difference of 0.15"-0.18".

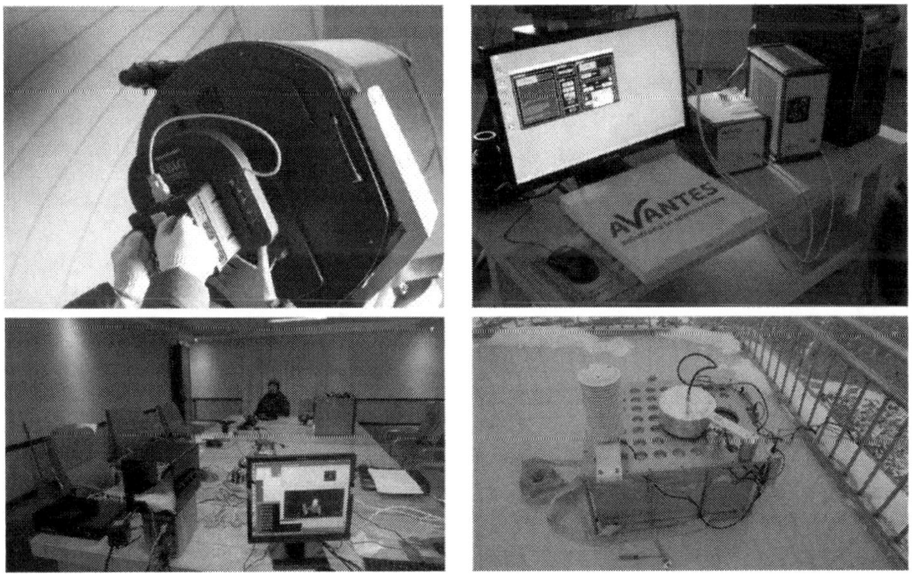

Figure 5. The 50cm telescope at Ali site equipped with large format CCD (top left) and optical-NIR spectrometer (top right). The setup of MIR imager for cloudiness and sky background emission (bottom left). The Fourier Transform Spectrometer measuring atmospheric opacity in the sub-mm bands(bottom right).

Figure 5 presents the configuration of various instruments at Ali site in 2014, including the 50cm telescope with optical image and spectrometer, the mid-infrared imager, and the Fourier Transform Spectrometer in the sub-mm bands, to show the effort to make real observations involving infrared and sub-mm wavebands. All the instruments have passed the test observations in the period of Jan.-March 2014.

4. The site development

The new Ali observatory is under construction for both detailed site characterization and the proposed small telescopes. The space of the current site seems not enough for an international level observing site. There are good possible sites on higher ridges located south and south-east directions, as shown in Figure 6. The higher ridge B seems to have good topographical location and much wider area. Also the peak C could be a good site for future large telescope. There is already a funded plan to construct the road to ridge B in 2015.

The best site, such as Ali area to reach a consensus, can be very helpful for general purpose astronomy and for cooperation in East Asia. A Site Committee will be organized internationally to decide individual telescope location plans, and to review and advise the site development.

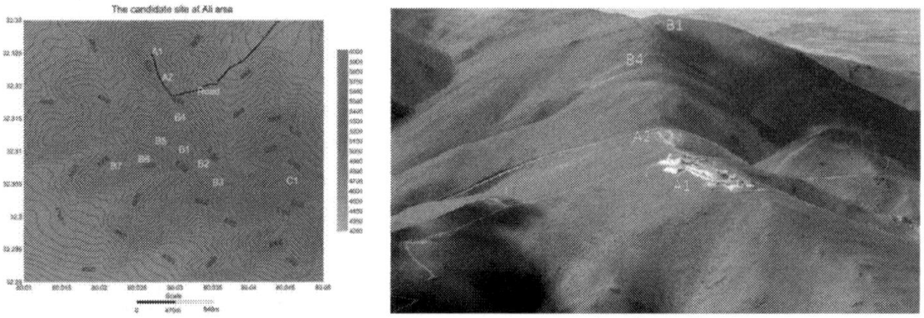

Figure 6. A topographical map of the Ali site, with marks A1, B1, C1 etc. for possible sites for large observing facilities, and the current site operated in point A1.

Acknowledgements

This work is supported by the National Natural Science Foundation of China (NSFC Grants 10903014, 11373043, 11103042, 11203044, 11303055 and 11473041).

References

[1] Yao Y 2005 *J. Korea Astron. Soc.* **38** 113
[2] Yao Y, Wang H, Liu L et al 2012 *Proc. SPIE* **8444** 84441K
[3] Stone R 2012 *Science* **337** 1156

Laser setup for measuring ground-level turbulence

L Yatcheva, R Barros, S Keary and S Gladysz
Fraunhofer Institute of Optronics, System Technologies and Image Exploitation IOSB
Gutleuthausstraße 1, 76275 Ettlingen, Germany

E-mail: lydia.yatcheva@iosb.fraunhofer.de

Abstract. Ground-level turbulence is usually strong and can thus be a limiting factor for many applications such as laser communications or astronomical observations. In this paper, the working principle of a setup for characterization of optical turbulence along a horizontal path in the lower boundary layer is described. The system includes a telescope focusing a laser beam onto a retro-reflector located at a distance of 410 meters as well as an optical test-bench comprising two cameras and a Shack-Hartmann wavefront sensor. Some of the challenges encountered when using this type of sensor for measurement of strong turbulence are discussed. Furthermore, two techniques for estimation of Fried's coherence length r_0 are presented and their results compared.

1. Introduction

Electro-optical systems, whether used for astronomical observations [1], remote-sensing and surveillance from space [2], tracking and high-resolution imaging of satellites [3], delivery of directed energy to space-based platforms [4], or horizontal-path imaging and laser communications [5], are always affected by atmospheric turbulence and in the majority of cases this turbulence imposes a fundamental limitation to their performance. Measurements of optical turbulence have a long tradition, from qualitative observations made by Newton [6] up to large scale measurement campaigns in preparation for the construction of future 30-m class optical telescopes [7].

It is well known that turbulence near the ground is usually the strongest. Still, the influence of turbulence on adaptive optics (AO) systems in this layer is rarely investigated. Our focus is laser propagation near the ground and for this application an experimental setup for turbulence measurement, characterization and testing of advanced wavefront sensing modalities was built [8]. The system includes a telescope to focus a laser beam onto a retro-reflector located at a distance of 410 meters, and an optical test-bench to measure wavefronts and focal-plane images, as well as pupil-plane intensity fluctuations of the reflected beam. The setup allows the collection of motion-corrected point-spread functions, individual wavefronts from the Shack-Hartmann sensor, and scintillation data, all simultaneously. In this paper the design of the experiment and the methods for determination of turbulence strength are discussed and some shortcomings of the Shack-Hartmann wavefront sensor under strong turbulence conditions are shown. In the future the setup will be used as the main optical system for testing of the holographic wavefront sensor [9].

2. Design of the setup

The concept of the system and its current implementation are shown in figure 1. The setup was designed and optimized to work in the visible spectrum (λ = 632.8 nm). However, for the purpose of free-space laser communications, further modifications will be made in order to operate in the near-infrared wave band (λ = 1.55 μm). In the present configuration, a Gaussian beam leaving the 30-cm aperture is focused on the retro-reflector while propagating through a random medium (air), and is reflected back to the transmitter through the same instance of the random medium (figure 1a).

a)
b)

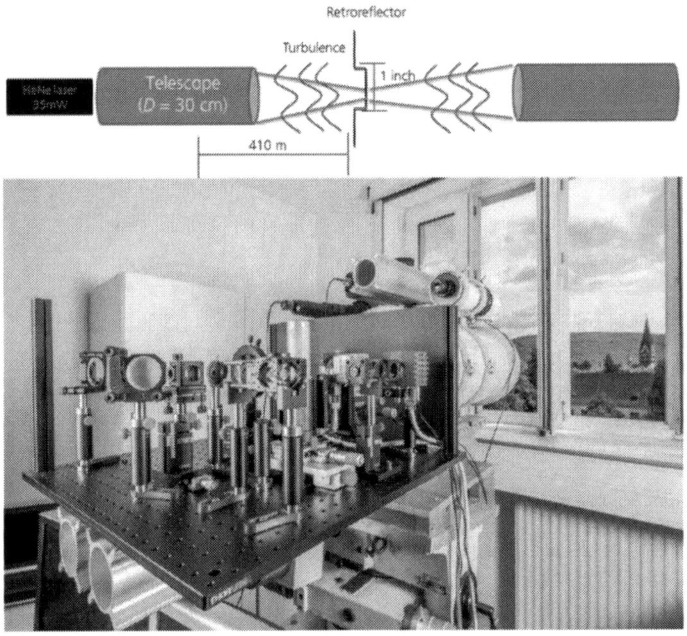

Figure 1. a) System concept: propagation geometry of the beam. b) Photograph of the experimental setup: test-bench, telescope and in the background the church tower where the reflector is located.

The corner cube retro-reflector rotates the wavefront by 180 degrees and sends it back in the direction of the transmitter. Thus, the setup is in principle a monostatic one albeit the area of the pupil where the coherence and subsequently intensity would be amplified by the double-pass propagation [10] is blocked by a large central obscuration of our Dall-Kirkham telescope. Since the peripheral parts of the wavefront are flipped horizontally and vertically upon reflection from the corner cube, it is assumed that statistics derived from these annular wavefronts will correspond to 820 m propagation of (almost) completely uncorrelated turbulence. Naturally, one must remember that in this particular case light is focused in the middle of the path and therefore the presented values of Fried's parameter correspond to such a propagation geometry only. Additionally, the retro-reflector sends light in the exact direction it originated from which makes the system insensitive to atmospheric tip and tilt.

Figure 2 depicts the optical setup schematically. The system is fed by a 35 mW linearly polarized, red laser beam that is focused by a converging lens (lens 1) on the image plane of the Dall-Kirkham telescope (Mewlon 300). The latter is conjugated with the retro-reflector plane located at the church tower 410 m away. Before leaving the system, the beam passes through a half-waveplate and is reflected by a polarizing beamsplitter (BS) through a quarter-waveplate. Thus the now circularly polarized beam travels to the reflector and back, and upon the second passage through the quarter-waveplate, has the opposite polarization with respect to the initial one. The incoming light can thus be collected by the telescope and re-collimated by a second lens (lens 2). A pellicle beam-splitter (BS 2) after the second lens produces two images of the entrance pupil of the telescope. One is dedicated to the Shack-Hartmann wavefront sensor (SHWFS) while the other is imaged again by a relay system (lenses 3 and 4) onto a high-speed camera for the purpose of scintillation measurements. Another beamsplitter (BS3) and a further detector placed at the focal plane of lens 3 allow concurrently measuring the focal plane image corresponding to a point spread function (PSF).

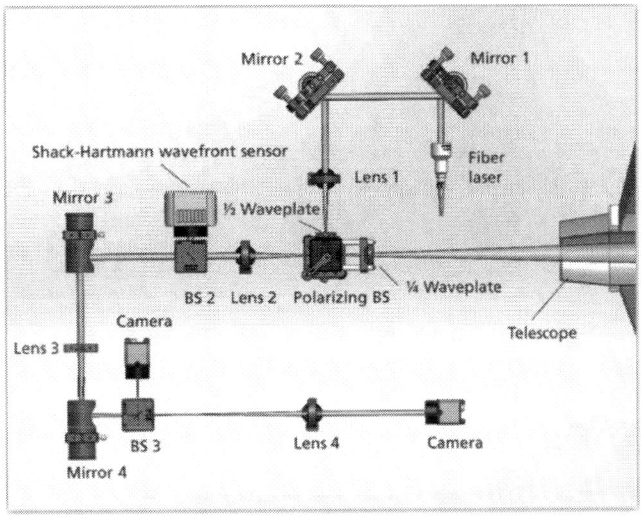

Figure 2. Detailed scheme of the optical test-bench.

The wavefront sensor used is SHSCAM from Optocraft with a 1.4-Megapixel, 8-bit greyscale Photonfocus sensor, with a frame rate set to 38 Hz and a square pattern lenslet array consisting of 42×31 lenslets with a pitch of 248 µm. PSF images are collected using a DCC3240M camera from Thorlabs, running at 100 fps. The exposure time for all cameras was set to 2 ms.

The measurements delivered by SHWFS and the PSF-camera are the subject of this paper. The scintillation index analysis is not described here as it has been presented in [8].

3. Turbulence strength estimation

For the purpose of this work, Fried's coherence length (r_0) was chosen to quantify turbulence strength under the propagation geometry shown in figure 1a. Two approaches for measurement estimation of r_0 are analyzed: Zernike mode variances fitting and fitting of the modulation transfer function (MTF). In the following, the methods are described in more detail.

3.1. Zernike variances

One approach for measuring turbulence strength involves estimation of r_0 based on the variance of Zernike modes [11].

For Kolmogorov turbulence, regardless of propagation geometry, it can be shown that the variance of any Zernike mode Z_n in radial order n can be represented as [12]:

$$Z_n = 1.095\,(n+1)\,\frac{\Gamma\left[-{}^{5}/_{6}+n\right]\Gamma\left[{}^{7}/_{3}\right]}{\Gamma\left[{}^{23}/_{6}+n\right]\Gamma\left[{}^{17}/_{6}\right]}\left(\frac{D}{r_0}\right)^{5/3} \quad (1)$$

where Γ is the gamma function and D is the telescope diameter.

Throughout the course of an experiment, the first 25 Zernike modes are recorded in regular intervals. Variances are computed for sets of 1000 measurements and compared to the corresponding theoretical values for different r_0 in accordance with the model. Thus, Fried's coherence length is obtained by the best fit to experimental data.

3.2. Fried's "short-exposure" MTF

The second method of estimating turbulence strength is based on Fourier optics.

Fried [13] derived a theoretical model for the tip-tilt-corrected ("short-exposure") MTF. In this theory, ensemble-averaged MTF, i.e., the Fourier transform of the PSF, is given by:

$$\mathbf{H}_S(\vec{u}) = \mathbf{H}_0(\vec{u}) \times \mathbf{H}_{SE}(\vec{u}) \tag{2}$$

where $\mathbf{H}_{SE}(\vec{u})$ is the average short-exposure MTF of the atmosphere only, and $\mathbf{H}_0(\vec{u})$ is the telescope's MTF ($\vec{u}$ is the two-dimensional spatial frequency). For a diffraction-limited circular aperture of diameter D we have:

$$\mathbf{H}_0(\vec{u}) = \frac{2}{\pi}\left[\arccos\left(\frac{\bar{\lambda} z u}{D}\right) - \frac{\bar{\lambda} z u}{D}\sqrt{1-\left(\frac{\bar{\lambda} z u}{D}\right)^2}\right] \tag{3}$$

where $u = |\vec{u}|$, $\bar{\lambda}$ is the average wavelength, and z is the distance from the exit pupil to the image plane. For the presented work, this approximation, although neglecting the central obscuration of the pupil and Gaussian apodization, was found to yield satisfactory results.

Furthermore, Fried postulated that:

$$\mathbf{H}_{SE}(\vec{u}) = \exp\left\{-3.44\left(\frac{\bar{\lambda} z u}{r_0}\right)^{5/3}\left[1-\alpha\left(\frac{\bar{\lambda} z u}{D}\right)^{1/3}\right]\right\} \tag{4}$$

where α is a parameter that varies between 1/2 when there are both intensity and phase variations across the collecting aperture and 1 when only phase distortions are present.

The method consists therefore of building up a temporal-average PSF, Fourier transforming it, and comparing it to a set of templates, based on equations (2) – (4). Since varying α was found not to affect the measurement result significantly, the only free parameter in these models is r_0 and thus the template that best fits the data automatically yields turbulence strength.

4. Efficacy of the Shack-Hartmann wavefront sensing in strong turbulence

The Shack-Hartmann wavefront sensor is a reliable and widely used solution in astronomical adaptive optics (AO). However, two fundamental characteristics can inhibit the application of this sensor to more challenging scenarios like laser propagation over long horizontal paths within extended-volume turbulence that causes scintillation and branch points in the wavefronts [14].

Firstly, due to the procedure of wavefront reconstruction the bandwidth of SHWFS is limited. This has consequences for deploying SHWFS-based AO systems on moving platforms and/or for satellite tracking. Secondly, SHWFS is highly sensitive to scintillation effects. Obscurations or saturations of parts of the sensor's pupil can lead to failure rate of the wavefront reconstruction process. We demonstrate the onset of this effect in figure 3 where circles denote either the errors signalled by the SHWFS software (gaps in the time series) or the ones which are easy to spot by eye but have not been flagged as errors by the software (sharp peaks and dips in the data).

Given this reservation, one needs to use datasets which were not corrupted by errors or such where errors could be quickly identified and removed.

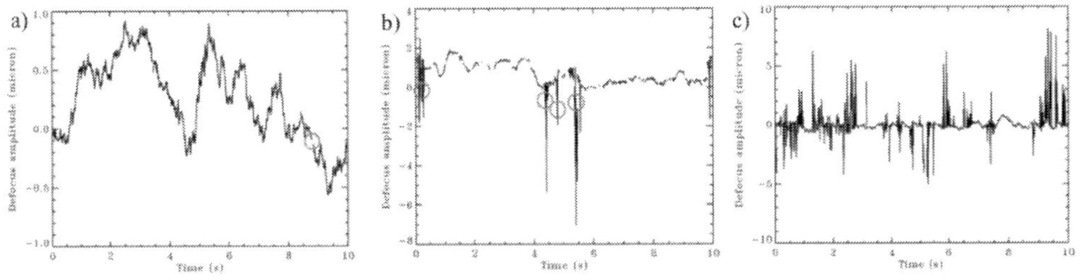

Figure 3. Values of defocus given by the Shack-Hartmann wavefront sensor at 9 am (a), 10 am (b) and 11 am (c) on the 21st of October 2014. In each panel 10 s of measurements are plotted. Please note the different scales. Turbulence is generally increasing from early morning until midday and then decreasing again. Consequently, errors can increase throughout the day as well. The presented data from 11 o'clock contains predominantly erroneous measurements and is thus difficult to draw conclusions from.

5. Experimental results

The two methods were numerically implemented and tested experimentally. Figure 4 shows a comparison of the yielded values throughout the course of one morning. A fairly good agreement between the values can be observed.

One should note, however, that turbulence did not vary significantly during this experiment, with an average r_0 of 7.1 mm and 7.8 mm for the MTF and Zernike fitting methods respectively. This was mainly due to the nearly constant meteorological conditions during the measurement time, with the temperature changing less than 2 °C.

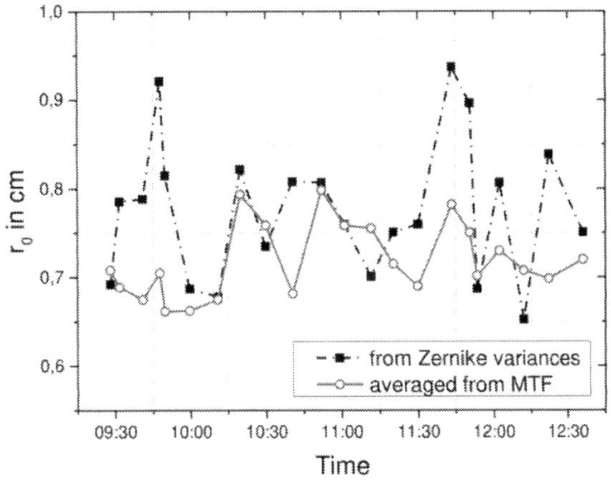

Figure 4. Comparison of the two estimation methods. Measurements from the 8th of December 2014. Dash-dotted line and square data points represent the values obtained from the Zernike variance fitting for sets of 1000 frames, while solid line and hollow circular data points depict values from averaging three subsequent sets of 3000 frames.

6. Conclusions and outlook

A laser setup capable of measuring turbulence on a horizontal urban path was successfully implemented and results from fitting of Zernike mode variances and modulation transfer functions were found to be in agreement despite some challenges of using the Shack-Hartmann sensor for this application. In the future, a portable measurement system, based on the same principles, would allow characterizing turbulence at different sites, e.g. in a telescope dome [15].

Furthermore, the assumption of Kolmogorov turbulence near the ground shall be investigated using the Shack-Hartmann angle-of-arrival structure functions [16]. Additionally, novel wavefront sensing modalities could be implemented [17]. Finally, the shortcomings of the Shack-Hartmann wavefront sensor could be compensated using a different device, such as e.g. the holographic wavefront sensor which is insensitive to scintillation [9].

References

[1] Davies R and Kasper M 2012 Adaptive optics for astronomy *Annual Review of Astronomy and Astrophysics* **50** 1-47

[2] Hajlaoui N, Chaux C, Perrin G, Falzon F and Benazza-Benyahia A 2010 Satellite image restoration in the context of a spatially varying point spread function *J. Opt. Soc. Am.* A **27** 1473-81

[3] Matson C L, Borelli K, Jefferies S, Beckner J C C, Hege E K and Lloyd-Hart M 2009 Fast and optimal multiframe blind deconvolution algorithm for high-resolution ground-based imaging of space objects *Appl. Opt.* **48** A75-A92

[4] Murphy D V 1992 Atmospheric-turbulence compensation experiments using cooperative beacons *Lincoln Laboratory Journal* **5** 25-44

[5] Vorontsov M, Riker J, Carhart G, Gudimetla V S R, Beresnev L, Weyrauch T and Roberts L C Jr 2009 Deep turbulence effects compensation experiments with a cascaded adaptive optics system using a 3.63 m telescope *Appl. Opt.* **48** A47-A57

[6] Newton I, 1952 Optics *Great Books of the Western World* vol 34 ed R M Hutchins (Chicago: Encyclopedia of Brittanica)

[7] Schoeck M *et al.* 2009 Thirty meter telescope site testing I: overview *Publications of the Astronomical Society of the Pacific* **121** no 878 384-395

[8] Barros R, Keary S, Yatcheva L, Toselli I and Gladysz S 2014 Experimental setup for investigation of laser beam propagation along horizontal urban path *Proc. of SPIE* **9242** 92421L

[9] Palomo P M, Zepp A and Gładysz S 2014 Characterization of the digital holographic wavefront sensor *Proc. of SPIE* **9242** 92421T

[10] Nelson W, Palastro J P, Wu C and Davis C C 2014 Enhanced backscatter of optical beams reflected in atmospheric turbulence *Proc. of SPIE* **9224** 922411

[11] Noll R J 1976 Zernike polynomials and atmospheric turbulence *J. Opt. Soc. Am.* **66** no 3 207-11

[12] Sasiela R J 2007 *Electromagnetic wave propagation in turbulence: evaluation and application of Mellin Transforms* 2nd ed (Bellingham, Washington USA: SPIE Press)

[13] Fried D L 1966 Optical resolution through a randomly inhomogeneous medium for very long and very short exposures *J. Opt. Soc. Am.* **56** 1372-79

[14] Zepp A, Gladysz S and Stein K 2013 Holographic wavefront sensor for fast defocus measurement *J. Adv. Opt. Technol.* **2** 433–37

[15] Berdja A, Guzman D, Saez N, David N, Dubost N, Sarazin M and Ziad A 2014 LOTUCE2: a dome-seeing instrument for the E-ELT *Proc. of SPIE* **9147** 91479R-1

[16] Nicholls T W, Boremann G D and Dainty J 1995 Use of a Shack-Hartmann wave-front sensor to measure deviations from a Kolmogorov phase spectrum *Optics letters* **20** no 24 2460-62

[17] Polans J, McNabb R P, Izatt J A and Farsiu S 2014 Compressed wavefront sensing *Optics letters* **39** no 5 1189-92

Cloud Cover Measurement from All-Sky Nighttime Images

Jia Yin, Yongqiang Yao, Liyong Liu, Xuan Qian and Hongshuai Wang

National Astronomical Observatories, Chinese Academy of Sciences, 20A Datun Road, Chaoyang District, Beijing 100012, China

E-mail: jyin@bao.ac.cn

Abstract. The site quality of astronomical observatory critically depends on cloud coverage, and the measurement of cloudiness is particularly important for site survey. A method to deal with all-sky images in no-moon nights is described. By identifying the positions of bright reference stars and making photometry for a set of all-sky images in clear nights, we can set up a reference image with median smoothing differential magnitude values. The standard image can be taken as the threshold for clear nights, and the detectivity of stars on other images can be utilize to reveal cloud coverage. Four types of all-sky images, clear night, icy lens, part of cloud, and full of cloud, are selected to check up the method. For the moonlight image, the formula of the CIE (Commission International de l'Éclairage) standard general sky model are used to fit the luminance distribution of the image, then to estimate the cloud cover of the image from the difference between the fitting image and the real image.

1. Introduction

The cloudiness is one of the most important parameters for evaluating the quality of the astronomical site. In recent years, the all-sky-camera is widely used to monitor the cloudiness in site testing campaign. The cloudiness image taken with the camera has a higher spatial resolution, and its analyzed result can accurately determine the local cloudiness and the characteristics of its variation. For the project of the astronomical site survey in Western China, the all-sky camera is actively adopted to make the monitoring and analysis of the cloudiness over the observatory and candidate site, such as Gaomeigu observatory, Karasu, Oma and Ali sites.

However, the methods for automatically processing the cloudiness images are quite few and remain to be deeply researched, and the cloudiness itself also lacks a clear and united quantization standard. As for the daytime cloudiness images, the luminance ratio of the blue wavelength to the red wavelength (B/R) is used to distinguish the cloud and clear sky[1], due to there exists the difference in the light scattering between cloud and atmosphere. As for the nighttime cloudiness images, the threshold cutting of B/R is not enough to distinguish the cloud and the night sky. By means of a more glancing method the images are made into the corresponding animated cartoon and the cloud cover is judged with the naked eyes[2], but the influence of the artificial factors can not be neglected. Smith et al. proposed a kind of more accurate method, the cloud cover is judged from the extinction of atmosphere or opacity of cloud, which obtained by the photometry of the bright background stars[3-4]. However, this method needs a lot of clear-night images to be accumulated as the standard image to compare.

Based on the cloudiness data of Gaomeigu observatory, we had developed the effective method to determine the cloudiness of night image. The photometry method of bright stars and the CIE model are adopted for the no-moon image and the moonlight image, respectively.

2. Cloud monitoring and image processing

The SBIG All-Sky-340 camera is adopted as the cloud monitor. The field of view (FOV) of the SBIG All-Sky-340 camera is $140° \times 185°$, of which the CCD system has a very high sensitivity. The exposure time is automatic adjustment, and the brightness variation of stars or the fluctuation of sky luminance with moonlight can be detected.

2.1. The no-moon image

The all-sky camera adopts the fixed-point photography mode, the coordinates of the bright stars in the image are correlated to the earth's rotation and geographical position, therefore, the astronomical position of a bright star can match its position in the image through a coordinate transformation. The matching method proposed by Pickering[5], which can be used to carry out the coordinate transformation between the image and the catalogue.

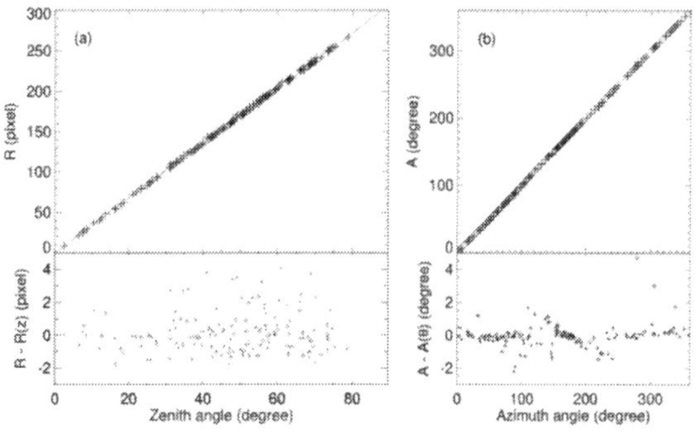

Figure 1. Comparison between the fitting position of the bright star in the image and its horizontal coordinates converted from the catalogue. (a) the comparison between the polar distance of the bright star and the zenith distance, (b) the comparison between the two groups of azimuthal angles.

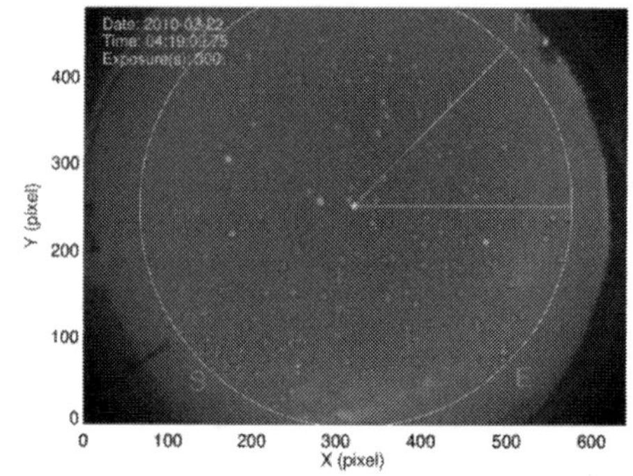

Figure 2. Comparison between the fitted result of the bright stars and the actual coordinates, where the blue square and red sign "+" show the actual coordinates of the bright stars and the fitted result, respectively.

The right ascension, declination, apparent magnitude and name of the bright stars are obtained from the Yale bright star catalogue. The bright stars selected from the catalogue are converted from the equatorial coordinates (α, δ) into the horizontal coordinates (A, h or z). In the image, the polar distance R from the center of the image has a linear relation to the zenith distance z, and a constant offset exists between the polar angle θ and the azimuthal angle A. Thus, the corresponding relation between the polar coordinates (R, θ) and the image coordinates (x, y) can be given as follow:

$$R = a \times z, \quad (1)$$
$$\theta = \theta_{\text{off}} - A, \quad (2)$$
$$x = x_0 + R \times \cos\theta, \quad (3)$$
$$y = y_0 + R \times \sin\theta. \quad (4)$$

(x_0, y_0) indicates the coordinates of the zenith in the image, $\theta_{\rm off}$ is the constant offset.

Fig.1(a) shows the comparison between the two sets of coordinates of the bright stars, the abscissa is the horizontal coordinate converted from the Yale catalogue and the ordinate is the polar coordinate converted from the image coordinate. In Fig.1(b) the ordinate is $A = \theta_{\rm off} - \theta$, which corresponds to the azimuthal angle of the bright stars. The comparison is also shown in Fig.2, with circle $z = 75°$ and the acute angle $\theta_{\rm off}$.

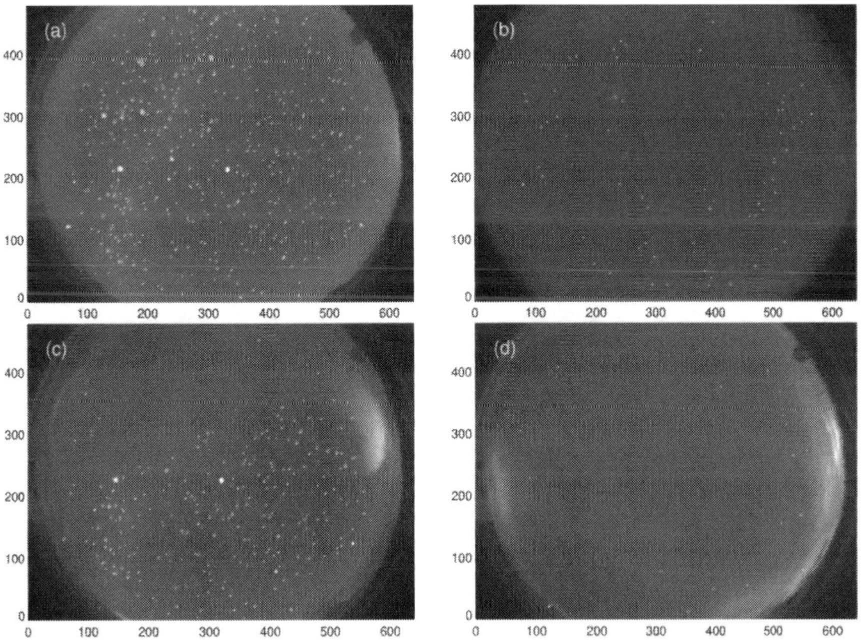

Figure 3. The test images with detected stars: (a) clear-night image, (b) the icing lens, (c) part clouds and (d) full of clouds, respectively.

Based on the matching method, most bright stars of the catalogue can be easily found in the all-sky image, and then be estimated the photometric magnitude ($m_{\rm phot}$) of the bright point at its position. The magnitude difference of the stars can be calculated as formula $\Delta m = m_{\rm phot} - m_{\rm tab}$, $m_{\rm tab}$ can be obtained from the catalogue. The magnitude difference Δm of a few clear night images are computed to construct the reference image, which reflects the atmospheric extinction on the position of the bright star for the clear image. The difference Δm of the other position (pixel) in the reference image is obtained from the median value of Δm at 30 adjacent position.

After the reference image is constructed, the absorption of clouds can be described as formula: $A_{\rm cloud} = \Delta m - \Delta m_{\rm std}$. $\Delta m_{\rm std}$ is the corresponding value of Δm in the reference image. For the clear-night images, the statistical distribution of $A_{\rm cloud}$ exhibits the standard normal distribution. The Gauss fitting on the $A_{\rm cloud}$ distributions of the clear-night images is carried out, and then the mean σ is calculated. For the images with clouds, their $A_{\rm cloud}$ distributions is slant to the right side, not a standard normal distribution. The 3σ value is taken as the threshold to distinguish the bright stars, which become dimmer for the absorption of cloud or not be affected ($N_{\rm good}$). Therefore, the cloudiness of one cloud image can be represented as $P_{\rm cloud} = 1 - N_{\rm good}/N_{\rm std}$, N_{std} is the average value of the $N_{\rm good}$, which is obtained from the clear night images.

There are 4 sorts of cloudiness images selected to test the usability of the above-mentioned method, including the clear image, the image with icing lens, the image with part clouds and the image full of clouds. Fig.3 shows the result of detected bright stars for 4 types of images, the cross sign "+" expresses the position of the detected star which satisfies the threshold condition.

2.2. The moonlight image

The night-time images with moon are seriously affected by the moonlight, especially for the full moon time. The bright stars can not easily to extract from the image with moon, and can not be used to estimate the cloudiness of the image. If one can get the luminance distribution of the sky with moonlight, the cloud cover can be determined by the difference between the luminance distribution and the cloudiness image. Krisciunas[6] proposed a model of brightness of moonlight, which is a function of the Moon's phase, the zenith distance of the Moon, the zenith distance of the sky position, the angular separation of the Moon and sky position, and the local extinction coefficient. The IDL software package is adopted to fit the model of the moonlight and a clear night image, but the difference between the fitting result and the night image is not good enough to distinguish the cloudiness.

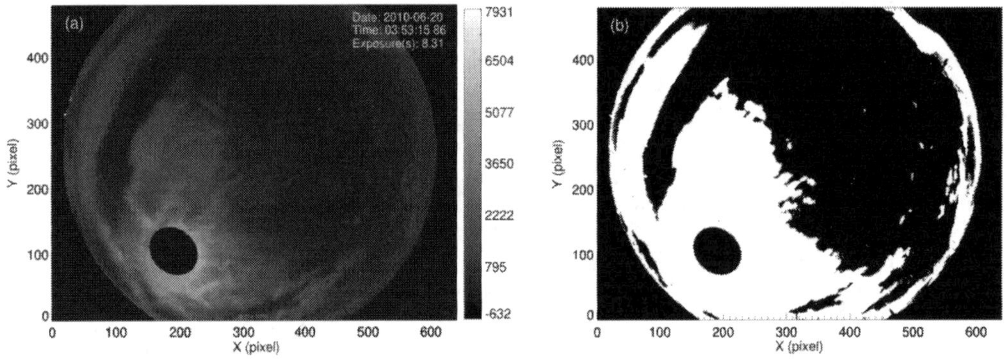

Figure 4. (a) The selected cloud image with moon, (b) the binary image obtained from the difference between the CIE model and the cloud image.

For the nighttime luminance of moon from the sunlight reflected, there is a similar luminance distribution between the moonlight and the daylight. The CIE standard general sky model is considered to carry out the moonlight distribution, like the moonlight model, to estimate the cloud cover by the difference between the fitting result and the cloudiness image [7-8]. The formula of CIE model as following :

$$L_s = \frac{(1 + C(e^{D\chi} - e^{D\pi/2}) + E\cos^2\chi)(1 + Ae^{B/\cos z_s})}{(1 + C(e^{Dz_\odot} - e^{D\pi/2}) + E\cos^2 z_\odot)(1 + Ae^{B})} \cdot L_z, \quad (5)$$

and,

$$\chi = \arccos(\cos z_\odot \cos z_s + \sin z_\odot \sin z_s \cos(|A_\odot - A_s|)). \quad (6)$$

where, χ is the distance of sky element from the sun, ($A_\odot$, $z_\odot$) and (A_s, z_s) indicates the horizontal coordinates of the sun and the sky element, respectively. The L_z is the zenith luminance of the clear night image. The parameters A, B, C, D, E of the CIE model is determined by the comparison between the fitting result and a clear image with moon, which similar to the parameters of CIE standard clear sky with low illuminance turbidity. With the fitting parameters, the reference image is computed from the formula of the CIE model for one moonlight image with cloud. Fig.4(a) shows the difference between the reference image and the moonlight image. A certain threshold is set to distinguish the cloudiness, and get the binary image (as shown in Fig.4(b)).

3. Conclusion and discussion

We develop a method to estimate the cloudiness of the all-sky nighttime image without moon, and the feasibility of this method is tested by four types images. The result of testing shows that the cloud cover in the no-moon image can be determined effectively by the method. For the image with moon, we try to use the CIE model to estimate the cloud cover of the image. The difference between the clear sky luminance of the CIE model and the cloud image mainly caused by the cloud cover, which can be used to estimate the cloudiness of the image.

The determination of the cloudiness still exists some deviation, such as the unstable atmospheric conditions. As the no-moon image, the photometry method is accurate enough to quantize the cloudiness, the simplification of atmospheric extinction will be better. As the image with moon, the CIE model can be used to estimate the cloudiness of image, while the zenith luminance of image need to been obtained, which is difficult to the image with cloud. Moreover, a reasonable approach should be develop to obtain the certain threshold to distinguish cloud cover.

Although the photometry method or sky models can be used to estimate the cloudiness of images, the processing is not easily and directly. If more characteristic of the cloud can be given in the image, including multi-band information, light polarization, absorption of cloud, etc, the statistics of the cloudiness will be easier and more accurate than just using the grey level information.

Acknowledgments

This work is supported by the National Natural Science Foundation of China (grant Nos. 11103042, 11203044, 11373043 and 11303055).

References

[1] Shi Y, Yao Y Q and Liu L Y 2008 *Astronomical Research and Technology* **5** 415
[2] Skidmore W, et al 2008 *Ground-based and Airborne Telescopes II* vol 7012, ed L M Stepp and R Gilmozzi *Proc. Society of Photo-Optical Instrumentation Engineers* June 23, 2008, Marseille France
[3] Smith R, Walker D and Schwarz H E 2004 *Scientific Detectors for Astronomy: The Beginning of a New Era* vol 300, ed P Amico, J W Beletic and J E Belectic (Dordrecht: Kluwer Academic Publishers) p 379
[4] Shamir L and Nemiroff R J 2005 *Publ. Astron. Soc. Pac.* **117** 972
[5] Pickering T E 2006 *Ground-based and Airborne Telescopes* vol 6267, ed L M Stepp *Proc. Society of Photo-Optical Instrumentation Engineers* May 24, 2006, Orlando Florida USA
[6] Krisciunas K and Schaefer B E 1991 *Publ. Astron. Soc. Pac.* **103** 1033
[7] Hosek L and Wilkie A 2012 *Association for Computing Machinery Transactions on Graphics* vol 31 (New York: ACM) p 1
[8] CIE DS 011.2/E:2002 *Spatial distribution of daylight - CIE standard general sky* Draft standard (Vienna: CIE Central Bureau)

Towards a fully automatic and robust DIMM (DIMMA)

A.M. Varela[1,2], C. Muñoz-Tuñón[1,2], A.M. del Olmo-García[1,2], L.F. Rodríguez[1], J.M. Delgado[1] & J.A. Castro-Almazán[1,2]
[1]Instituto de Astrofísica de Canarias, E-38200, La Laguna, Spain
[2]Dept. Astrofísica, Universidad de La Laguna, E-38200, La Laguna, Spain
E-mail: skyteam@iac.es

Abstract. Quantitative seeing measurements have been provided at the Canarian Observatories since 1990 by differential image motion monitors (DIMMs). Image quality needs to be studied in long term (routine) measurements. This is important, for instance, in deciding on the siting of large telescopes or in the development of adaptive optics programmes, not to mention the development and design of new instruments. On the other hand, the continuous real time monitoring is essential in the day-to-day operation of telescopes.

These routine measurements have to be carried out by standard, easy-to-operate and cross-calibrated instruments that required to be be operational with minimum intervention over many years. The DIMMA (Automatic Differential Image Motion Monitor) is the next step, a fully automated seeing monitor that is capable of providing data without manual operation and in remote locations. Currently, the IAC has two DIMMs working at Roque de los Muchachos Observatory (ORM) and Teide Observatory (OT). They are robotic and require an operator to start and initialize the program, focus the telescope, change the star when needed and turn off at the end of the night, all of which is done remotely. With a view to automation, we have designed a code for monitoring image quality (avoiding spurious data) and a program for autofocus, which is presented here. The data quality control protocol is also given.

1. Introduction

The optical conditions are crucial both for site characterization and selection, and for feasibility studies for adaptive optics. The DIMM provides the full atmospheric seeing with an accuracy better than 0.1, with a sampling rate better than 1 datum/minute, and produces accurate, absolute and reproducible data [1]. Control tests are important and have to be implemented routinely (e.g. [2], [3]). The DIMM is the ideal instrument for long term monitoring of image quality and has been adopted at the Canarian Observatories [4].

The basis of the DIMM is to separate two light beams coming from the same star but received in two separate zones of the primary mirror (telescope aperture). This task is accomplished by completely covering the telescope aperture except for two small apertures. The light going through one of the apertures will be slightly deflected by a prism, thus obtaining two images of the same star in the focal plane. Seeing data can be computed from the variance of the distance between the two images, in both the horizontal and vertical axes (more details in [5] and [1]).

Routine measurements require instruments to be working with minimum operational life-span of many years. In this sense, for the optimum performance of the DIMM, we have designed and

Journal of Physics: Conference Series **595** (2015) 012041 doi:10.1088/1742-6596/595/1/012041

implemented new applications towards obtaining a fully automated DIMM (hereafter DIMMA).

A number of systems based on this scheme are at present installed in most astrophysical observatories and are being routinely used for site surveys. There are two DIMMs installed at the Canary Islands Observatories (Roque de los Muchachos Observatory - La Palma - and Teide Observatory - Tenerife; see Figs 1 and 2); however, nowadays operators (IAC Telescopic Operation Technicians at Teide Observatory or TOTs) are needed.

The DIMMA system (Differential Image Motion Monitor *Automático*) is a step forward in telescope-based equipment used for site evaluation in astrophysics. Using the experience and know-how of the DIMM systems [1] regularly operated since 1995 at the IAC by the Sky Quality Team, the DIMM provides a comprehensive set of data without the need for manned operation, in close linkage with the rest of the world using internet.

Figure 1. Image of the DIMMA station at the ORM (Las Moradas site), 300 m east of the Gran Telescopio Canarias (GTC).

Figure 2. Image of the DIMMA station at Teide Observatory, close to the ESA Optical Ground Station (OGS).

Currently, the IAC DIMMs are robotic and require an operator to start and initialize the program, focus the telescope, change the star when needed and turn off at the end of the night or in bad weather conditions, all remotely. With a view to complete automation, we have designed a code for monitoring the image quality (avoiding spurious data) and a program for autofocus, including software and web cameras for surveillance, alerts and protection.

The second phase for full automation will incorporate automatic scheduling for observations and appropriate response in bad weather conditions. Figure 3 shows the configuration of the DIMMA. The dome control is completely independent. All components and sensor status are controlled with a common interface. The system is designed and implemented in LabVIEW.

2. Autofocus algorithm

The autofocus algorithm is summarized in Figure 4. The cluster focus configuration, i.e. the input parameters used, is based on the hardware (CCD, mechanics of the focus motor, etc.) of our DIMM system.

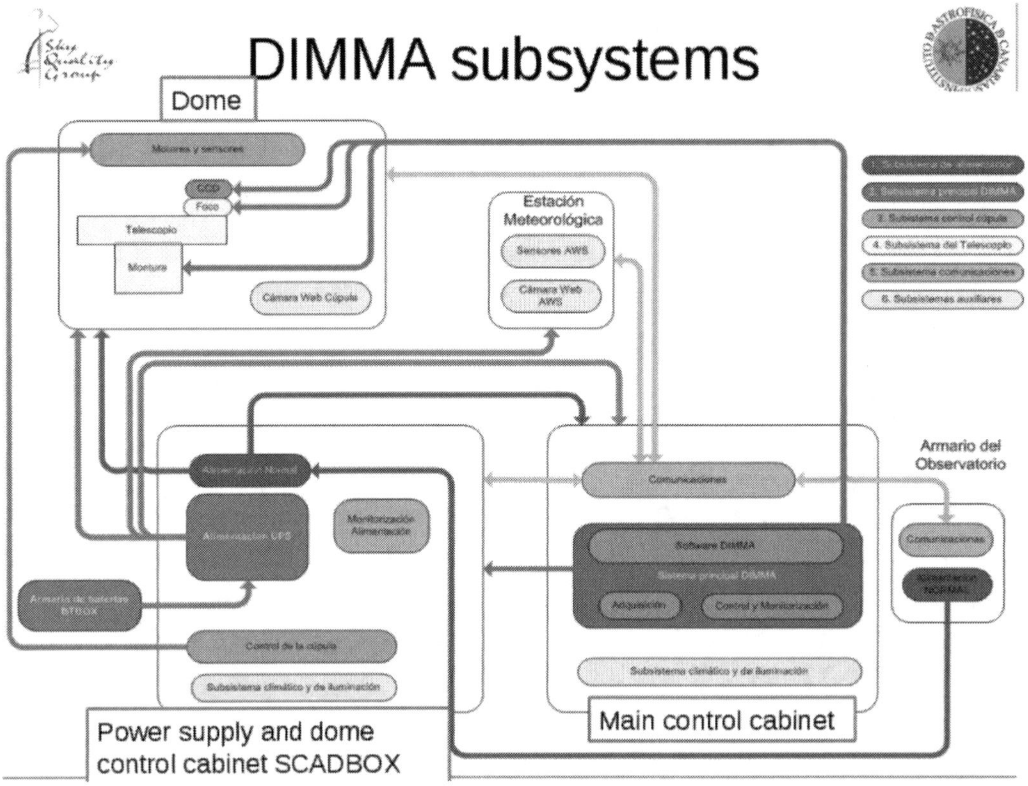

Figure 3. Configuration of the DIMMA: power supply (red); main DIMMA subsystem - software for acquisition and control of data - (blue); dome control (orange); telescope subsystem (yellow); communication (green) and auxiliary subsystems - web cameras and automatic weather station sensors - (white).

The initial sampling position ($PosIni$) or distance between the centroids of spots is suggested to be 25 pixels (corresponding in our case with 18180 counts, using a focus change or Step = 10 counts). The final position $PosEnd = PosIni + (Step \times Nsample)$, must be set up at 40 pixels. N images (20) are scanned, stored and processed, the mean and the standard deviation (σ) of the mean of the distance of the centroids are calculated. The σ threshold is 3 pixels and the maximum allowed σ is 10 pixels. This process is repeated 30 times ($Nsample$), and finally we plot the relationship between the distance in pixels and the focus position to obtain the best linear regression (see Fig. 5). We want to obtain a focus position precision better than 1 pixel, therefore, the scan must be smaller than the counts per pixel ratio, but it is limited by the mechanics of the motor. The focus position from 25 to 40 pixels is summarized in Table 1. The theoretical focus values for DIMMA-ORM and DIMMA-OT are 35 and 33 pixels respectively.

3. Quality control of the data

The first control routine concerns the number of accepted and rejected data. Reasons of rejection are mainly associated with a discrepancy greater than 20% between the fwmh longitudinal (fwhml) and the fwhm transversal (fwhmt). This could be an indication of focus problems and/or conditions where the technique cannot be employed (see [1] for further details). Other

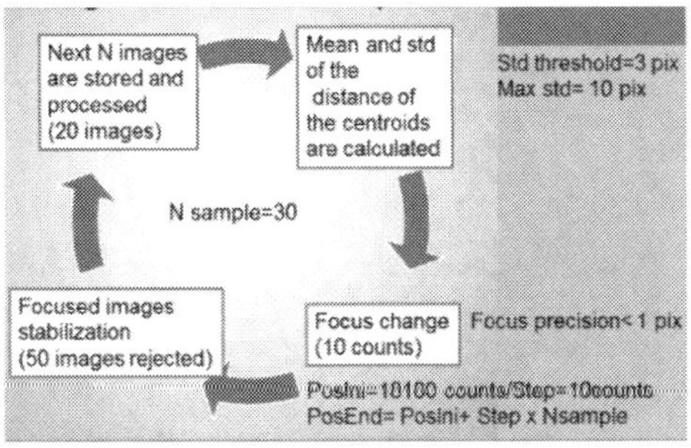

Figure 4. Autofocus algorithm for the DIMMA.

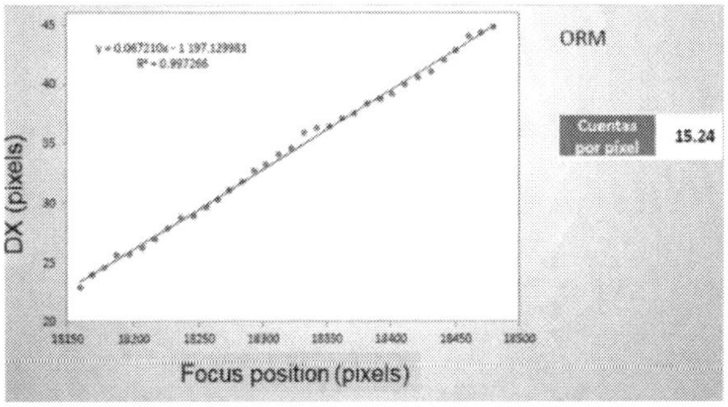

Figure 5. Relationship between the distance in pixels and the focus position.

reasons for the rejection of data (automatically identified with an error code) and the percentages of occurrence are indicated in Table 2, corresponding to 551 observing nights with the DIMM at Teide Observatory. The statistics of the rejected measurements in a given time-lapse is an indicator of the robustness of the accepted data. The percentages of rejected data are represented in Fig. 6, the total number of rejected data being 13% and 11.9% for the DIMMA–OT and DIMMA–ORM respectively. The main cause of lost data is the discrepancy between the fwhml and the fwhmt, being 66.4% and 62.3% of the total rejected data for the DIMMA–OT and DIMM–ORM respectively, followed by the low signal to noise ratio of the images and unfocused images (this study was performed without autofocus implementation).

Besides the IAC equipment, there are two more DIMMs installed at the ORM summit

Table 1. Focus position results for DIMMA-ORM and DIMMA-OT.

pixels	DIMMA ORM (counts)	DIMMA OT (counts)
25	18181	12815
33	–	12646
35	18344	–
25 – 40	229	-273

Table 2. Causes and percentajes of rejected data for a sample of 551 and 304 observing nights at the OT and ORM respectively.

Code #	Diagnostic	Rel. Freq	Abs. Freq
0	**No errors**	**86.98** %	**377661**
9	Full discrepancy	0.00 %	9
11	Distorted spots	0.16 %	696
12	Unfocused images	1.04 %	4517
13	Low SNR	2.32 %	10084
14	No maximum	0.01 %	48
16	Lost images	0.00 %	18
17	Offset image	0.34 %	1494
31	Wrong airmasses	0.23 %	995
50	fwhml/fwhmt > 20 %	8.64 %	37520
99	User's stopping	0.25 %	1103
1001	Source of images disconnected	0.00 %	6
-10	Negative codes	0.01 %	22
Total:		100.00 %	434173

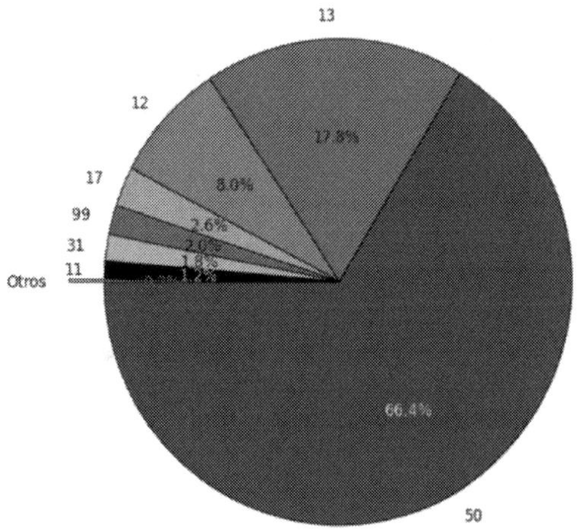

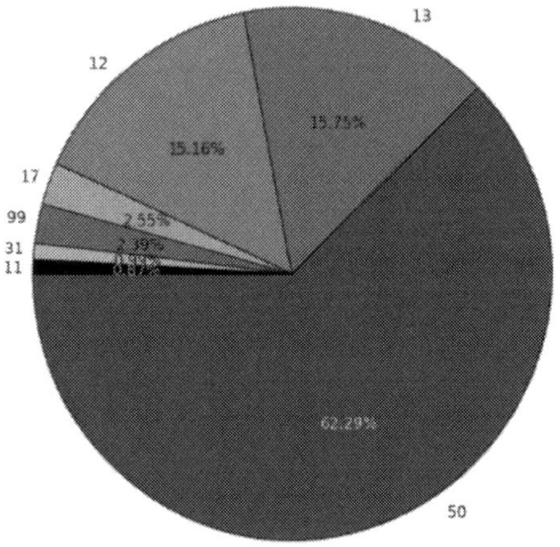

Figure 6. Diagram of the causes and percentages of rejected data for the DIMMA-OT (left) and the DIMMA-ORM (right). The total rejected data are 13.0% and 11.9% for OT (551 nights) and ORM (304 nights), respectively. See rejection codes in Table 2.

operated by the Isaac Newton Group (ING) and Telescopio Nazionale Galileo (TNG) teams. We perform routine comparisons (the most recent was published in Canarian Observatories Updates (CUps) [6] The statistical results are consistent to within the errors. The difference in the median seeing is less than 0.1 arcsec.

4. Remarks

- Two cross-calibrated DIMMs have been in operation at the OT and ORM since March 2011 and April 2012 respectively.
- A robust and reliable automated DIMMA capable of providing data without manual operation and in remote locations is being developed at the IAC.
- Automatic focus and data quality control has been satisfactorily implemented.

The future is the implementation of the automatic scheduling and weather monitoring.

Acknowledgments
This work has been funded by the IAC. We thank the IAC Telescopic Operation Technicians at Teide Observatory (TOTs) for the operation of the DIMMs and the Instrumental Maintenance and Civil Maintenance teams for their work with the DIMMs. Finally, we are grateful to Terry Mahoney for the language correction of the manuscript.

References

[1] Vernin J and Muñoz-Tuñón C 1995 *PASP* **107** 265–272
[2] Varela A M, Muñoz-Tuñón C and Vernin J 2004 *Second Backaskog Workshop on Extremely Large Telescopes* (*SPIE Conference Series* vol 5382) ed Ardeberg A L and Andersen T pp 656–660
[3] Tokovinin A and Kornilov V 2007 *MNRAS* **381** 1179–1189
[4] Muñoz-Tuñón C, Varela A M and Castro-Almazán J A 2015 *These Proceedings*
[5] Sarazin M and Roddier F 1990 *A&A* **227** 294–300
[6] Varela A M, Muñoz-Tuñón C and Castro-Almazán J *Canarian Observatories Updates* –`www.iac.es/cups`– *CUps* **2-2014**

Astronomical Site Characterization at the Canarian Observatories

C. Muñoz-Tuñón,[1,2] A.M. Varela[1,2] and J.A. Castro-Almazán,[1,2]

[1]Instituto de Astrofísica de Canarias, E-38200, La Laguna, Spain

[2]Dept. Astrofísica, Universidad de La Laguna, E-38200, La Laguna, Spain

E-mail: skyteam@iac.es

Abstract. Roque de los Muchachos Observatory (La Palma) and Teide Observatory (Tenerife) are prime astronomical sites, as confirmed by more than 30 years of intensive site-testing campaigns. The IAC has long been aware of the importance of promoting initiatives for the characterization and protection of the Canarian Observatories. For this purpose, in the late '80s a Sky Team was created to measure the atmospheric parameters relating to astronomical observations, to design and develop new instruments and techniques for astronomical site testing, and to improve and maintain a high level of instrumentation in site characterization. New instruments and techniques are welcomed by the Observatories.

1. Introduction

More than 60 Institutions have facilities at the Canarian Observatories and also contribute to the better characterization of the sites. SUCOSIP[1] is the sub-committee that brings together the work carried out by different countries. The most updated databases and models (NCEP/NCAR, WRF, satellites, etc.) are also explored and used [1], [2], [3], [4]. We present a summary of the instruments, results and projects for monitoring the relevant parameters for site characterization: weather conditions, clear/useful time, precipitable water vapor content, sky brightness, image quality and atmospheric turbulence [5], [6], [7] and [8]. The data are also useful for telescope operation, for nightly observing scheduling and forecasting [9], and for feasibility studies for adaptive optics [10], [11], [12], [13].

2. The locations

In Figs 1 and 2 we show the locations at the Roque de los Muchachos (ORM) and at Teide (OT) Observatories where site-testing instruments are currently installed. The Observatories are located in La Palma and Tenerife, respectively (Canary Is., Spain).

3. Seeing

Seeing measurements are taken with instruments based on the Differential Image Motion technique [14]. In Figs 3 and 4 we show the Differential Image Motion Motion Monitors –

[1] The SUb-COmmittee on SIte Properties (SUCOSIP) has been established within the International Scientific Committee (CCI; `http://www.iac.es/eno.php?op1=5&lang=en`) to guarantee the preservation of the astronomical conditions at the Canarian Observatories. This committee reviews all new developments at an Observatory to evaluate their possible impact on the existing telescopes.

Figure 1. Map of Roque de los Muchachos Observatory, indicating locations of various instruments routinely operating for site testing.

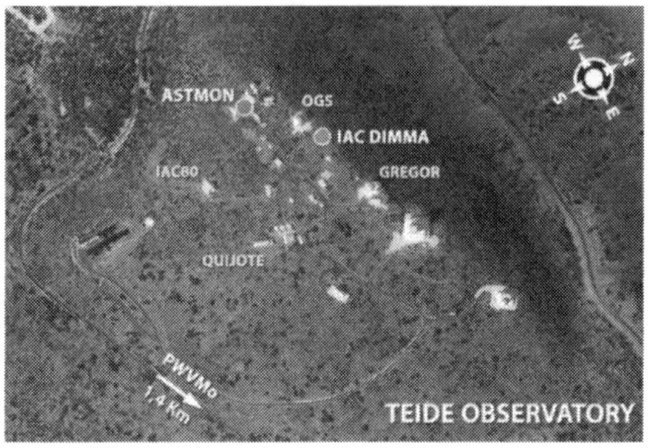

Figure 2. Map of Teide Observatory, indicating locations of various routinely operating instruments for site testing.

or DIMMs– located at the ORM and the OT. Both are installed on a 5 m tower in order to avoid surface-layer effects. The DIMMs are remotely operated by the IAC Telescopic Operation Technicians (TOTs) and supported by the Instrument Maintenance Department of the IAC.

There are other instruments for seeing monitoring at the ORM belonging to the Telecopio Nazionale Galileo (TNG DIMM) and to the Isaac Newton Group (ING RoboDIMM), shown in Figures 5 and 6 respectively (see Appendix A for a compilation of websites). Systematic comparison among measurements are carried out (see [15] for the last one). Seeing data gathered at the ORM and at the OT (see Fig. 7) are provided in real time in the web page of the Sky Quality Team for the telescopes operation (see Appendix A, web 13). The data are automatically stored for statistical analysis and available upon request.

Figure 3. IAC DIMMA at Las Moradas site (ORM), 300 m east of the Gran telscopio Canarias (GTC). Colour code in Figure 1.

Figure 4. IAC DIMMA at the OT, installed close to the Optical Ground Station (OGS). Colour code in Figure 2.

Figure 5. DIMM installed close to the TNG telecope (ORM). Colour code in Figure 1.

Figure 6. DIMM installed close to the William Herschel Telescope of the ING (RoboDIMM). Colour code in Figure 1.

4. Meteorology

Meteorological parameters are continuously recorded by Automatic Weather Stations (AWSs) belonging to various institutions operating at the Canarian Observatories. The IAC automatic weather stations consist of a metal framework holding the data acquisition unit (DAU), a combined hygrometer and air-temperature sensor, a solarimeter, a surface-wetness probe and a temperature-compensated barometer. A 10 m lattice tower supporting an anemometer and a wind vane is erected near the station. The system is supplied with a solar power generator and a radio communication system. The data logger can be programmed either to take instantaneous data or time-averaged data. Wind speed is measured every 10 s; the average during a minute is the wind speed and the maximum instantaneous measurement during the minute is the gust value. More details in [16].

Real time meteorological parameters provided by different weather stations at the ORM are compiled in a common webpage managed by the Isaac Newton Gropup (Appendix A, web 5). Weather information at the OT is provided by the IAC (Appendix A, web 7).

5. Precipitable Water Vapour content

Precipitable Water Vapour (PWV) content is crucial for infrared atmospheric characterization, and also for for the operational use of the CanaryCam instrument at GTC, and for optimizing queue-scheduled observations. The PWV monitoring is based on GPS data processing from a permanent GPS installation [17]. An estimation process is run every 30 minutes using fixed orbits to produce the latest PWV estimation for the ORM. The real-time data are published on the Sky Quality Team website (Appendix A, web 16) and the interface is in Figure 8. The PWV monitor has been extended to the OT and is going to be released briefly.

6. Sky brightness

One of the most damaging effects for astronomy is brightness or glare in the night sky caused by artificial light reflecting off gases and particles in the air. The Sky Quality Protection Technical Office (OTPC) was set up by the IAC in January 1992 to provide advice on the application

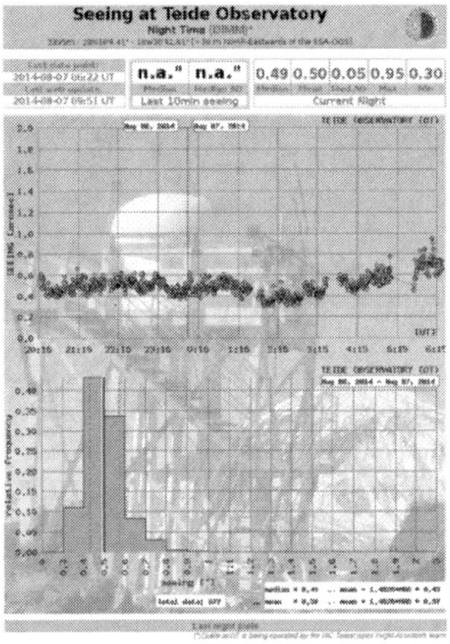

Figure 7. Webpage of the on-line seeing data retrieved at the Canarian Observatories by the DIMMAs (Appendix A, webs 14 and 15). The upper plot shows the seeing values in UT, while the lower plot shows the histogram of the seeing values for that night.

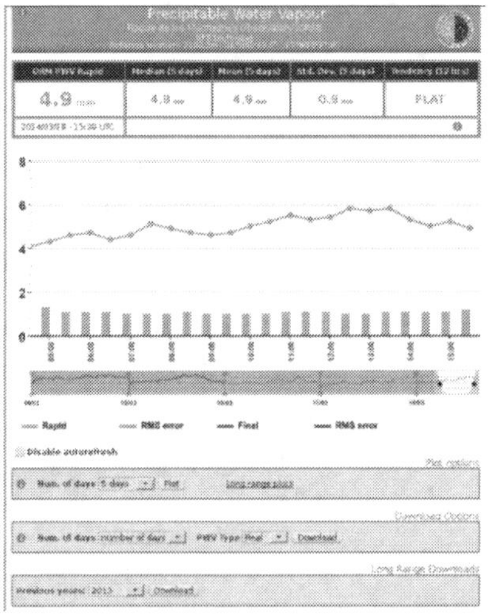

Figure 8. Webpage showing the on-line precipitable water vapour content gathered at the Canarian Observatories (Appendix A, webs 16 and 17).

of the Sky Law (Law 31/1988), which protects the astronomical quality of observatories in the Canaries from light pollution, radioelectric and atmospheric pollution and aviation routes (see Appendix A, web 8). The OTPC also develops important tasks concerning the application of the law and advising in the implementations of lights in the cities of the Islands [18].

ASTMON (All-Sky Transmission Monitor) and SQM (Sky Quality Meter) are the most recommended instruments for monitoring the sky brightness. Fig. 9 shows the ASTMON located at the OT. Software to provide the clear night fraction is in preparation in collaboration between IAC, Complutense University of Madrid and GTC. SQM at the ORM and at the OT provide on-line data through the IAC's webpage (Appendix A, webs 9 and 10).

7. Atmospheric extinction

Sky transparency is a key parameter, as it defines the quality of the photometry to be acquired in astronomical observations. Extinction is associated with the absorption/scattering of incoming photons by the atmosphere of the Earth and is characterized by the extinction coefficient, K. Sources of sky transparency degradation are clouds (water vapour) and aerosols (dust particles included). This coefficient is wavelength-dependent and can be determined by making observations of a star at different airmasses ([2] and [19]). The Carlsberg Meridian Telescope (CMT, formerly the Carlsberg Automatic Meridian Circle), a 17.8 cm diameter telescope, started operations in 1984 May and has since then provided continuous automated measurements of atmosphere extinction (see Fig. 10). the longest and most complete and homogeneous in situ database available for any observatory. Relative monthly frequency of weather downtime at the CMT from the log for 5 yr compared with that of the WHT for 18 yr is shown in Fig. 11. The

Figure 9. ASTMON installed at the OT. Colour code in Fig. 2. There is also an ASTMON at Las Moradas (ORM; Fig. 1).

Figure 10. The Carlsberg Meridian Telescope has been operative at the ORM since 1984. Colour code in Fig. 1.

average monthly weather downtime estimated from the CMT data log is 20.7%, in reasonable agreement with earlier studies [20]. Monthly K_V medians are shown in Figure 12.

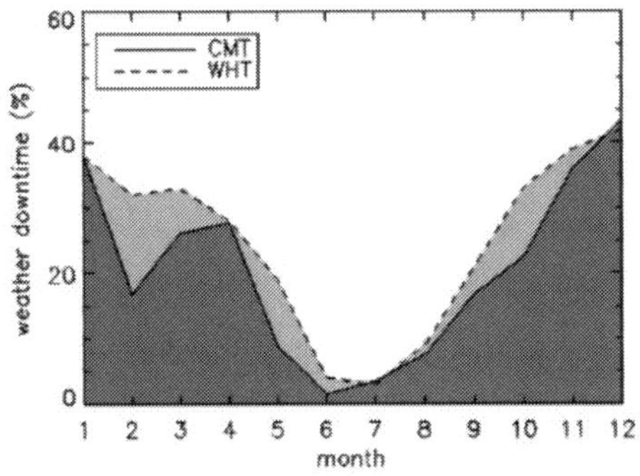

Figure 11. Relative monthly frequency of weather downtime at the CMT from the log for 5 yr compared with that of the WHT for 18 yr.

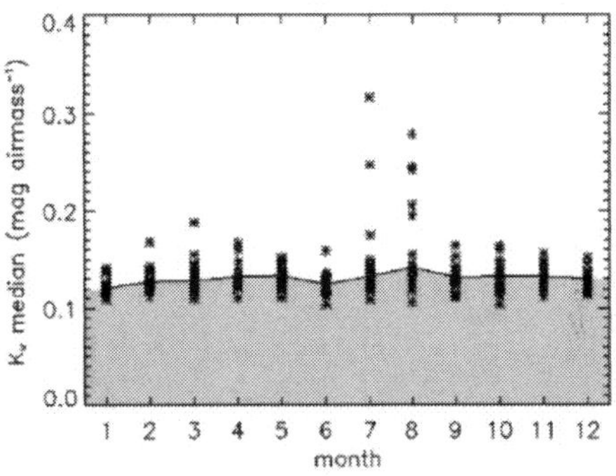

Figure 12. Monthly K_V medians for 20 years. The solid line joins the median K_V for each month [19].

Acknowledgments

This work has been funded by the IAC. The IAC manages the ORM and OT that currently home to facilities belonging to 60 institutions. We thank all who actively contribute to the characterization of the Sites. In particular the ING, TNG and Nordic Optical Telescope (NOT). We also thank the IAC TOTs, the Instrumental Maintenance team and L.F. Rodríguez and J.M. Delgado, of the Istrumental Area of the IAC. We acknowledge the Spanish Instituto Geográfico Nacional (IGN) for maintaining the GPS antennas at ORM and OT that make possible the PWV estimations and to Canary Advanced Solutions (SAC) for programming the PWV monitor. Finally, we are grateful to Terry Mahoney for the language correction of the manuscript.

Appendix A. Miscellaneous compilation of webs

[1]	Telescopio Nazionale Galileo Galilei	`www.tng.iac.es`
[2]	Weather and seeing panel	`.../weather`
[3]	Isaac Newton Group of Telescopes	`www.ing.iac.es`
[4]	Weather and seeing panel	`catserver.ing.iac.es/weather`
[5]	Weather site view	`.../index.php?view=site`
[6]	Instituto de Astrofísica de Canarias	`www.iac.es`
[7]	Weather panel for OT	`.../telescopes/tiempo/weather.html`
[8]	Sky Quality Protection Technical Office	`.../servicios.php?op1=28&lang=en`
[9]	Sky Quality Monitor at ORM	`.../telescopes/pages/es/inicio/meteo /sqm-orm.php`
[10]	Sky Quality Monitor at OT	`.../telescopes/pages/es/inicio/meteo /sqm-ot.php`
[11]	Canarian Observatories Updates CUps – *electronic journal*	`.../cups`
[12]	IAC site characterization (Sky Team)	`www.iac.es/site-testing`
[13]	Real time panel	`.../realtime`
[14]	DIMM at ORM	`.../DIMMA_ORM`
[15]	DIMM at OT	`.../DIMMA_OT`
[16]	PWV Monitor at ORM	`.../PWV_ORM`
[17]	PWV Monitor at OT	`.../PWV_OT`

References

[1] García-Lorenzo B, Fuensalida J J, Muñoz-Tuñón C and Mendizabal E 2005 *MNRAS* **356** 849–858
[2] Varela A M, Bertolin C, Muñoz-Tuñón C, Ortolani S and Fuensalida J 2008 *MNRAS* **391** 507–520
[3] Castro-Almazán J A, García-Lorenzo B and Fuensalida J J 2009 *Optical Turbulance: Astronomy Meets Meteorology* ed Masciadri E and Sarazin M pp 350–357
[4] Pérez-Jordán G, Castro-Almazán J, Muñoz-Tuñón C, Codina-Sánchez B and Vernin J 2015 *MNRAS (submitted)*
[5] Muñoz-Tuñón C 1998 *New Astronomy Reviews* **42**
[6] Muñoz-Tuñón C, Varela A M and Fuensalida J J 2007 *Revista Mex. de A. y A. Conf. Series* **31** 36–46
[7] Vernin J, Benkhaldoun Z and Muñoz-Tuñón C (eds) 2002 *Astronomical site evaluation in the visible and radio range : IAU technical workshop* (*Astronomical Soc. of the Pacific Conf. Series* vol 266)
[8] Vernin J, Muñoz-Tuñón C, Sarazin M, Vázquez Ramió H, Varela A M, Trinquet H, Delgado J M, Jiménez Fuensalida J, Reyes M, Benhida A, Benkhaldoun Z, García Lambas D, Hach Y, Lazrek M, Lombardi G, Navarrete J, Recabarren P, Renzi V, Sabil M and Vrech R 2011 *PASP* **123** 1334–1346
[9] Giordano C, Vernin J, Vázquez Ramió H, Muñoz-Tuñón C, Varela A M and Trinquet H 2013 *MNRAS* **430** 3102–3111
[10] Vernin J and Muñoz-Tuñón C 1992 *A&A* **257** 811–816
[11] Vernin J and Muñoz-Tuñón C 1994 *A&A* **284** 311–318
[12] Muñoz-Tuñón C, Vernin J and Varela A M 1997 *A&A Supplement Series* **125** 183–193
[13] García-Lorenzo B, Eff-Darwich A, Fuensalida J J and Castro-Almazán J 2009 *MNRAS* **397** 1633–1646
[14] Vernin J and Muñoz-Tuñón C 1995 *PASP* **107** 265–272
[15] Varela A M, Muñoz-Tuñón C and Castro-Almazán J *Canarian Observatories Updates* –`www.iac.es/cups`– *CUps* **2-2014**
[16] Varela A M, Vázquez Ramió H, Vernin J, Muñoz-Tuñón C, Sarazin M, Trinquet H, Delgado J M, Jiménez Fuensalida J J, Reyes M, Benhida A, Benkhaldoun Z, García Lambas D, Hach Y, Lazrek M, Lombardi G, Navarrete J, Recabarren P, Renzi V, Sabil M and Vrech R 2014 *PASP* **126** 412–431
[17] García-Lorenzo B, Eff-Darwich A, Castro-Almazán J, Pinilla-Alonso N, Muñoz-Tuñón C and Rodríguez-Espinosa J M 2010 *MNRAS* **405** 2683–2696
[18] Díaz-Castro J *Canarian Observatories Updates* –`www.iac.es/cups`– *CUps* **1-2013**
[19] García-Gil A, Muñoz-Tuñón C and Varela A M 2010 *PASP* **122** 1109–1121
[20] Erasmus D A and van Rooyen R 2006 *SPIE Conference Series* vol 6267 p 1

AUTHOR INDEX

AUTHOR INDEX